AF595553

Remote Sensing of the Oceans

Remote Sensing of the Oceans: Blue Economy and Marine Pollution

Editors

Andrea Buono
Yu Li
Rafael Lemos Paes

MDPI • Basel • Beijing • Wuhan • Barcelona • Belgrade • Manchester • Tokyo • Cluj • Tianjin

Editors
Andrea Buono
University of Naples "Parthenope"
Italy

Yu Li
Beijing University of Technology
China

Rafael Lemos Paes
Institute of Advanced Studies
Brazil

Editorial Office
MDPI
St. Alban-Anlage 66
4052 Basel, Switzerland

This is a reprint of articles from the Special Issue published online in the open access journal *Remote Sensing* (ISSN 2072-4292) (available at: https://www.mdpi.com/journal/remotesensing/special_issues/blue_economy).

For citation purposes, cite each article independently as indicated on the article page online and as indicated below:

LastName, A.A.; LastName, B.B.; LastName, C.C. Article Title. *Journal Name* **Year**, *Volume Number*, Page Range.

ISBN 978-3-0365-1272-3 (Hbk)
ISBN 978-3-0365-1273-0 (PDF)

Contents

About the Editors

Andrea Buono was born in Napoli, Italy, in 1984. In 2017, he received the Ph.D. degree in Information Engineering from the Università degli Studi di Napoli "Parthenope" (Napoli, Italy), where he has been an Assistant Professor at the Engineering department since 2018. He was author/co-author of more than 70 works among publications on peer-reviewed international journals, book chapters and refereed proceedings presented at international conferences. He was involved in several national and international scientific collaborations with space agencies and research centers as ESA, ASI, NOAA. He was the recipient of several national and international awards received for his research and reviewer activities. His main research activities deal with applied electromagnetics, including electromagnetic modeling, wave and radar polarimetry, SAR ocean applications and electromagnetic compatibility.

Yu Li was born in Beijing, China in 1986. He received his B.S. degree in Electronics and Information Engineering in 2009, and the M.S. degree in Information and Communication Engineering from Beihang University in Beijing, China in 2012. He got his Ph.D. degree in Microwave Remote Sensing from the Chinese University of Hongkong, Hong Kong, China, in 2015. Since 2016, he has been a Lecturer and Associate Professor at Department of Information and Telecommunication Engineering, Faculty of Information Technology, Beijing University of Technology, Beijing, China. He is the author of more than 20 peer-reviewed articles and three book chapters. His major research interests include synthetic aperture radar signal and image processing, machine learning and data fusion. He is a guest editor of the journal Remote Sensing.

Rafael Lemos Paes was born in Brazil in 1980. He received the B.Sc. degree in Aeronautical Sciences from the Brazilian Air Force Academy, Pirassununga, Brazil, in 2001 and the M.Sc. degree in remote sensing from the National Institute for Space Research (INPE), São José dos Campos, Brazil, in 2009, where he is currently working toward the Ph.D. degree in remote sensing. Since 2007, he has been a Researcher in SAR remote sensing, focusing on target detection, with the Institute of Advanced Studies (IEAv), São José dos Campos, Brazil. His main research interests deal with SAR target detection and classification, including electromagnetic modeling, single-polarization and multipolarization sea surface scattering, oceanic environment monitoring, sea ice monitoring, pattern recognition, and computer intelligence.

 remote sensing

Editorial

Editorial for the Special Issue "Remote Sensing of the Oceans: Blue Economy and Marine Pollution"

Andrea Buono [1,*], Yu Li [2] and Rafael Lemos Paes [3]

1 Engineering Department, University of Naples "Parthenope", Centro Direzionale Isola C4, 80143 Naples, Italy
2 Faculty of Information Technology, Beijing University of Technology, No. 100 PingLeYuan Road, Chaoyang District, Beijing 100124, China; yuli@bjut.edu.cn
3 Remote Sensing Department, Institute of Advanced Studies, Trevo Coronel Aviador Josè Alberto Albano do Amarante, 1–Putim, Sao Josè Dos Campos 12228-001, Brazil; rafaelpaes@ieav.cta.br
* Correspondence: andrea.buono@uniparthenope.it; Tel.: +39-081-5676706

Citation: Buono, A.; Li, Y.; Paes, R.L. Editorial for the Special Issue "Remote Sensing of the Oceans: Blue Economy and Marine Pollution". *Remote Sens.* **2021**, *13*, 1522. https://doi.org/10.3390/rs13081522

Received: 7 April 2021
Accepted: 13 April 2021
Published: 15 April 2021

Publisher's Note: MDPI stays neutral with regard to jurisdictional claims in published maps and institutional affiliations.

Oceans represent an extraordinary source of resources that needs to be preserved while being exploited. The blue economy lies at the basis of the future of human society because it aims at developing a sustainable and renewable economy, getting benefits from the ocean while reducing pollution and waste. Hence, improving our understanding of ocean processes and their changes, as well as how ocean resources are affected by anthropogenic activities is crucial.

Within this framework, the continuous, updated, and synoptic monitoring capabilities provided by Earth observation instruments play a key role. Nowadays, an unprecedented amount of large-scale and long-term information is available that can support decision makers, environmental agencies, business companies, and local authorities in the management of ocean resources. Remote sensing tools operating on different platforms (e.g., satellite, airborne, unmanned aerial vehicle (UAV), shore-based) at different frequencies (e.g., microwaves, infrared, visible) provide the unique chance of generating added-value products, retrieving geophysical parameters of interest, and boosting the knowledge of ocean processes and marine awareness.

In this special issue, several topics have been addressed that deal with the remote sensing of the ocean for blue-economy-supporting and marine-pollution-monitoring purposes. The articles published in this special issue cover:

- Detection of targets such as marine raft aquaculture, moving vessels, and the shoreline [1–3];
- Observation of spatio-temporal pattern of oil spills and coastal marine litter [4–6];
- Study of natural sea processes, including typhoon-induced storm surges, sub-mesoscale eddies and migration of the along-slope counter-flow [7–9];
- Analysis of scattering and spectral properties of the sea surface [10,11].

Those goals have been pursued using multi-platform and multi-frequency remote sensing tools together with theoretical models, numerical simulations, and in-situ measurements. Most of the study exploited satellite data, including microwave–synthetic aperture radar (SAR) imagery collected in single-, dual- and quad-polarimetric imaging modes, radar altimeters [1–5,7,11], and optical–spin-scanning radiometers and spectroradiometers [7,8]. Other studies used airborne or shore-based sensors, including UAV cameras and high-frequency (HF) coastal radars [6,9]. Numerical tools and simulations were also considered in [3,4,8–10], while in-situ information was exploited in [5,7,8].

Further details on material and methods addressed in the articles published in this special issue, together with the main outcomes those studies achieved in the context of blue economy and marine pollution, are presented as follows:

In [1], single-polarization (under vertical transmit) C-band Sentinel-1 SAR satellite data are exploited to detect marine raft aquaculture in coastal areas. To this aim, a segmentation network combined with a non-subsampled contourlet transform is proposed to extract the sea area covered by the raft aquacultures. It was pointed out that interferences due to significant sea waves can reduce the effectiveness of the proposed method.

In [2], the problem of shoreline extraction is addressed by means of single-polarization (under horizontal transmit) C-band spaceborne Radarsat-2 SAR images. An improved geometric active contour model is proposed, which resulted in a fast, stable and accurate extraction of the land/sea boundary.

In [3], a new method to improve the refocusing of moving vessels under high sea state conditions is proposed. Experimental results, performed on C-band Gaofen-3 SAR satellite imagery, showed that the adaptive time-frequency analysis based on the particle swarm optimization results in an increased and faster global convergence and better processing effectiveness and robustness.

In [4], the monitoring from space of marine pollution due to oil spills is addressed. Simulated compact-polarimetric SAR data are considered for the analysis and were collected from Alos PalSAR-1 (L-band), Radarsat-2 and SIR-C/X (C-band) satellites over ocean slicks of known origin. A set of polarimetric parameters is investigated to identify actual oil spills of natural origin and to distinguish them from oil look-alikes as biogenic films. It was shown that scattering-based features are effective in oil spill detection and that, even though the slant linear compact-polarimetric mode results in better detection performance, the circular compact-polarimetric architecture is to be preferred to preserve the integrity of the detected oil spill.

In [5], the problem of oil pollution is also considered. A novel approach, based on a convolutional neural network and simple linear iterative clustering superpixel, is proposed to classify sea oil spills from quad-polarimetric SAR measurements collected on C-band by Radarsat-2 and SIR-C/X satellite missions. It was found that the simple linear iterative clustering superpixel method significantly improves the classification accuracy, especially for oil emulsion, and that, among the polarimetric features considered in the study, the scattering model-based parameters derived from the four-component Yamaguchi decomposition results in the highest classification performance.

In [6], the pollution of marine coastal areas due to anthropogenic debris, including plastic and metal objects, is investigated. The spatial and temporal patterns of marine debris accumulation along the beaches are analyzed by means of cameras on-board a UAV. Results showed that the equilibrium of the accumulation process depends on the season and on the size of the debris and that it can be significantly affected by extreme events, such as floods. A fairly good agreement between the UAV observations and the standard manual counting is found for medium-/large-size litters, while discrepancies were found for small-size objects, which is likely attributed to the transparent, buried, or hidden nature of such debris.

In [7], a study on the spatiotemporal variations of the along-slope counter-flow off northeastern Taiwan is investigated by means of satellite data and in-situ observations. A synergistic approach is followed, which integrates geostrophic velocity from radar altimeter data from the Archiving, Validation, and Interpretation of Satellite Oceanographic data, sea surface temperature measurements from the moderate-resolution imaging spectroradiometer, the re-analysis ocean data from the assimilative global Hybrid Coordinate Ocean Model, and horizontal velocity records from a mooring acoustic Doppler current profiler. It was observed that the along-slope counter-flow in the subsurface layer was remarkably uplifted and lowered with this phenomenon that was closely linked with the Kuroshio intrusion.

In [8], the impact of tropical cyclone size on storm surges in semi-enclosed areas is addressed. Typhoon information from meteorological satellites, data from tide stations, and simulations performed according to a finite-volume coastal ocean model were considered for the study. It was found that the size of the tropical cyclones is a key parameter that must be accounted for when predicting marine-economic effects and risk assessment. The highest storm surges occur at maximum wind speeds of 40–45 m/s, while the radius of maximum wind only affects the inner area of the typhoon. The peak surge values have been found to approximately follow a linear trend with respect to the seven-level wind circle range.

In [9], HF coastal radar observations of the sea surface current velocity field are used to detect sub-mesoscale eddies in an unsupervised way. A novel algorithm is proposed to overcome the drawbacks due to the high non-geostrophic winds of the observed sea surface currents, therefore resulting in the detection of eddies characterized by significant asymmetry. It was shown that the proposed method allows estimating the eddy boundary profiles and spatial distribution effectively.

In [10], theoretical advancements on the scattering mechanisms of the sea surface when observed by HF and very HF airborne radars are presented. Once the sea surface height has been expressed as the superposition on linear and non-linear wave heights, numerical models are used to simulate the sea surface normalized radar cross section according to the small perturbation method under different environmental and radar imaging parameters and to derive the first- and second-order sea-echo Doppler spectra. The proposed model gives further insights on the sea surface scattering and the wave height spectrum, also providing a theoretical baseline to design a potential airborne radar for ocean surface remote sensing.

In [11], C-band Sentinel-1 satellite SAR measurements are exploited to analyze the spectral signatures of low-backscattering sea areas. The latter can be due to several natural and anthropogenic phenomena, such as oil spills, algal blooms, and low-wind areas. A physically-based approach that relies on the inherent synthetic aperture radar imaging characteristics of the sea surface with and without slicks is proposed to evaluate the signatures of low-backscattering sea areas in terms of the auto-correlation function estimated along the azimuth direction. Results showed that the presence of a low-backscattering area at sea modifies the shape and the width of the azimuth auto-correlation function with respect to the reference sea surface, and that oil spills result in the largest departure.

Acknowledgments: We want to thank all the authors who contributed towards this Special Issue on "Remote Sensing of the Oceans: Blue Economy and Marine Pollution", as well as all the kind reviewers who provided constructive comments and useful suggestions to the authors.

Conflicts of Interest: The authors declare no conflict of interest.

References

1. Zhang, Y.; Wang, C.; Ji, Y.; Chen, J.; Deng, Y.; Chen, J.; Jie, Y. Combining Segmentation Network and Nonsubsampled Contourlet Transform for Automatic Marine Raft Aquaculture Area Extraction from Sentinel-1 Images. *Remote Sens.* **2020**, *12*, 4182. [CrossRef]
2. Wei, X.; Zheng, W.; Xi, C.; Shang, S. Shoreline Extraction in SAR Image Based on Advanced Geometric Active Contour Model. *Remote Sens.* **2021**, *13*, 642. [CrossRef]
3. Yu, L.; Li, C.; Chen, J.; Wang, P.; Men, Z. Refocusing High-Resolution SAR Images of Complex Moving Vessels Using Co-Evolutionary Particle Swarm Optimization. *Remote Sens.* **2020**, *12*, 3302. [CrossRef]
4. Yin, J.; Yang, J.; Zhou, L.; Xu, L. Oil Spill Discrimination by Using General Compact Polarimetric SAR Features. *Remote Sens.* **2020**, *12*, 479. [CrossRef]
5. Zhang, J.; Feng, H.; Luo, Q.; Li, Y.; Wei, J.; Li, J. Oil Spill Detection in Quad-Polarimetric SAR Images Using an Advanced Convolutional Neural Network Based on SuperPixel Model. *Remote Sens.* **2020**, *12*, 944. [CrossRef]
6. Merlino, S.; Paterni, M.; Berton, A.; Massetti, L. Unmanned Aerial Vehicles for Debris Survey in Coastal Areas: Long-Term Monitoring Programme to Study Spatial and Temporal Accumulation of the Dynamics of Beached Marine Litter. *Remote Sens.* **2020**, *12*, 1260. [CrossRef]
7. He, Y.; Hu, P.; Yin, Y.; Liu, Z.; Liu, Y.; Hou, Y.; Zhang, Y. Vertical Migration of the Along-Slope Counter-Flow and Its Relation with the Kuroshio Intrusion off Northeastern Taiwan. *Remote Sens.* **2019**, *11*, 2624. [CrossRef]
8. Li, J.; Hou, Y.; Mo, D.; Liu, Q.; Zhang, Y. Influence of Tropical Cyclone Intensity and Size on Storm Surge in the Northern East China Sea. *Remote Sens.* **2019**, *11*, 3033. [CrossRef]
9. Bagaglini, L.; Falco, P.; Zambianchi, E. Eddy Detection in HF Radar-Derived Surface Currents in the Gulf of Naples. *Remote Sens.* **2020**, *12*, 97. [CrossRef]
10. Ding, F.; Zhao, C.; Chen, Z.; Li, J. Sea Echoes for Airborne HF/VHF Radar: Mathematical Model and Simulation. *Remote Sens.* **2020**, *12*, 3755. [CrossRef]
11. Corcione, V.; Buono, A.; Nunziata, F.; Migliaccio, M. A Sensitivity Analysis on the Spectral Signatures of Low-Backscattering Sea Areas in Sentinel-1 SAR Images. *Remote Sens.* **2021**, *13*, 1183. [CrossRef]

Article

Unmanned Aerial Vehicles for Debris Survey in Coastal Areas: Long-Term Monitoring Programme to Study Spatial and Temporal Accumulation of the Dynamics of Beached Marine Litter

Silvia Merlino [1,*], Marco Paterni [2], Andrea Berton [2] and Luciano Massetti [3]

1 Istituto di Scienze Marine del Consiglio Nazionale delle Ricerche, ISMAR – CNR, 19032 Lerici (SP), Italy
2 Istituto di Fisiologia Clinica del Consiglio Nazionale delle Ricerche, IFC – CNR, 56124 Pisa (PI), Italy; marco.paterni@ifc.cnr.it (M.P.); andrea.berton@ifc.cnr.it (A.B.)
3 Istituto per la Bioeconomia del Consiglio Nazionale delle Ricerche, IBE – CNR, 50145 Firenze (FI), Italy; luciano.massetti@ibe.cnr.it
* Correspondence: silvia.merlino@sp.ismar.cnr.it; Tel.: +39-0187-178-8902

Received: 10 March 2020; Accepted: 13 April 2020; Published: 16 April 2020

Abstract: Unmanned aerial vehicles (UAVs) are becoming increasingly accessible tools with widespread use as environmental monitoring systems. They can be used for anthropogenic marine debris survey, a recently growing research field. In fact, while the increasing efforts for offshore investigations lead to a considerable collection of data on this type of pollution in the open sea, there is still little knowledge of the materials deposited along the coasts and the mechanism that leads to their accumulation pattern. UAVs can be effective in bridging this gap by increasing the amount of data acquired to study coastal deposits, while also limiting the anthropogenic impact in protected areas. In this study, UAVs have been used to acquire geo-referenced RGB images in a selected zone of a protected marine area (the Migliarino, Massacciuccoli, and San Rossore park near Pisa, Italy), during a long-term (ten months) monitoring programme. A post processing system based on visual interpretation of the images allows the localization and identification of the anthropogenic marine debris within the scanned area, and the estimation of their spatial and temporal distribution in different zones of the beach. These results provide an opportunity to investigate the dynamics of accumulation over time, suggesting that our approach might be appropriate for monitoring and collecting such data in isolated, and especially in protected, areas with significant benefits for different types of stakeholders.

Keywords: unmanned-aerial-vehicles; UAVs; anthropogenic-marine-debris; AMD; beached-marine-litter; BML; marine-protected-areas; MPA; ortho-photo; marine-pollution; accumulation-rate

---

1. Introduction

Interactions between geosphere and anthroposphere, in sensitive areas such as the land–sea interface, are constantly evolving due to population growth and exploitation of natural resources. Therefore, the growing problem of the accumulation of anthropogenic marine debris (AMDs, or marine litter–ML), especially in isolated/protected coastal areas, is one of the emerging problems of recent decades. The interest in AMDs pollution in recent years has led to a significant increase in data related to such material in oceans [1]. In the Mediterranean area, the increasing knowledge of the concentration and type of ML [2–7] and the increasing efforts to survey off-shore areas, have not been accompanied by an equally increasing knowledge of the sources, composition and distribution of materials deposited along the coast (beached marine litter, BML), and the mechanism through which they accumulate in particular coastal areas. A long stay on the coast of BML can cause considerable damage. Especially plastic objects left on the beach for months/years are subject to photodegradation at a higher rate than expected if they were at sea [8,9]. Rapidly fragmented, reduced to meso-plastics (5 mm–2.5 cm) and micro-plastics (MPs, ≤ 5 mm long), they mix with the substrate and can produce a stream of particles flowing into the sea [10–12] adding to those directly released by rivers, which are key agents in the release of macro and micro marine litter in oceans [13,14]. There is an urgent need to develop new methods of spatial and temporal mapping of beaches to identify the areas of greatest accumulation, quantify the abundance and types of material, and trace their origin, in line with the protocol and standard monitoring strategies [15–19]. Numerous studies have reported that 80% of waste present in the sea is probably of terrestrial origin and rivers also seem to play a key role in the transport of debris from land to oceans [20]. Therefore, it becomes important to estimate the flow of material transported by water courses and its impact on the areas surrounding the river mouths. So far, few studies have investigated the transport, deposition, and accumulation of AMDs through internal water. To better understand this problem, monitoring actions are needed to verify how rivers transport AMDs and how they affect coastal deposits. These surveys should collect data in a consistent manner throughout the investigated territory, in one or more seasons, and with different replicas, trying to correlate the data obtained with the anthropic impact (urbanization, presence of parks and protected areas) and with the morphological characteristics (rivers, ports, types of beaches) of the area. Particularly at risk are the MPAs, which often suffer from a large influx of AMDs, as they are located in or near densely populated and industrialized areas. AMDs in protected coastal areas are often difficult to clean from waste due to the inherent difficulty of reaching these isolated areas that are not served by roads or facilities, and also due to regulations that limit human intervention. In this context, the use of aerial survey could be a valuable aid. To get the best results from aerial survey processing it is important to choose the right scale [21]. Since the highest resolution of commercial satellite images is about 0.3 m (WorldView-4), this platform is not the most suitable for observing beach waste [22]—a spatial resolution below decimeter is required and UAVs, especially commercial UAVs, have proven to be effective in this respect. The longer battery life, the ability to plan automatic flights with easy-to-use ground station software, and their small size are real advantages, and the structure from motion algorithms (SFM) allow accurate digital elevation models (DEM) and ortho-mosaic terrain models over large areas. Today UAVs are increasingly accessible and have widespread applications, such as in environmental monitoring systems for agroforestry, structural geology, archaeology, marine habitats, supervised hazards, and accidents [23–39], and recently also in monitoring ML on the coast [40–44] or that floating in rivers [45]. These studies are not uniform with regard to the data processing procedures, ranging from visual interpretation of images [42] and analysis of the spectral profile of litter [46], to the use of machine learning methods [43,44]. Moreover, since this is an "emerging field of study", there is no single standardized protocol for data acquisition and processing, but only a few suggested protocols [42,43]. The difficulty of developing scalable procedures that do not depend on local environmental constraints, is also due to the different objectives to be achieved. In any case, most of the studies carried out focused on the detection of BML stocks, especially in isolated areas. The advantages offered by UAVs, in terms of survey resolution and repeatability, are particularly suitable for the purpose we are interested in, that is to study the model of aggregation and distribution of BML in such remote areas, and are particularly useful to monitor the most sensitive areas, such as protected areas. So far, little attention has been paid to the long-term study of a particular area in order to understand the dynamics of BML deposition, and to obtain the rate of accumulation and the variability of spatial distribution over time. The vertical spatial distribution (cross shore) of debris on a beach has its own dynamics, which is strongly influenced by the physical processes determined by the wind and waves on the beach profile. Therefore, to understand this phenomenon, it is important to monitor it over a long period, with frequent sampling [46,47]. The "manual" collections and cataloguing of BML are the usual way of carrying out such monitoring, but they take time and involve many people. UAVs can reduce both monitoring time and human staffing requirements. For this reason, starting from April 2019, we have implemented a pilot monitoring program through UAVs in the Migliarino,

Massacciuccoli, and San Rossore (SRPRK) park, a marine protected area with 34 km of protected coastline, north of the mouth of the Arno river. We planned to use UAVs to acquire geo-referenced RGB images in a selected area about 100 meters long, from the dune crest to the low-tide terrace (shoreline base), during a one-year monitoring program (about two recognition flights per month). For each monitoring and image acquisition date, a post-processing system allowed the localization and identification of BML within the area scanned by ortho-photos. A specially created software for pattern recognition (based on visual interpretation of the images, [42]) provided the estimation of typology, quantity, density, and position of the identified items (see Material and Methods).

The UAVs used in this study were in the 'multicopter' category, and in the 'light' and 'very light' classes, and had an autonomy of about 30 minutes of flight, sufficient for the purposes of our current survey, as suggested in previous studies [42,43]. In our case, each flight allowed us to cover the entire selected area (100 m × 15 m), even considering that it used a conservative approach with high overlap between images and a "stop and go" shooting method that increased the overall flight time.

In this specific area, where there are airworthiness constraints due to the presence of a control traffic region (CTR), flight missions were allowed at specific heights provided for by the regulations; in any case, a resolution of at least 2/2.5 cm/pixel (and even higher) would be guaranteed, which would be sufficient to recognize even the cotton buds or caps, i.e., the small BML typically present on beaches.

2. Materials and Methods

2.1. The Study Area

The target area was the afitoic backshore of a stretch of sandy beach inside the marine protected area of SRPRK. This area, located between the two rivers Arno (N 43°40′47.408″, E 10°16′40.466″) and Serchio (N 43°47′1.704″, E 10°16′0.016″) is affected by the marine current that goes from the mouth of Arno to the north, with a considerable transport of fluvial material. The Arno River is, in fact, an important Italian water course that crosses the Tuscany region, running through large cities like Florence and Pisa, and industrial and production centers such as the province of Prato and Pontedera. The limitation to tourism in this area of the Park allows the study of the dispersion of marine debris and its accumulation on the coasts, caused by natural and meteorological events, as there is no direct contribution to such accumulation by human presence. This site is in fact located within the area "A" of SRPRK, which means there is an absence of tourism throughout the year. Access to this area is forbidden from both land and sea, and access is only allowed for research purposes. In summer, some excursions are made with environmental guides, but only on a few specific paths and never beyond them. The beach is a "natural" beach, with a dune cordon parallel to the coastline that delimits the hinterland. In our case the foreshore is very small, given the weakness of the tidal phenomena, as on most of the coasts of the Italian peninsula. We have taken into account the maximum extent of the tide in this area when choosing the points of our stretch of beach (10°16′40.70″ E 43°42′55.07″ N; 10°16′42.12″ E 43°42′51.86″ N; 10°16′41.68″ E 43°42′51.74″ N; 10°16′40.23″ N. E 43°42′54.93″ N), such that it started at the edge of the "swash zone" (wave run-off zone). The size of the selected area was about 100 meters long and 15 meters wide, with a south-west exposure (Figure 1).

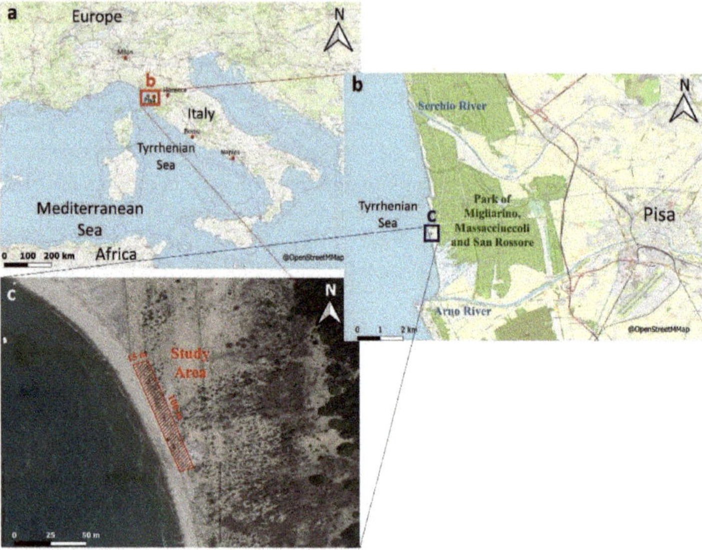

Figure 1. Geographical location of the stretch of beach studied, on the Northern Italian coast (coordinates: 10°16′40.70″ E 43°42′55.07″ N; 10°16′42.12″ E 43°42′51.86″ N; 43°42′51.86″ N; 10°16′41.68″ E 43°42′51.74″ N; 10°16′40.23″E 43°42′54.93″N). Maps created by using QGIS 3.12 [48], @OpenStreetMap Contributors [49] and Geoscopio WMS service by Regione Toscana [50].

2.2. Characteristics of the Used UAV

In the present study the Phantom 4 PRO v2 quadcopter [51] was used. It is a commercial UAV suitable for this type of application, thanks to the good resolution of the camera (5472 × 3078 pixels, which, flying at 6 meters above ground level allowed us to reach the theoretical value of 0.16 cm/pixel—with perfectly flat ground and in the best conditions—and, in our case of not perfectly flat ground, 0.18 cm/pixel), the compactness of the aircraft, and flight stability. It had a titanium and magnesium alloy structure, increasing the strength of its frame and reducing its weight; together with the good battery capacity (5870 mA), this gave a flight time of up to approximately 30 minutes. It had a gimbal three-axis stabilized camera with a 1-inch 20-megapixel CMOS sensor (Figure 2), capable of shooting up to 4K/60 fps video and photo bursts, at up to 14 fps. The gimbal was set to −90° to look at nadir. This allowed it to capture photos perpendicular to the direction of flight. It was equipped with HD video transmission capable of reaching a maximum range of 7 km. The correct position management was obtained thanks to two satellite tracking systems: GPS and GLONASS. The use of UAVs for 3D mapping of the terrain or sites has the advantage to access utilities, like waypoint mapping for identifying the surveyed area and flight path planning and control, provided by third-party applications. Three sets of dual vision sensors formed a 6-camera navigation system that worked constantly to calculate the relative speed and distance between the UAV and any object; this system allowed it to fly more safely and avoid obstacles along the way. A remote controller allowed a pilot to control the flight of the UAV; a smartphone (or tablet) could be connected to the remote controller to view the camera, read the telemetry, and enable automatic functions. The maximum speed was 72 km/h and the maximum control range was 7 km from the driver. The UAV operated automatically using the Drone Harmony (DH) ground station software (see Section 2.4). The mapping and modeling of the aerial photography area was selected during the configuration process, and the flight plans were selected. The automatic process of the mission included take-off and landing, route planning and calculation of the corresponding spatial resolution of the flight altitude, which were displayed on the screen. Establishment of flight altitude depended on the spatial resolution we wanted. As the number of pixel per item depended on the type of object (our test gave 20–30 pixel for a bottle cup, 18–22 pixel for cotton-buds, and 40–50 pixel for a spoon) with best light condition and since these conditions were not always present, we tried to increase the resolution by acting on the flight altitude, so as to guarantee at least twice the pixel/item values previously tested.

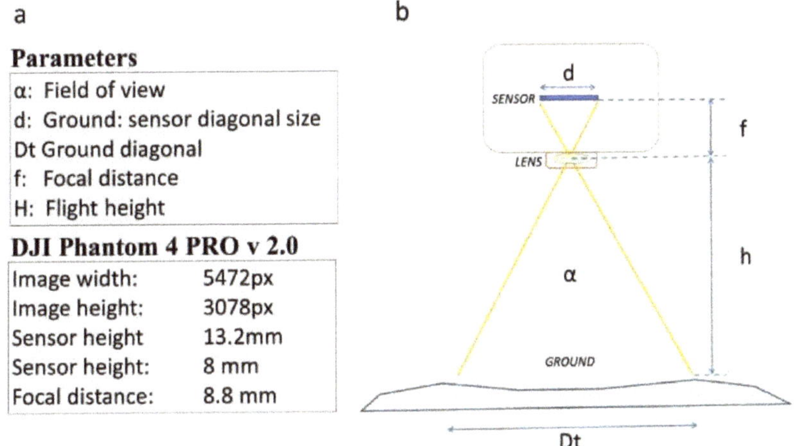

Figure 2. (**a**) Sensor characteristics of the DJI Phantom 4 PRO v 2.0; (**b**) parameters for calculating the distance of the sample to the ground: sensor diagonal size (**d**), focal distance (**f**), field of view (α), ground diagonal (Dt), and vertical ground distance or flight height (**h**). With this choice, a spatial resolution of 0.01–0.03 m was obtained, which was sufficient to recognize the smallest macro beached marine litter (BML).

Once the flight altitude of the UAV was established, the ground sample distance (GSD) was calculated using the parameters in Figure 1 and Equations (1), (2), and (3):

$$\alpha = 2 \times \text{atan}(d2/f), \quad (1)$$

where α is the field of the view angle, d2 is half the diagonal of the sensor area, and f is the focal distance,

$$Dt = h \times \tan(\alpha/2) \times 2, \quad (2)$$

where Dt is the ground diagonal, h is the vertical ground distance, and α is the field of view angle,

$$GSD = Dt/n, \quad (3)$$

where GSD is the ground sample distance, Dt is ground diagonal, and n is number of pixels.

2.3. Survey Realization

After having identified and delimited the coastal profile of the study (a stretch of about 100 meters in length and 15 meters in depth, starting from the coastline to the dune area), we established the flight plans and the take-off/landing points for the UAV (Figure 3). These plans must have guaranteed the necessary resolution, the respect of the autonomy of the aircraft, the compliance with the regulations in force, and the acquisition of photos with overlap of at least 70%–80%. This overlap level ensured a greater precision during the preparation of the ortho-mosaic. In fact, by using a low altitude flight, the scanning area for each photogram was limited (8.76 m × 5.84 m) and an error of even 1 or 2 meters could have had a big impact on the photogrammetric reconstruction. A high overlap, therefore, compensated for this fact, and also allowed us to get a better 3D reconstruction.

UAVs operating in the multicopter category and in the "very light" classes had an autonomy of about 25/30 minutes of flight. In order to have the maximum resolution in the photos, a flight altitude of 6 m was chosen. Using this configuration, we got a ground sampling distance (GSD) of 0.18 cm/pixel. In addition, specific precautions were used to ensure optimal image capture: 1 m/s speed with a "stop and go" mode for each shot, to ensure shooting in a stationary position due to low flight altitude and to avoid blurred photos; with the manual focus set to infinity (i.e., autofocus disabled) to avoid variations in focus; initial exposure setting of autoexposure (AE) was disabled to avoid variations in brightness. In the specific study area, several scans were performed using UAVs. For each scan, the specific software (Agisoft Photoscan Professional) provided the alignment and creation of an orthophoto for the entire area. The memorization of the flight plans allowed the replication of the scans at a later time, keeping the same areas of interest and the established take-off/landing points. We used the Litchi application to manage the storage of the flight plan; if the conditions required landing before the end (for example due to the sudden occupation of the airspace), the Litchi application allowed you to resume the flight from the point where it was interrupted.

After the initial scan of the monitoring program (12 April 2019), the removal of anthropic material from the stretch of beach considered was carried out, excluding only objects with a linear dimension less than 2.5 cm (OSPAR protocol). The collected material was subsequently catalogued and counted according to a protocol previously adopted in this type of monitoring; this protocol integrated the Marine Strategy Framework Directives (MSFD) survey procedures [16], the OSPAR guideline [15] for size and type classification of BML and citizen science contribution, involving volunteers, researchers, and university students during beach cleaning operations, classification, and counting of objects (SeaCleaner protocol [52,53]). As a result, we could compare and match the ortho-photo data with those collected by standard manual surveys. A second survey was then carried out by acquiring images immediately after the cleaning of the beach. From this date, every 10/15 days and for a period of about 4 months, surveys were carried out with UAVs. The scanned images/videos were transferred to the servers. A post-processing system that uses a pattern recognition software, located and identified the different BML within the scanned area and estimated the accumulation rate of the different classes of objects and dimensions and other parameters of interest. On 13 July 2019, the beach was cleaned for the second time, the BML were catalogued and counted, and the procedure described was started again for the second period of study.

(a) (b)

Figure 3. (**a**) An example of a transect performed by unmanned aerial vehicles (UAVs) to obtain the necessary images for the entire coverage of the area of interest, with an overlapping of the shooting areas of the adjacent frames. (**b**) The stretch of beach investigated before and after the complete cleaning carried out by researchers and volunteers.

2.4. Image Acquisition and Processing

The software used for the automatic flight of UAVs was "DRONE Harmony", a commercial software (free to use only for one month) that can perform several operations (to create the flight plan necessary, to capture the photos, etc.) in a simple and accurate way [54]. The total acquisition time of all images of the studied area, on each date, was about 21 minutes. The speed of the flight was 2 m/s, but the total acquisition time increased because the "stop and go" mode option was used at each selected point for the photographic acquisition. The photogrammetry technique was used to define the position, shape, and size of the objects on the ground, using the information contained in appropriate photographic images of the same objects, taken from different points. The photos were in fact taken such that there was an overlap between the adjacent frames, with a coverage of about 70% or 80% (see Section 2.3). This technique can be used both at the ground level and in aerial mode, and allows the obtainment of a 3D reconstruction of the objects (Figure 4), whose potential, even if not exploited in the present study, could be useful for several applications (Supplementary Material B).

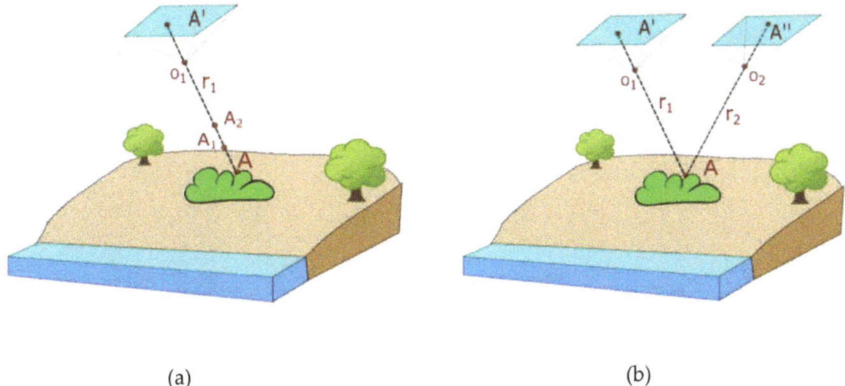

(a) (b)

Figure 4. Photogrammetry technique allowing the 3D reconstruction of the objects. By precisely knowing the position of the homologous points A' and A" on the two photographs, and the spatial position of the two sectors and the two perspective centers O1 and O2, the point A remains geometrically defined, since it is the intersection point of the two projecting rays r1 and r2 connecting the two homologous points with the perspective centers (**b**). This does not happen with a single photo shoot (**a**).

In our specific application, we used Agisoft Photoscan, a standalone software product that performs photogrammetric processing of digital images and generates 3D spatial data for use in GIS applications, cultural heritage documentation, and visual effects production, as well as for indirect measurements of objects of various scales. The use of this software allows us to obtain ortho-mosaic, a calibrated image that constitutes the ortho-rectified mosaic of the entire area covered by the scan. We also obtain the digital elevation model (DEM), with the aim of estimating the height variations of the ground that must be used to correct the dimensional measures of the objects (see Figure 5).

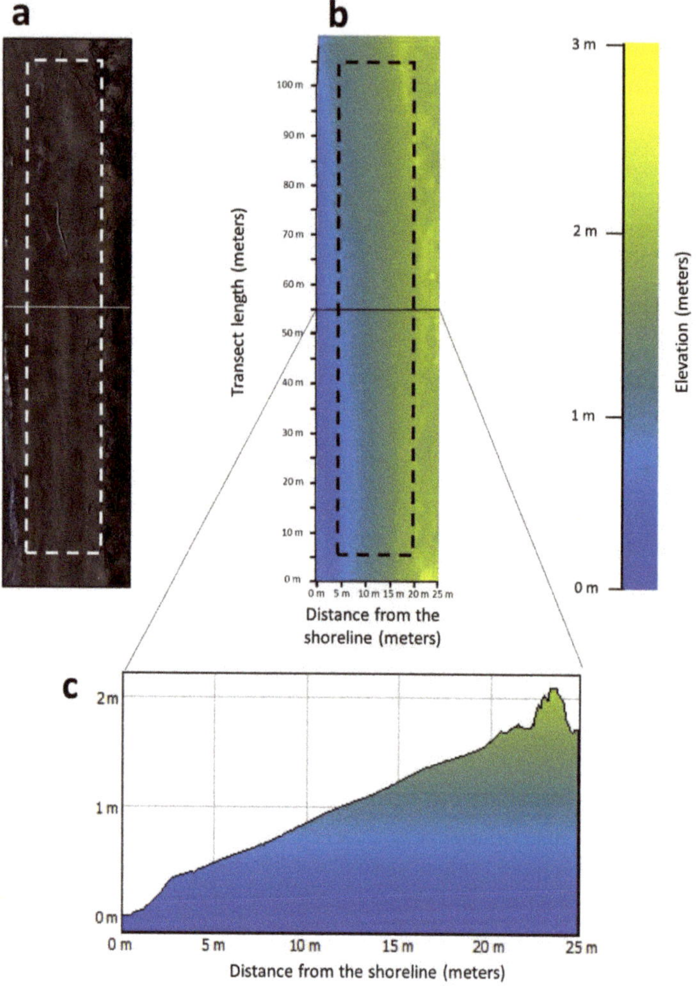

Figure 5. Orthophoto (**a**) and digital elevation models (DEM) (**b**) of the study area, with the profile of the beach corresponding to the central point of the sampled transect (**c**). Distances and elevation are indicated in the image, with the dotted line that delimits the study area inside the beach.

2.5. Data Acquisition from Images and Data Analysis

After data collection, the image sets were processed by Agisoft Photoscan for the generation of dense points cloud and Digital Terrain Models. GPS information extracted from the EXIF (EXchangeable Image File) information of each image file was used to create a georeferenced ortho-photo map (ortho-mosaic), with a resolution of 0.18 cm/pixel. Once the ortho-mosaic was obtained, it was possible to extract the data of interest for this study. We were mainly interested in obtaining an estimate of the spatial coverage of the monitored stretch of beach, over time, by the BML. This involved the knowledge of the surface occupied by the surveyed objects, and not only their possible type identification (material), the standard linear dimension and the numerical estimate [15,16]. For this reason, we decided to develop a semi-automatic software (waste mapping, WM) to quantify the waste detected with the ortho-photos acquired with aerial survey. This software (currently only available for internal use, but we plan to share it within the concerned scientific community when we have concluded the necessary updates) was developed for the most popular operating systems (Windows, OSX, Linux); it could load an ortho-rectified image of the analyzed area and also offered analysis tools on it. The image must contain some acquisition data as ground sample distance and geographic coordinates; if not available, these data could also be entered manually. The user had a cursor which identified the presence of the object; subsequently, the user could have an automatic drawing of the shape of the object, but, if it did not work (overlapping objects, unclear image, noise, etc.), he could manually carry out the drawing; the user then associated the object with a class (plastic, metal, multi-material, etc.). Once these operations were performed on the whole image, the software estimated the number of objects, GPS position, area, and principal linear dimension of each object, and then calculated the properties of the objects: total number for each material type (plastic, glass, metal, etc.), total number for each size category, both in terms of standard linear dimensions and measured area, also expressed as a pseudo-color map. This visualization of the local density of BML (local percentage of area covered by the objects) was obtained in the following way—the whole stretch of the beach was divided into 32 × 6 rectangles, each measuring 3.125 m × 2.5 m = 7.8125 m^2: = TA (see Section 3.1 for the results of this visualization methods). Then, for any rectangle of beach, the local density surface area (DA) would be DA = $\sum OA_i/TA$, where OA_i is the object coverage area, with i varying from 1 to n; n = total number of detected objects inside the considered rectangle.

All information obtained through the WM software could be exported to a CSV (Comma Separated Values) file and managed by other processing software. The effectiveness of this evaluation method was validated by distributing a defined number of objects of a known size (surface area and linear dimension) on the same stretch of beach (Figure 6a,b, and Figure 7c,d) and then analyzing the WM errors in identifying the sizes and the number of the same objects. Thus, we evaluated the errors in percentage terms:

- PE_S = (Sm-Sa)/Sa × 100 where PE_S is the percentage error on size (object surface area) estimation, Sm is the measured object size, Sa is the actual object size;
- PE_N = (Nm-Na)/Na × 100 where PE_N is the percentage error on the number of objects estimation, Nm is the measured object number, Na is the actual object number.

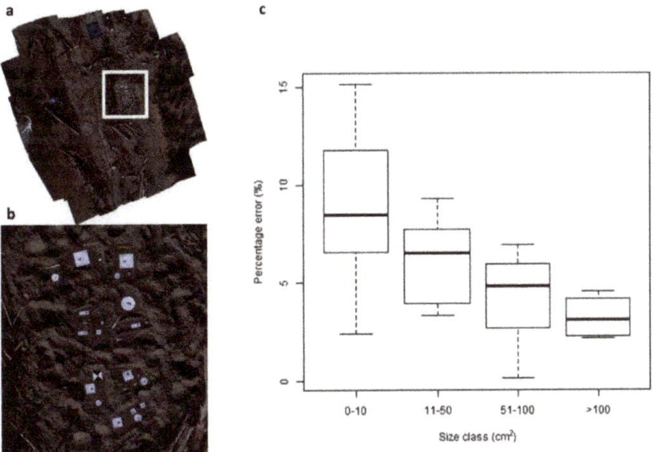

Figure 6. Error evaluation on items size (object surface) estimation. (**a**) Part of the whole study area; (**b**) zoomed-in view on the portion of beach used for validation, of about 18 m^2; (**c**) boxplot of percentage error (PE_S) per size class (in m^2). PE_S was calculated using Microsoft Excel and boxplot were made with R 3.4.1 software for windows (https://www.r-project.org/). Defining PDs the "percentage of detected objects by UAVs", we obtain that PDs is 100–PE_S.

Concerning the object surface area estimation, in Figure 6c we can see how the relative error grows with a decreasing size class, but it is still quite small for all classes. For item counting, the relative errors were classified according to the linear size of the objects, because it was the same classification used during the standard manual monitoring (see Section 3.3). Additionally, in this case, the percentage error increased while the size class decreased (Figure 7c). The validation of this method was done in the best possible weather conditions (Figure 7a), but we also showed, in Figure 7b, how the ortho-photo image could be in the worst weather conditions (wind, change of brightness, etc.). In this case the relative error increased, especially for smaller objects.

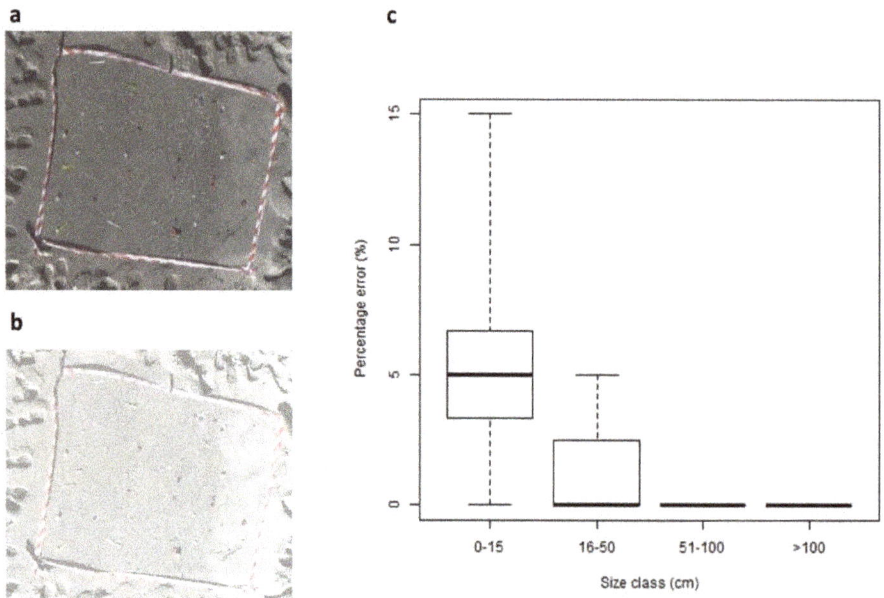

Figure 7. Error evaluation on items counting estimation in (**a**) good weather condition; (**b**) a case of possible bad weather/not optimal visual condition; and (**c**) boxplot of percentage error in items counting (PE_N) per size (linear dimension, following OSPAR guidelines prescriptions). PE_N was calculated with Microsoft Excel and the boxplots were made with the R 3.4.1 software for windows (https://www.r-project.org/). PD_N indicates the percentage of detected objects by UAVs; $PD_N = 100 - PE_N$.

3. Results

From April 2019 to January 2020 we carried out 17 total recognition flights of the studied area, within the SRPRK, near Pisa, Italy. For each flight, we have elaborated and realized an ortho-mosaic. The visual screening of each ortho-mosaic took about 60–80 minutes, and was carried out using our WM software, which allowed us to obtain different types of information.

3.1. *Two-Dimensional Distribution of BML on the Beach*

WM enabled us to estimate the size of the classified objects in terms of the surface area they occupied, to visualize the amount and the distribution of BML over time in the different zones of the beach (Figures 8 and 9), and to investigate the dynamics of their accumulation over time (Figure 10). This is an aspect that, to our knowledge, has not been previously investigated using UAVs.

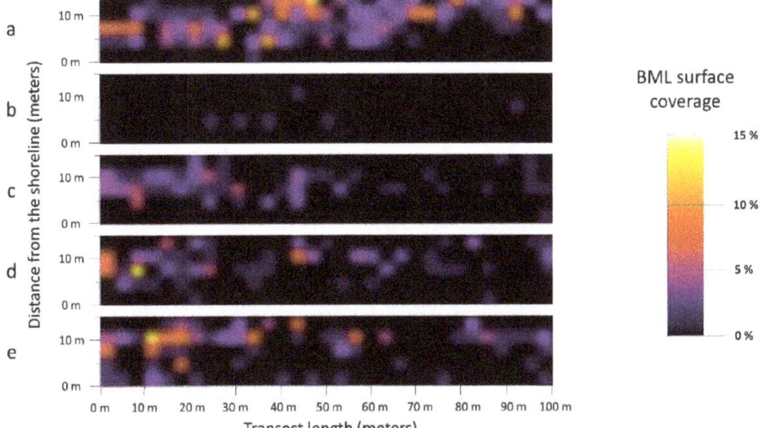

Figure 8. Variation in beached marine litter surface coverage of the monitored area, over time, for the first temporal period (from April 2019 to July 2019). The sequence of strips, from (**a**) to (**e**), shows the beach before the first total cleaning in (**a**), immediately after the first cleaning in (**b**), and at the following dates of our monitoring in (**c**), (**d**), and finally (**e**), after 90 days. The correspondent dates were: (**a**) 15 May 2019; (**b**) 24 May 2019; (**c**) 04 June 2019; (**d**) 25 June 2019; and (**e**) 13 July 2019. The accumulation of waste is displayed both qualitatively and quantitatively using a color gradation, which corresponded to the percentage of the surface area covered by the objects in relation to the total area. The sequence shows the pattern of spatial and temporal distribution, with a clear increase in waste accumulation in the upper part of the strips (the dune zone).

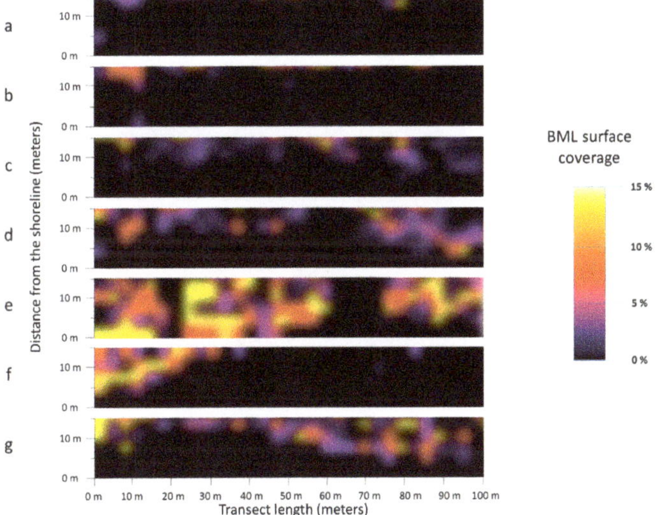

Figure 9. Variation in beached marine litter surface coverage of the monitored area, over time, for the second temporal period (from half July 2019 to January 2020). The correspondent dates were: (**a**) 19 July 2019; (**b**) 21 August 2019; (**c**) 18 September 2019, (**d**) 03 October 2019, (**e**) 20 November 2019, (**f**) 11 December 2019; and (**g**) 17 January 2020. As in Figure 8, the pattern of spatial and temporal distribution highlights the accumulation of waste increasing in the upper part of the strips (dune zones). In strip (**e**) the "footprint" of the Arno flooding, that occurred during the correspondent time period, is evident.

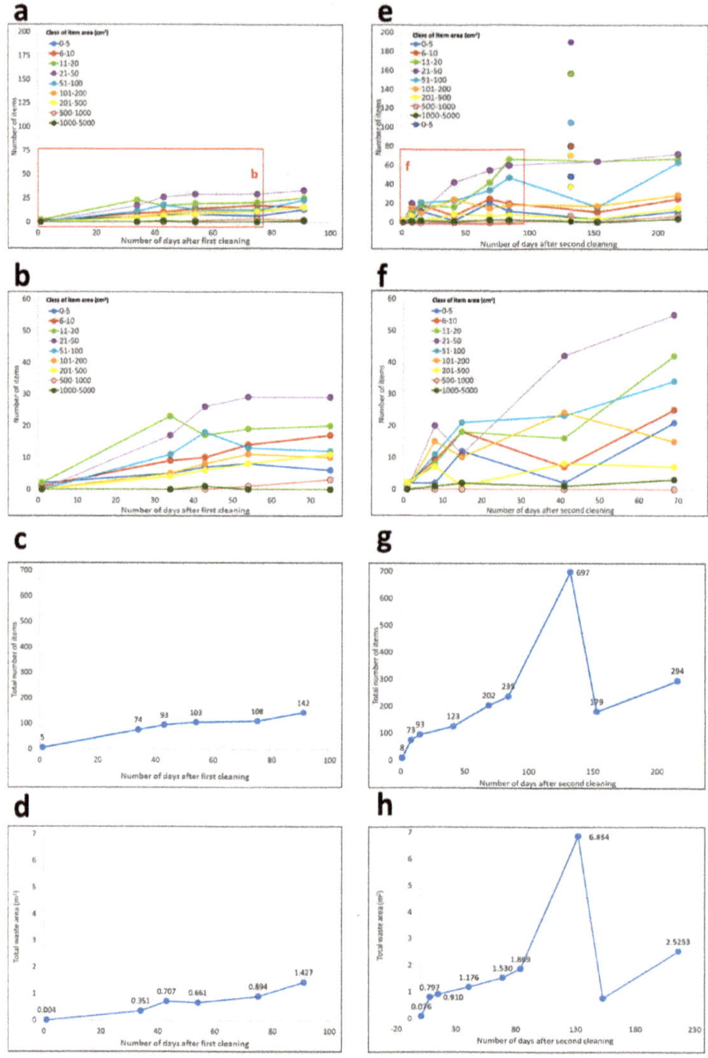

Figure 10. Accumulation trend of the BML with time for both time periods studied, starting from a "completely clean beach" situation, for the different size classes (**a**,**b**,**e**,**f**), in terms of the total number of elements accumulated (**c**,**g**) and the total area covered (**d**,**h**). The time, in the x- axes, for both periods starts from '0', which represents the date of total beach cleaning. The points correspond to the acquisition dates of the orthophotos (graphs on the left: 12 April 2019; 15 May 2019; 24 May 2019; 04 June 2019; 25 June 2019; 13 July 2019; graphs on the right: 13 July 2019; 19 July 2019; 03 August 2019, 21 August 2019; 18 August 2019, 03 September 2019, 20 November 2019, 11 December 2019, 17 January 2020). The "burst" detected on 20 November 2019, following the flood of the 15 November 2019 is well visible. Data collected before the first cleaning (12 April 2019) of the monitoring programme, that were not included in the graph, are the total number of items (203 items) and the total waste area (2.141 m^2).

Figures 8 and 9 show the distribution of the density (percentage of covered area) of beach waste in the studied area and its variation over time, from April 2019 to July 2019 and from July 2019 to January 2020, respectively. By using a color gradation to represent the amount of covered surface (see

Section 2.5), it was possible to highlight both qualitatively (visually) and quantitatively (percentage of covered surface) how the accumulation changed with time. In Figures 8 and 9, therefore, for both temporal periods, the initial "random" pattern of BML assumed, and with time, an increasingly precise connotation with a greater accumulation (red/yellow) mainly located near the dunes clearly emerges. The fact that in this type of beach most of the debris is in the area towards the dune was already highlighted by other studies, including those carried out using UAVs [43]. The new results that emerged from our long-term programme were the time-series that allowed us to obtain the trend of BML accumulation over three or four months, starting from a cleaned-up beach. In our opinion, these data are important because they can contribute to the understanding of the dynamic mechanisms that determine the "vertical" (cross-shore) distribution of BML observed on beaches.

3.2. Quantity, Typology, and Accumulation Rate of BML

Another important result obtained from our long-term monitoring programme is the estimation of the quantity of waste accumulated over time and its rate of accumulation, for different size classes of BML and for two different periods (spring–summer and summer–autumn).

In Figure 10, the graphs on the left side refer to the first period (spring–summer). Figure 10a,b show the number of objects deposited on the studied stretch of beach with time, for different size (surface area) classes (Figure 10b is just a zoomed-in view of Figure 10a). Figure 10c shows the total number of objects accumulated with time, while Figure 10d shows the total surface covered by BML, with time. Figure 10e–h show the same graphs but for the second period (summer–autumn). Looking at graphs (a)–(e), we note that, starting from the date of the total cleaning of the beach (date "0" for both periods), there was a progressive increase in waste on the beach for all size classes in the first months. Unfortunately, during the first period (spring–summer) we could not monitor frequently during the first month and, therefore, we did not have any data in the first 30 days, as can be seen in Figure 10b. On the contrary, the more frequent data-acquisition flights that were performed in the second period (summer–autumn) highlighted a clear general fast growth of items of all size classes in the first ten days (Figure 10f). Then, up to 40/50 days from the cleaning, this initial common steep growth changed to a less fast growth or, for some size classes, to a decrease. Finally, from the third month (around 60–80 days), the accumulation seemed to be in an "almost-flat" phase, with no or very low growth in the total quantity of items, for almost all size classes. The range of 10–40 days after the full beach cleaning was the one in which major changes in the accumulation occurred, depending on the size class. After 40 days, the growth continued, but at a much lesser rate, especially in the first period (spring–summer), both for the total number of objects and for the total area covered (Figure 10c,d). The second period (summer–autumn) was characterized, in its final phase, by an anomalous accumulation due to a flood in the Arno river, which occurred on 15 November 2019. During this extraordinary event, there was a large increase in the flow rate of the river, with a peak of 1473.75 m^3/s compared to the previous period (mean flow of 33.44 m^3/s, with a minimum of 8.7 m^3/s, and a maximum of 121.1 m^3/s). In addition, a strong south-west wind (Libeccio) was recorded, with a maximum peak of 86.4 Km/h and an average daily value of 50.1 Km/h, just in the direction of the coast [55]. Parallel to the increase in the transport of solids from the river and their discharge into the sea [20,45], the effect of the wind must also be taken into account, which prevents their dispersion offshore but, instead, helps to push them towards the coast where they accumulate on the beaches. Our monitoring, which took place on 20 November 2019, i.e., immediately after this flood, showed a huge increase in the number of stranded materials, highlighted both in Figure 10 (dotted line) and Figure 11, and through the high density and uniform spatial distribution of BML displayed in Figure 9e. In the following monitoring dates, we could observe a return to the pre-flooding values, both of the quantity and of the spatial distribution of the BML, as evidenced again in Figures 10 and 11 and in the last strip of Figure 9, corresponding to our last survey (17 January 2020). The data concerning the number of BML, obtained during this last survey (performed in January 2020), are reported in Figure 10 and included in Tables 1 and 2; from the Tables, and by comparing graphs (c) and (g) of

Figure 10, it can be noted that the amount of BML in the autumn and winter season was higher than that found in the previous seasons (spring–summer). The survey of 17 January 2020 thus concluded the second studied time-period (summer–autumn) with a considerable delay, as it was impossible for us, after the last monitoring on 11 December 2019, to carry out other surveys with UAVs, due to the continuous bad weather conditions, which lasted until mid-January 2020.

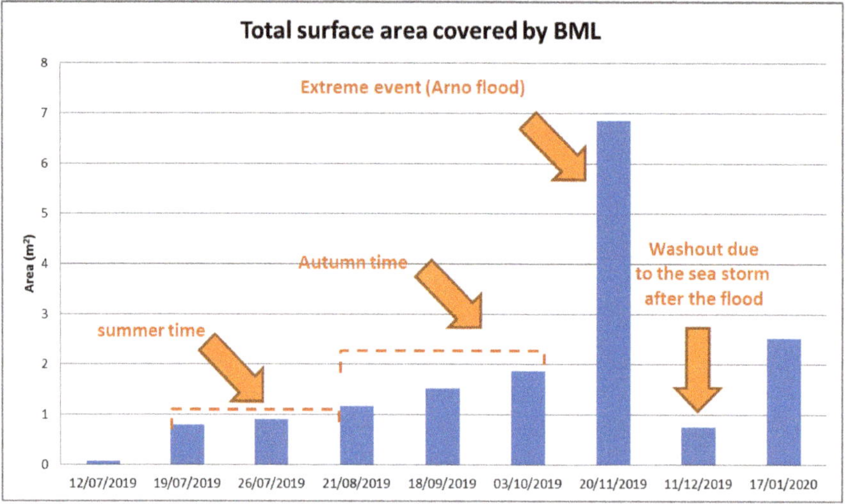

Figure 11. Histogram of the total surface area covered by BML for the different survey dates of the temporal period summer–autumn. Note that the time gap between different dates is not constant.

3.3. Comparison with "Standard" Survey Results

Standard monitoring campaigns were carried out during our long-term monitoring program with the help of volunteers, citizens, researchers, and students, following the SeaCleaner protocol [52,53] (which meets the MFSD survey procedures [16] and the OSPAR guideline for size and type classification of BML [15]). To date, three beach cleaning operations have been carried out—at the beginning of the first period (12 April 2019), at the beginning of the second period (13 July 2019), and at the beginning of the third, for which data acquisition has not yet been completed (17 January 2020). The cleaning and cataloguing operations take about one full day for each date of monitoring. Tables 1 and 2 show the results from the manual standard survey of the studied beach, compared with the one obtained from ortho-mosaic.

Table 1. Results from manual standard survey of the studied beach, compared with the one obtained from ortho-mosaic (material type classification).

	12 April2019			13 July 2019			17 January 2020		
	Standard Survey	**UAV Results**	**UAV vs.* Standard(in Percentage)**	**Standard Survey**	**UAV Results**	**UAV vs.* Standard(in Percentage)**	**Standard Survey**	**UAV Results**	**UAV vs.* Standard(in Percentage)**
MATERIAL	Number of items			Number of items			Number of items		
Plastic**	879	182	20.71%	741	142	19.16%	1503	277	18.43%
multimaterial	42	6	14.29%	28	3	10.71%	70	2	2.86%
Glass	17	12	70.59%	3	2	66.67%	10	5	50.00%
Metal	4	3	75.00%	3	2	66.67%	17	10	58.82%
Other (Clothes . . . .)	1	0	0.00%	2	2	100.00%	1	0	0.00%
Total items	943	203	21.53%	768	151	19.66%	1599	294	18.39%
	Density (items·/m^2)			Density (items·/m^2)			Density (items·/m^2)		
Total density	0.63	0.13	20.63%	0.51	0.1	19.61%	1.07	0.2	18.69%

Note: *Percentage of litter identification compared to that derived using the standard terrain assessment. **including EPS, rubber, foam.

Table 2. Results from a manual standard survey of the studied beach, compared with the one obtained from ortho-mosaic (dimensional classification).

Date	12 April2019			13 July 2019			17 January 2020		
	Standard Survey (Number of Items)	**UAV Results (Number of Items)**	**UAV vs.* Standard(in Percentage)**	**Standard Survey**	**UAV Results (Number of Items)**	**UAV vs.* Standard(in Percentage)**	**Standard Survey**	**UAV Results (Number of Items)**	**UAV vs.* Standard(in Percentage)**
Small (2.5–15 cm)	859	124	14.44%	716	103	14.39%	1526	240	15.73%
Medium (15–50 cm)	67	64	95.52%	49	46	93.88%	60	45	75.00%
Large (>50 cm)	17	15	88.24%	3	2	66.67%	13	9	69.23%
Total	943	203	21.53%	768	151	19.66%	1599	294	18.39%

Notes: *Percentage of litter identification compared to that derived using the standard terrain assessment.

In a 1500 m^2 transect on 12 April 2019, we detected a total of 943 objects through a standard manual census, while the visual screening of UAV orthophotos, for the same area, reported 203 objects. Thus, the percentage of litter identification, from images at an altitude of 6 m, compared to that derived using the standard terrain assessment, was about 21%. At the end of the first monitoring period, on 13 July 2019, 768 items with manual census and 151 items from UAV ortho-photos screening were counted. Therefore, the probability of litter identification compared to the standard manual census, was about 20%. At the end of the third monitoring period, on 17 January 2020, 1599 items with manual census and 294 items from UAV ortho-photos screening were counted—which gave an 18% probability of litter identification, compared to the standard manual census. From Table 2 we can see that the main differences between the standard manual and the UAV concern the small objects, this is not surprising and it agrees with what was observed in [43]. We must point out that the objects belonging to our "small" category had dimensions between 2.5–15 cm, in line with the OSPAR prescription. Although through the use of UAV and the other settings adopted, we would be able to detect even smaller objects (up to 1 cm of average linear dimension), during data extraction using the WM software we only counted BML included in the size of the "small" category, to be able to make the comparison with data extracted from the manual census.

In Table 1, the major differences between UAVs and the standards results were mainly found for the plastics and the multimaterial categories. This is because most plastics are small objects and multimaterials are inherently difficult to identify, as compared to other well-classified BML.

4. Discussion

The visualization of the orthographic maps allowed the study of the two-dimensional distribution of the accumulations (Figures 8 and 9), which was not feasible with the standard monitoring approach (manual collection and classification of the objects). The information we got in this way confirmed some previous results, such as the fact that BML accumulated prevalently in the dune zone [41,45], and was a cause for concern as, in this area, there were semi-permanent structures (trunks, clumps of plants, etc.) that contributed to retaining anthropogenic debris in the long-term, by hindering their return to the sea, even in conditions of heavy storms. This favored their photo-degradation, which was faster on the rather than the sea [8] and led to the consequent formation of meso and micro-plastics, [9]. However, with our monitoring approach we were also able to evaluate the dynamics of the accumulation process, and its dependence on the object sizes. In fact, Figure 10 shows that in the first ten days (the first time range) there was a fast growth of the number of objects of any sizes. This was understandable, because we started from a cleaned beach. Then, during the second time range (from 10 to about 60 days), the dynamics was more influenced by the size of the objects (Figure 10b and especially Figure 10f). Probably, the great variability that characterized this time interval was due to the different dynamic equilibrium times, between deposition and removal, for objects of different size classes. From the third month, the general trend was a regular growth with a dynamic that decreased a lot, for all size classes, towards the equilibrium between deposition and removal, possibly disturbed by the occurrence of sporadic events such as, in our case, the flooding of the river (Figure 10g,h). At the end of this last period, it was not unusual to see a large prevalence of small objects: this agreed with the results of previous manual surveys, which considered the size classes, carried out in the same area [53,56]; more generally this agrees with the fact that the number of macro-AMDs was higher when the size was smaller, both at sea and on the coast. [55,57].

The predominance of smaller objects led to some issues concerning the counting method using an aerial survey. In fact, despite the expected accuracy in counting beach objects (85%–100%, Figure 7) with UAV, for small objects, the figure obtained when compared to the standard (manual) counting, was quite different—about 15% (Section 3.3, Tables 1 and 2). This result agreed with that of Martin et al. [43]; in particular that result shown in Table 2 They pointed out that "smaller items < 4 cm in average linear dimension, were those that were not mainly recorded through aerial surveying (small fragments, bottle caps, plastic rings, etc.)". It is interesting to note that litter density estimates by

Martin et al. [43] through aerial survey and manual image processing (0.27 items/m2) was in line, even if slightly higher, with what we observed (Table 1). However, we must point out that their surveys were carried out in a very different place from ours, so this agreement between the two measures could only be fortuitous. As far as counts performed manually using the SeaCleaner protocol are concerned, the numerical density values (items/m^2) found (0.51–1.07, see Table 1) agreed well with those found with the same protocol during previous surveys, in the same geographical area [53,56].

In any case, our work supported the conclusion of others [42,43], regarding the difficulty in identifying the smallest objects through UAV. The reasons for this gap between manual and UAV counting of small objects could be the following:

1. Hidden BML (for example under trunks or other objects) could be easily identified by human inspection, while this was almost impossible for UAV;
2. Almost completely buried BML could be extracted by humans from sand, identified and then counted, while UAVs cannot do the same, obviously;
3. Some transparent BML, especially fragments of plastic bags and thin films, are often not detected by the UAV camera;
4. Small BMLs can be overestimated by manual counting. In fact, while for the protocol suggested by the OSPAR macro-waste guidelines (to which our specific SeaCleaner local monitoring protocol refers), objects smaller than 2.5 cm should not be considered [15,52,53,55], however, during manual counting such small objects were often counted equally, contrary to the recognition and counting by orthophotos.

In the course of this work, we have noted the importance of taking into account the occurrence of extreme events, as evidenced by the "burst" in Figures 10 and 11, where the increase in the flow river heavily affected the transport of the solid bodies by the rivers, and so influenced the accumulation of BML [20]. It is interesting to note that, after this extreme event, the situation reverted back to that similar to previous ones (Figures 10–12). The decrease in detected BML after the flood event was due to the successive swells, i.e., the dynamics of the waves, which, even in "standard" conditions, not only tend to accumulate, but also remove objects from the shoreline, tending towards a situation of equilibrium. As a result of this, in normal conditions most of the waste is found in the dune area, where the presence of trunks and plant material tend to prevent it from returning to the sea (Figure 12a). During exceptional events, such as that of 15 November 2019 (Figure 12b), the enormous amount of material transported and washed up on the beach produced situations such as that in Figure 12c, i.e., a significant increase in the number of waste, evenly distributed on the beach (Figure 10e). The waves (and even more so the strong swells) that act in the period following this event, partly removed this material, bringing the beach back to a situation similar to the previous ones with regards to the number and density of BML, with the trash mainly accumulated in the dune area (Figures 9g and 11). In our particular case, the heavy swells following the flood also contributed to a change in the conformation of the beach, removing a part of the sand from the ordinary berm zone (the upper part of the beach, with depositional features due to the accumulation of sediment caused by the waves, Figure 12d). Moreover, the removal time of the plant material accumulated in the dune area was quite long—it was in fact present, in a higher than normal quantity, even during our last monitoring, which was 32 days after the occurrence of the exceptional event when the beach had recovered its standard conformation. This had probably caused a greater difficulty in detecting the BML with aerial devices, with even relatively large BML sometimes being hidden from aerial detection (see Point 1 above), as observed in our latest census that showed a worse agreement between UAV and standard manual survey, also for "large" and, especially, for "medium" objects, with respect to previous dates. A part of this small difference, the percentage of BML identification with UAV compared to that derived using the standard terrain assessment was very similar for all three dates reported in Tables 1 and 2, ranging from 18.39% to 21.53% for total item number, and from 18.69% to 20.63% for total density of items, indicating that the bias described above (points 1, 2, and 3) always affect in the same way. However, our study was limited to a small stretch of beach (100 meters long and 15 meters deep), while extreme events also affected the shape of all coastlines of SRPRK. Therefore, an effective understanding of the consequences of these strong events should imply the study of a large portion of the coastline, also taking into account the erosion phenomena.

Even if not quantitatively reported in this work, we have observed an effective role of the wind influencing the special distribution of BML. This applies, in particular, to expanded polystyrene (EPS), which is easily fragmented and, because of its very low density, it was conveyed by the wind more than by the sea. Not surprisingly, from Figure 10 and also Tables 1 and 2, we noted that the amount of BML accumulated during spring and summer was smaller than that during late autumn and in the winter, which were characterized by stronger winds and larger swells.

The main task of the present work was to test a possible methodology for studying the coastal dynamics of waste accumulation through aerial survey devices. Thus, our attention was more focused on the study of accumulation dynamics in the coastal area rather than on the precision and detailed cataloguing of the found objects, as done by other works [43,44]. However, we realized that the high precision in object recognition could lead to a high accuracy even for its size (surface area) estimation, a priority target for understanding the dynamic behavior of different classes of AMDs on the beaches (Figure 10). Actually, the aerial survey size evaluation was intrinsically influenced by some errors that led to an underestimation of the values. For example, as discussed in Point 2 of the list above, and due to the fact that the area of a "flat" BML positioned in an almost vertical direction was highly underestimated by UAVs, because of the almost fixed direction of view. The estimation of this bias required a dedicated study.

Aerial surveys save time, compared to standard manual approaches to the BML study, even after accounting for the time needed, in addition to mere monitoring, for image processing, labeling, and imaging. In fact, covering even larger areas requires the work of only one person, and data extraction by orthophotos requires no more than 2 hours of visual census, for each monitoring campaign. Moreover, geo-referenced images can provide useful information that standard counting cannot provide. The study of accumulation dynamics, like ours, requires two/three monthly monitoring campaigns, repeated for 3/4 months. Many people should be involved in standard manual procedures. This is probably the reason why data on the accumulation rate of BML are scarce in the literature. The use of UAVs can, in our opinion, help to fill this gap. In addition to the pure scientific aspect of the phenomenon, the knowledge of the accumulation behavior, possibly in different areas of the coast, was a useful information for marine parks/protected areas (MPAs). Presently UAV technology, compared to some years ago, is sufficiently low-cost and it can be foreseen that parks and administrations of MPAs are equipped with such devices, with the patents/permits related to their use. The knowledge of the time scale for the formation of the maximum stock of BML after the beach cleaning operation, could help the MPAs and the local authorities to optimize the planning of the cleaning campaigns, minimize the effort, and maximize the result, thus, preventing the degradation and fragmentation of most of the material and the production of microplastics.

Figure 12. (**a**) The beach before the sea storm of 20/11/2019 (normal conditions). (**b**) The beach during the storm. (**c**) The beach 3 days after the storm. (**d**) The beach 10 days after the storm. At the time of the last monitoring (17 January 20120, i.e., 32 days after the storm) the beach was returned almost to its normal condition and conformation.

5. Conclusions and Further Improvement

The UAVs flying at a low altitude provided high resolution data, which was useful in detecting plastic, metal, and other type of beached objects. Moreover, UAV allowed the repeatability of the surveys in a short time, which was essential for the study of accumulation dynamics. Other important advantages were the reduced anthropogenic impact (just one person for the survey) and the possibility of obtaining 3D and 2D characterization of the monitored areas. Our pilot test to use UAVs for monitoring spatial and temporal dynamic of BML accumulation in coastal areas that started in April 2019, proved to be a useful procedure. To our knowledge, this was the first case of using the aerial survey methodology through UAVs to monitor the presence of BML on Italian beaches (Figure 7 of [40]) and the first case of using UAVs to estimate the accumulation rates of BML on beaches in general. The results of our work showed that the dynamics and the equilibrium of the accumulation process depended, in general, on season, but also on the size of the specific BML. Moreover, extreme events lead to strong fluctuations, but the normal situation was quickly restored, for both the dynamics and the equilibrium features. Observational evidence of this phenomenon, as well as that visible in the peak accumulation of BML recorded by us (Figures 9–11), are available in our short film "Before&3-10dayafter.avi" (Supplementary Material), which shows the situation of the beach before the flood, three days after this extreme event and, finally, ten days after it.

The comparison between the UAV and standard manual counting (made according to the SeaCleaner protocol [52,53], which met the MFSD survey procedures [16] and the OSPAR guideline for size and type classification of BML [15]) showed a good agreement for "medium" and "large" size objects (~67%–95%), while this was not the case for the "small" ones (~15%). The possible causes of this discrepancy are analysed in the Discussion section.

Our study focused on the distribution of BML in the different areas of a stretch of beach, from the coastline to the dunes. In a next work, we would like to analyze, with the same techniques, the dynamics of accumulation on several stretches of beach of the entire coastline of the park. In this way, we could provide information on the importance of both, the distance from the mouths of rivers (Arno and Serchio) and the presence of possible obstacles in the process of accumulation of debris. To this end, we are currently testing different types of UAVs and increasing the flight height (requesting the necessary clearance in advance). The objective was to increase the size of the scanned area for each individual flight, losing a little of the detail, in the process. Ideally, we should do a lower resolution scan on the whole coastal area of the park, and do a detailed analysis, like the one in this paper, on a few randomly sampled areas.

As shown in Tables 1 and 2, besides the fact that UAV census counts are generally underestimated for all categories of objects, it seemed that the recognition of some particular types was more difficult and in our case was the "multimaterial". As already pointed out in the Discussion section, although the recognition of the object typology through UAV did not play a central role in the present work, its importance in improving the effectiveness of the monitoring system (estimation of the count and the surface/volume of the objects) had emerged. Therefore, we are trying to apply automatic systems based on machine learning, as already tested by other BML recognition studies [40,41,43,44]. In order to carry out a more accurate study of how coastal dynamics affect the accumulation of BML, it would also be necessary to cross-reference the data obtained from AMD monitoring campaigns with those related to sea and wind weather conditions, while accounting for extreme events, as far as possible.

The main difficulties encountered in this type of monitoring are, in our opinion and personal experience, those common in autumn and winter due to adverse weather conditions; windy or rainy days or heavy swells prevent the surveys from being carried out (see Figure 12) and might, therefore, compromise the planned surveys (as for our last late survey of the summer–autumn period). In particular, the flight operation must be suspended if it rains, the wind exceeds 10 m/sec and the temperature exceeds the range of 0–40 °C. Operating near the shoreline, the force of the sea can produce aerosol in the area–low aerosol levels can be managed with a simple cleaning, while high values (present with rough seas and strong wind conditions) can damage the UAV and require extraordinary assistance. Take-off and landing can raise the sand, which can damage the moving parts; it is therefore, important to use an appropriate drone landing pad. Even sub-optimal light conditions (Figure 7) and ground shading can affect the results–during the scan it is preferable to have a constant brightness, therefore, the fast passage of clouds could produce variations of light that needs to be corrected in post-processing. Another operational limit is the flight autonomy of the UAV, especially if one wants to analyze large areas; in this case the APP Drone Harmony was very useful, because it managed the interruption of the flight, the replacement of the battery, and the resumption of the flight starting from the last position.

To date, the use of UAVs for BML monitoring had only just begun, and we think this work could help to highlight its great potential. It is, therefore, foreseeable that UAVs would be widely used in the future and would allow us to considerably increase the knowledge of the dynamics of accumulation of BML on beaches, especially in coastal areas with difficult access and MPAs. The understanding of the characteristics of this process and the possibility for acquiring a large amount of data, even in real-time, combined with the relatively modest costs of these methodologies, would help allow, through integrated programs, the different stakeholders involved in mitigation actions, citizencience, and research activities, to work together to optimize beach cleaning actions.

Supplementary Materials: The following are available online at http://www.mdpi.com/2072-4292/12/8/1260/s1, Figure S1: The figure summarizes the potential of the three-dimensional orthophoto reconstruction. In (**a**) a relatively large section of the beach is shown, with the largest BML in evidence, while in (**b**) a smaller area of the selected beach section has been enlarged. The resolution of this method is high enough to identify even small objects in a large area.

Author Contributions: Contribution by different authors can be described as follows: Conceptualization S.M. and M.P.; methodology: S.M., M.P. and L.M.; software: M.P. and L.M.; validation: S.M., M.P. and L.M.; formal analysis: M.P. and L.M.; investigation: S.M. and M.P.; resources: A.B. and M.P.; data curation: M.P.; writing—original draft preparation: S.M.; writing—review and editing: S.M. and L.M.; visualization: M.P. and L.M.; supervision: A.B. and L.M.; All authors have read and agreed to the published version of the manuscript.

Funding: This research received no external funding.

Acknowledgments: We would like to thank the environmental associations "Acchiapparifiuti" and "Legambiente" (www.legambiente.it), and the working group Win on Waste (WOW—interdepartmental group of the National Research Council) for the support given to us during the manual operations of cleaning and classification of waste, as well as for the outreach activities carried out in order to further spread awareness of the problem of Marine Litter, with the production of video and educational material (UAVSanRossoreENG.m4v, Supplementary Material).

Conflicts of Interest: The authors declare no conflict of interest.

References

1. Cózar, A.; Echevarría, F.; González-Gordillo, J.I.; Irigoien, X.; Úbeda, B.; Hernández-León, S.; Palma, A.T.; Navarro, S.; Garcia-de-Lomas, J.; Ruiz, A.; et al. Plastic debris in the open ocean. *Proc. Natl. Acad. Sci. USA* **2014**, *111*, 10239–10244. [CrossRef] [PubMed]
2. Cannizzaro, L.; Garofalo, G.; Giusto, G.; Rizzo, P.; Levi, D. Qualitative and quantitative estimate of solid waste in the channel of Sicily. In Proceedings of the Second International Conforence on the Mediterranean Coastal Environment, MED-COAST, Tarragona, Spain, 24–27 October 1995.
3. Galgani, F.; Jaunet, S.; Campillo, A.; Guenegen, X.; His, E. Distribution and abundance of debris on the continental shelf of the north-western Mediterranean Sea. *Mar. Pollut. Bull.* **1995**, *30*, 713–717. [CrossRef]
4. Cózar, A.; Sanz-Martín, M.; Martí, E.; González-Gordillo, J.I.; Ubeda, B.; Gálvez, J.Á.; Irigoien, X.; Duarte, C.M. Plastic accumulation in the mediterranean sea. *PLoS ONE* **2015**, *10*, 0121762. [CrossRef] [PubMed]
5. Alomar, C.; Estarellas, F.; Deudero, S. Microplastics in the Mediterranean Sea: Deposition in coastal shallow sediments, spatial variation and preferential grain size. *Mar. Environ. Res.* **2016**, *115*, 1–10. [CrossRef]
6. Suaria, G.; Avio, C.G.; Mineo, A.; Lattin, G.L.; Magaldi, M.G.; Belmonte, G.; Moore, C.J.; Regoli, F.; Aliani, S. The Mediterranean plastic soup: Synthetic polymers in Mediterranean surface waters. *Sci. Rep.* **2016**, *6*, 37551. [CrossRef]
7. MedSeaLitter EU Project. Available online: https://medsealitter.interreg-med.eu (accessed on 2 February 2020).
8. Andrady, A.L. Microplastics in the marine environment. *Mar. Pollut. Bull.* **2011**, *62*, 1596–1605. [CrossRef]
9. Andrady, A.L. The plastic in microplastics: A review. *Mar. Pollut. Bull.* **2017**, *119*, 12–22. [CrossRef]
10. Saliu, F.; Montano, S.; Garavaglia, M.G.; Lasagni, M.; Seveso, D.; Galli, P. Microplastic and charred microplastic in the Faafu Atoll, Maldives. *Mar. Pollut. Bull.* **2018**, *136*, 464–471. [CrossRef]
11. Saliu, F.; Montano, S.; Leoni, B.; Lasagni, M.; Galli, P. Microplastics as a threat to coral reef environments: Detection of phthalate esters in neuston and scleractinian corals from the Faafu Atoll, Maldives. *Mar. Pollut. Bull.* **2019**, *142*, 234–241. [CrossRef]
12. Olivelli, A.; Hardesty, D.; Wilcox, C. Coastal margins and backshores represent a major sink for marine debris: Insights from a continental-scale analysis. *Environ. Res. Lett.* **2020**, in press. [CrossRef]
13. Lebreton, L.C.M.; van der Zwet, J.; Damsteeg, J.-W.; Slat, B.; Andrady, A.; Reisser, J. River plastic emissions to the world's oceans. *Nat. Commun.* **2017**. [CrossRef] [PubMed]
14. Pierdomenico, M.; Casalbore, D.; Chiocci, F.L. Massive benthic litter funnelled to deep sea by flash- flood generated hyperpycnal flows. *Sci. Rep.* **2019**, *9*, 5330. [CrossRef] [PubMed]
15. OSPAR Commission. *Guideline for Monitoring Marine Litter on the Beaches in the OSPAR Maritime Area*; OSPAR Commission: London, UK, 2010.
16. Galgani, F.; Hanke, G.; Werner, S.; Oosterbaan, L.; Nilsson, P.; Fleet, D.; Kinsey, S.; Thompson, R.C.; van Franeker, J.; Vlachogianni, T.; et al. *Guidance on Monitoring of Marine Litter in European Seas*; Publications Office of the European Union: Brussels, Belgium, 2013. [CrossRef]
17. UN Environment. *Combating Marine Plastic Litter and Microplastics: An Assessment of the Effectiveness of Relevant International, Regional and Subregional Governance Strategies and Approaches*; UN Environment: Nairobi, Kenya, 2017.
18. GESAMP. *Guidelines on the Monitoring and Assessment of Plastic Litter and Microplastics in the Ocean*; Kershaw, P.J., Turra, A., Galgani, F., Eds.; (IMO/FAO/ UNESCO IOC/UNIDO/WMO/IAEA/UN/UNEP/UNDP/ISA Joint Group of Experts on the Scientic Aspects of Marine Environmental Protection). Rep. Stud. GESAMP No. 99 130p; GESAMP: London, UK, 2019.
19. Ryan, P.G.; Turra, A. *Guidelines for the Monitoring & Assessment of Plastic Litter in the Ocean Reports & Studies 99*; Kershaw, P.J., Turra, A., Galgani, F., Eds.; GESAMP: London, UK, 2019.
20. van Emmerik, T.; Schwarz, A. Plastic debris in rivers. *Wires Water* **2020**, *7*. [CrossRef]

21. Yao, H.; Qin, R.; Chen, X. Unmanned Aerial Vehicle for Remote Sensing Applications—A Review. *Remote Sens.* **2019**, *11*, 1443. [CrossRef]
22. Topouzelis, K.; Papakonstantinou, A.; Garaba, S.P. Detection of floating plastics from satellite and unmanned aerial systems (Plastic Litter Project 2018). *Int. J. Appl. Earth Obs. Geoinf.* **2019**, *79*, 175–183. [CrossRef]
23. Capolupo, A.; Pindozzi, S.; Okello, C.; Fiorentino, N.; Boccia, L. Photogrammetry for environmental monitoring: The use of drones and hydrological models for detection of soil contaminated by copper. *Sci. Total Environ.* **2015**, *514*, 298–306. [CrossRef]
24. Gonçalves, G.R.; Pérez, J.A.; Duarte, J. Accuracy and effectiveness of low cost UASs and open source photogrammetric software for foredunes mapping. *Int. J. Remote Sens.* **2018**, *39*, 5059–5077. [CrossRef]
25. Holman, R.A.; Holland, K.T.; Lalejini, D.M.; Spansel, S.D. Surf zone characterization from unmanned aerial vehicle imagery. *Ocean Dyn.* **2011**, *61*, 1927–1935. [CrossRef]
26. Manfreda, S.; Mccabe Matthew, F.; Miller Pauline, E.; Lucas, R.; Pajuelo, V.M.; Mallinis, G.; Dor EBen Helman David Estes, L.; Ciraolo, G.; Müllerová, J.; Tauro, F.; et al. Use of unmanned aerial systems for environmental monitoring. *Remote Sens.* **2018**, *10*, 641. [CrossRef]
27. Turner, I.L.; Harley, M.D.; Drummond, C.D. UAVs for coastal surveying. *Coast. Eng.* **2016**, *114*, 19–24. [CrossRef]
28. Pádua, L.; Vanko, J.; HruÊka, J.; Adão, T.; Sousa, J.J.; Peres, E.; Morais, R. UAS, sensors, and data processing in agroforestry: A review towards practical applications. *Int. J. Remote Sens.* **2017**, *38*, 2349–2391. [CrossRef]
29. Sankey, T.; Donager, J.; McVay, J.; Sankey, J.B. UAV lidar and hyperspectral fusion for forest monitoring in the southwestern USA. *Remote Sens. Environ.* **2017**, *195*, 30–43. [CrossRef]
30. Torresan, C.; Berton, A.; Carotenuto, F.; Di Gennaro, S.F.; Gioli, B.; Matese, A.; Miglietta, F.; Vagnoli, C.; Zaldei, A.; Wallace, L. Forestry applications of UAVs in Europe: A re- view. *Int. J. Remote Sens.* **2017**, *38*, 2427–2447. [CrossRef]
31. Baron, J.; Hill, D.J.; Elmiligi, H. Combining image processing and machine learning to identify invasive plants in high-resolution images. *Int. J. Remote Sens.* **2018**, *39*, 5099–5118. [CrossRef]
32. Bonali, F.L.; Tibaldi, A.; Marchese, F.; Fallati, L.; Russo, E.; Corselli, C.; Savini, A. UAV-based surveying in volcano-tectonics: An example from the Iceland rift. *J. Struct. Geol.* **2019**. [CrossRef]
33. Mlambo, R.; Woodhouse, I.H.; Gerard, F.; Anderson, K. Structure from motion (SfM) photogrammetry with drone data: A low cost method for monitoring greenhouse gas emissions from forests in developing countries. *Forests* **2017**, *8*, 68. [CrossRef]
34. Pérez-Alvárez, J.A.; Gonçalves, G.R.; Cerrillo-Cuenca, E. A protocol for mapping ar- chaeological sites through aerial 4k videos. *Digit. Appl. Archaeol. Cult. Herit.* **2019**, 13. [CrossRef]
35. Pérez, J.A.; Gonçalves, G.R.; Rangel, J.M.G.; Ortega, P.F. Accuracy and effectiveness of orthophotos obtained from low cost UASs video imagery for traf!c accident scenes documentation. *Adv. Eng. Softw.* **2019**, *132*, 47–54. [CrossRef]
36. Ventura, D.; Bonifazi, A.; Gravina, M.F.; Belluscio, A.; Ardizzone, G. Mapping and classification of ecologically sensitive marine habitats using unmanned aerial vehicle (UAV) imagery and Object-Based Image Analysis (OBIA). *Remote Sens.* **2018**, *10*, 1331. [CrossRef]
37. Colefax, A.P.; Butcher, P.A.; Kelaher, B.P. The Potential for Unmanned Aerial Vehicles (UAVs) to Conduct Marine Fauna Surveys in Place of Manned Aircraft. *Ices, J. Mar. Sci.* **2018**, *75*, 1–8. [CrossRef]
38. Kiszka, J.J.; Mourier, J.; Gastrich, K.; Heithau, M.R. Using unmanned aerial vehicles (UAVs) to investigate shark and ray densities in a shallow coral lagoon. *Mar. Ecol. Prog. Ser.* **2016**, *560*, 237–242. [CrossRef]
39. Levy, J.; Hunter, C.; Lukacazyk, T.; Franklin, E.C. Assessing the spatial distribution of coral bleaching using small unmanned aerial systems. *Coral Reefs* **2018**, *37*, 373–387. [CrossRef]
40. Fallati, L.; Polidori, A.; Salvatore, C.; Saponari, L.; Savini, A.; Galli, P. Anthropogenic Marine Debris assessment with Unmanned Aerial Vehicle imagery and deeplearning: A case study along the beaches of the Republic of Maldives. *Sci. Total Environ.* **2019**, *693*. [CrossRef]
41. Bao, Z.; Sodango, T.H.; Shifaw, E.; Li, X.; Sha, J. Monitoring of beach litter by automatic interpretation of unmanned aerial vehicle images using the segmentation threshold method. *Mar. Pollut. Bull.* **2018**, *137*, 388–398. [CrossRef] [PubMed]
42. Deidun, A.; Gauci, A.; Lagorio, S.; Galgani, F. Optimising beached litter monitoring protocols through aerial imagery. *Mar. Poll. Bull.* **2018**, *131*, 212–217. [CrossRef] [PubMed]

43. Martin, C.; Parkes, S.; Zhang, Q.; Zhan XMcCabe, M.F.; Duarte, C.M. Use of unmanned aerial vehicles for efcient beach litter monitoring. *Mar. Poll. Bull.* **2018**, *131*, 662–673. [CrossRef]
44. Gonçalves, G.; Andriolo, U.; Pinto, L.; Bessa, F. Mapping marine litter using UAS on a beach-dune system: A multidisciplinary approach. *Sci. Total Environ.* **2019**, in press. [CrossRef]
45. Geraeds, M.; van Emmerik, T.; de Vries, R.; bin Ab Razak, M.S. Riverine plastic litter monitoring using unmanned aerial vehicles (UAVs). *Remote Sens.* **2019**, *11*, 2045. [CrossRef]
46. Thornton, L.; Jackson, N.L. Spatial and temporal variations in debris accumulation and composition on an estuarine shoreline, Cliffwood Beach, New Jersey, USA. *Mar. Pollut. Bull.* **1998**, *36*, 705–711. [CrossRef]
47. Bowman, D.; Manor-Samsonov, N.; Golik, A. Dynamics of litter pollution on Israeli Mediterranean beaches: A budgetary, litter flux approach. *J. Coast. Res.* **1998**, *14*, 418–432.
48. QGIS Development Team (2020). QGIS Geographic Information System. Open Source Geospatial Foundation Project. Available online: http://qgis.osgeo.org (accessed on 30 March 2020).
49. OpenStreetMap contributors, licensed under the Creative Commons Attribution-ShareAlike 2.0 license (CC BY-SA). Available online: https://www.openstreetmap.org/copyright/en (accessed on 30 March 2020).
50. Regione Toscana. Geoscopio WMS, licensed under the Creative Commons (CC). Available online: https://dronedj.com/2020/01/08/dji-phantom-4-pro-v2-0-is-finally-back/ (accessed on 30 March 2020).
51. Merlino, S.; Locritani, M.; Stroobant, M.; Mioni, E.; Tosi, D. SeaCleaner: Focusing citizen science and environment education on unraveling the marine litter problem. *Mar. Technol. Soc. J.* **2015**, *49*, 99–118. [CrossRef]
52. Giovacchini, A.; Merlino, S.; Locritani, M.; Stroobant, M. Spatial distribution of marine litter along italian coastal areas in the Pelagos sanctuary (Ligurian Sea—NW Mediterranean Sea): A focus on natural and urban beaches. *Mar. Poll. Bull.* **2018**, *130*, 140–152. [CrossRef] [PubMed]
53. DRONE Harmony: flight planner and data capture tool, designed to collect drone data (photos or video) for a variety of applications. Available online: https://droneharmony.com (accessed on 30 March 2020).
54. Merlino, S.; Abbate, M.; Pietrelli, L.; Canepa, P.; Varella, P. Marine litter detection and correlation with the seabird nest content. *Rend. Lincei Sci. Fis. Nat.* **2018**, *29*, 867–875. [CrossRef]
55. Tuscany region. Regional Hydrological and Geological Sector Archive. Available online: http://www.sir.toscana.it/ (accessed on 30 March 2020).
56. Vlachogianni, T.; Anastasopoulou, A.; Fortibuoni, T.; Ronchi, F.; Zeri, C. *Marine Litter Assessment in the Adriatic and Ionian Seas*; IPA-Adriatic DeFishGear Project: Athens, Greece, 2017; p. 168, ISBN 978-960-6793-25-7.
57. Suaria, G.; Aliani, S. Floating debris in the Mediterranean Sea. *Mar. Pollut. Bull.* **2014**, *86*, 494–504. [CrossRef]

Article

Oil Spill Detection in Quad-Polarimetric SAR Images Using an Advanced Convolutional Neural Network Based on SuperPixel Model

Jin Zhang [1], Hao Feng [1,2], Qingli Luo [1,2,*], Yu Li [3], Jujie Wei [4] and Jian Li [1,2]

1 State Key Laboratory of Precision Measuring Technology and Instruments, Tianjin University, Tianjin 300072, China; zhangjin2018@tju.edu.cn (J.Z.); fenghao@tju.edu.cn (H.F.); tjupipe@tju.edu.cn (J.L.)

2 Binhai International Advanced Structural Integrity Research Centre, Tianjin 300072, China

3 Faculty of Information Technology, Beijing University of Technology, No. 100 PingLeYuan, Chaoyang District, Beijing 100124, China; yuli@bjut.edu.cn

4 Institute of Photogrammetry and Remote Sensing, Chinese Academy of Surveying and Mapping, Beijing 100036, China; weijj@casm.ac.cn

* Correspondence: luoqingli@tju.edu.cn

Received: 9 February 2020; Accepted: 12 March 2020; Published: 14 March 2020

Abstract: Oil spill detection plays an important role in marine environment protection. Quad-polarimetric Synthetic Aperture Radar (SAR) has been proved to have great potential for this task, and different SAR polarimetric features have the advantages to recognize oil spill areas from other look-alikes. In this paper we proposed an oil spill detection method based on convolutional neural network (CNN) and Simple Linear Iterative Clustering (SLIC) superpixel. Experiments were conducted on three Single Look Complex (SLC) quad-polarimetric SAR images obtained by Radarsat-2 and Spaceborne Imaging Radar-C/X-Band Synthetic Aperture Radar (SIR-C/X-SAR). Several groups of polarized parameters, including H/A/Alpha decomposition, Single-Bounce Eigenvalue Relative Difference (SERD), correlation coefficients, conformity coefficients, Freeman 3-component decomposition, Yamaguchi 4-component decomposition were extracted as feature sets. Among all considered polarimetric features, Yamaguchi parameters achieved the highest performance with total Mean Intersection over Union (MIoU) of 90.5%. It is proved that the SLIC superpixel method significantly improved the oil spill classification accuracy on all the polarimetric feature sets. The classification accuracy of all kinds of targets types were improved, and the largest increase on mean MIoU of all features sets was on emulsions by 21.9%.

Keywords: oil spill; Synthetic Aperture Radar; polarimetric decomposition; superpixel; convolutional neural networks

1. Introduction

Marine environment plays a crucial part in global ecosystems. Oil spill is one of the main marine pollution, which will cause serious damage to ocean ecology and resources. In 2010, the accident of the Gulf of Mexico oil spill lasted for about three months. Beaches and wetlands in many states of the United States were destroyed and local marine organism was devastated [1]. Thus it is necessary to monitor sea surface and detect oil spill. Remote sensing plays a crucial role in achieving this goal, and relevant methods have been effectively applied to oil spill detection.

Space-borne Synthetic Aperture Radar (SAR) is widely applied for oil spill detection due to its all-weather and all-time ability and wide area coverage. Full polarization SAR data provides four channels according to the transmit and receive mode of radar signal and they are HH, HV, VH and VV channels. The clean sea can be regarded as a rough surface, while the smooth oil spill layer usually floats on the water surface, existing as dark spots since it dampens capillary waves, short gravity waves and Bragg scattering [2]. The general steps for oil spill detection are divided into:(1) spot extraction, (2) feature extraction, (3) classification [3]. Early researches mainly focused on textual information of dark spots area. Several textual features including first invariant planar moment and power-to-mean ratio were extracted from SAR data, supplemented by statistical model or machine learning, to perform oil spill detection [4,5]. Some experiments were carried out to perform oil spill detection on different band SAR images [6,7]. Yongcun Cheng et al. used VV channel data acquired by COSMO-SkyMed to monitor oil spill and simulate a model [8]. M. Migliaccio et al. proposed a multi-frequency polarimetric SAR processing chain to detect oil spill in the Gulf of Mexico, and has been applied successfully [1]. These methods could successfully distinguish oil spill area from sea surface, and they are known as mature classification algorithms.

However, several environmental factors including low-speed winds, internal wave and biogenic films also appear as dark spots in SAR images [9], and they are called look-alikes. The most challenging thing of oil spill detection from SAR images is to distinguish the oil spill area from these look-alikes. That is the main obstacle of the early researches of texture analysis focused on single pol-SAR data. Oil spill area may experience complex deformation on the sea surface, which is easy to be confused with look-alikes. Moreover, a large amount of data are required for texture analysis. These problems become major hindrances for high-accuracy oil spill detection. With the new development of SAR satellites in recent years, the research focus of oil spill detection began to incline to dual-pol or quad-pol SAR image, and the derived compact SAR [10], which not only retain texture characteristics of dark spots area, but also provide a lot of polarized information. Polarimetric decomposition essentially reflects the scattering modes of microwave on the sea surface, which highlights the subtle differences of different ocean objects [11]. Many polarized parameters extracted from different SAR channels has been proved to have great ability to perform high accuracy oil spill detection [12–15]. On the perspective of polarized feature, S. Skrunes used k-means classification method to detect oil spill area on several polarized parameters [16]. With the rise of machine learning algorithms in recent years, neural networks have also been applied into oil spill detection. Yu Li et al. performed several comparative experiments between different machine learning classifiers based on multi polarized parameters [17], and the differences between fully and compact polarimetric SAR images [18] were explored.

Meanwhile, as a classical feedforward neural network, convolutional neural network (CNN) is widely used in image classification and recognition. Since it was proposed in 1989, CNN has experienced many improvements and changes, and derived several classic network structures such as Inception, Resnet and Cliquenet [19–21]. Min Lin et al. used the global average pooling (GAP) layer to replace the fully connection layer to reduce the parameter amount in 2014 [22]; Andrew Howard et al. put forward the depthwise separable convolution in MobileNet [23,24], which can maintain high accuracy even when the amount of parameters and calculation is reduced.

In 2015, Jonathan Long et al. proposed a fully convolutional network (FCN) with transposed convolution for image semantic segmentation [25]. The end-to-end operation is implemented with an encoder-decoder structure, and the classification prediction is given for each pixel on the image. The concept of dilated convolution into semantic segmentation in 2016, and it greatly improved classification accuracy [26]. Follow by that, advanced models have been developed for high precision segmentation, and they are Unet, Linknet and Deeplab series [27,28]. The encoder-decoder structure of semantic segmentation model based on CNN has been used in oil spill detection in recent studies [29,30], and achieved high accuracy. With the application of TerraSAR-X and other SAR satellites, dual-polarized SAR images are also introduced into oil spill detection. Daeseong Kim et al. extracted polarized parameters from dual-pol TerraSAR-X images, successfully mapped oil spill area with artificial neural networks [31].

The concept of superpixel is an image segmentation technology proposed in 2003 [32]. It refers to an irregular pixel block with certain visual significance. It is composed of adjacent pixels with similar texture, color brightness and other characteristics. The similarity of features between pixels are used to form a group of pixels, and image are expressed by a small number of superpixels. Superpixel greatly reduces the complexity of image post-processing, and it is used as a pre-processing step for image segmentation algorithm. Simple Linear Iterative Clustering (SLIC) is a widely used superpixel segmentation method [33], and has been introduced into some SAR scenes. Some researchers also use multi-chromatic analysis to perform target detection and analysis on SAR images [34,35].

Many current oil spill detection methods only divided images into oil or non-oil areas, which may cause false alarm and cannot recognize every target on the sea surface. The classification method combining neural networks also could not distinguish oil spill and look-alikes well, while the flexible structure of CNN provides the possibility to solve these problems. It allows a variety of parameters input, can simultaneously take into account the task of dark spots extraction and classification and identify every target on the sea surface. In this paper we proposed an oil spill detection method using SLIC superpixel and semantic segmentation algorithm based on CNN, combining several convolution kernels including dilated and depthwise separable convolution. It allows multiple-parameters input and realize pixel-level oil spill area classification, which finally realize further accuracy improvement. We carried out experiments on five groups of polarized parameters extracted from SLC quad-polarimetric SAR data of Radarsat2 and Spaceborne Imaging Radar-C/X-Band Synthetic Aperture Radar(SIR-C/X-SAR), and evaluated the classification results of superpixel segmentation combined with polarized parameters. The experiments results show that our method could effectively extract and classify dark spots on a SAR image. SLIC superpixel could further improve classification accuracy of oil spill area, and Yamaguchi 4-component decomposition combined with SLIC superpixel classification is considered the most suitable parameters for oil spill detection in our case.

2. Materials and Methods

2.1. Overall Framework

The flowchart of our oil spill detection method is illustrated as Figure 1. The four channels data was processed by Lee refined filter firstly, and different polarized parameters were then extracted from these four channels. All polarized parameters used in our experiments are divided into five groups according to different scattering principles and calculation methods. For monostatic SAR system, reciprocity always holds, which means that the complex scattering coefficient obeys HV = VH. For this reason, HV is considered for cross polarization channel in the analysis. Three channel data should be used to generate image in CIElab color space for SLIC superpixel model. We choose data of HH, HV and VV channel to do it, since co-polarized channels (HH/VV) data contains more polarized information than cross-polarized channels (VH/HV) [36]. The HH, HV and VV data were also used to calculate SLIC superpixels. Sections 2.2 and 2.3 explains the method used to extract polarized parameters and they are both set as the input of neural network models. The neural network is composed of an encoder and a decoder section. The output of the neural network is segmentation results of oil spill detection.

We designed a semantic segmentation model based on CNN as the classifier, as shown in Figure 2. The dims in the diagram represents the number of polarization parameters, since multiple polarization modes are applied in our work, the network parameters will be adjusted according to the parameter of dims. The depthwise separable convolution and dilated convolution was used in several layers in the bottom parts. The subsequent encode task is completed by standard convolution layer. Green blocks represent a skip connection structure similar to residual learning [19], in order to make top layers accessible to the information from bottom layers and help train the network easier. The feature maps extracted by encoder is decoded by progressive transposed convolution layers, and skip connection is also applied to absorb more features. The specific principle and implementation of network are explained in Section 2.4 in detail.

The different polarized parameters groups and superpixel segmentation results will be combined to input into the neural networks for training. The output is the result of oil spill detection. Finally,

Mean Intersection over Union (MIoU) was calculated between segmentation result and annotation images to evaluate the accuracy.

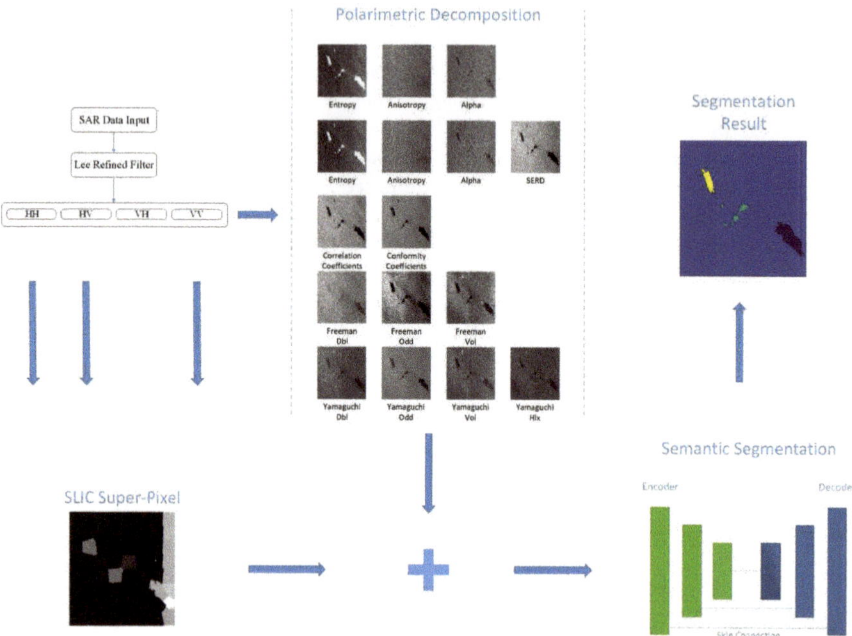

Figure 1. Flow chart of our oil spill detection method.

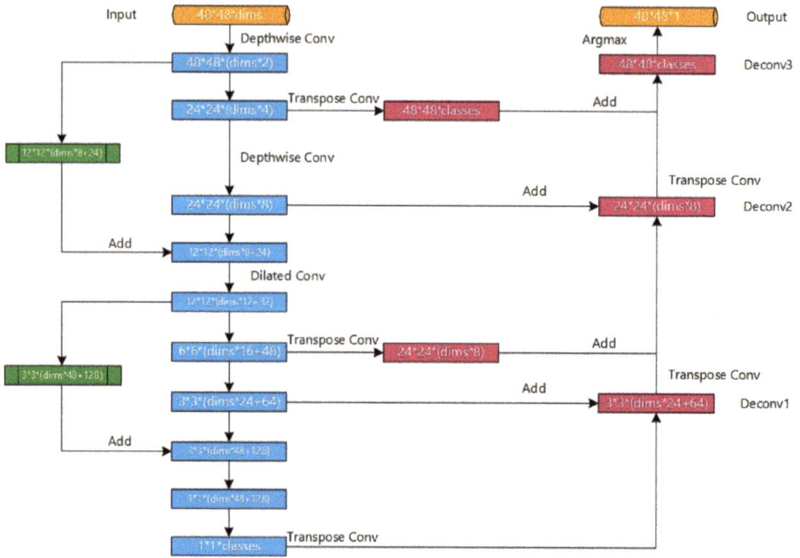

Figure 2. Structure of neural networks in our segmentation method. Blue and green blocks donate encoder parts, which consist of multi convolution layers, here we used depthwise separable convolution, dilated convolution and standard convolution as filter kernel, respectively. Purple-red blocks constitute the decoder part, it outputs a classification map with the same size as original image.

2.2. Polarimetric Decomposition

The whole process of extracting polarimetric parameters is shown in Figure 3. The boxes are different polarized parameter combinations for classification. This section will explain the calculation method of different polarized parameters in the followings.

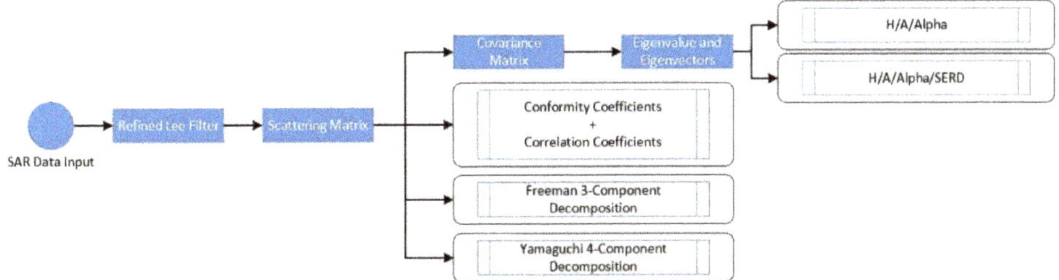

Figure 3. Polarized parameters extraction. We extracted 13 polarized parameters in total. They are divided into five groups as the position of the box in the figure, each group was input into neural network for classification.

2.2.1. H/A/Alpha Decomposition

The scattering matrix of a fully polarimetric SAR image can be expressed as

$$S = \begin{bmatrix} S_{HH} & S_{HV} \\ S_{VH} & S_{VV} \end{bmatrix} = \begin{bmatrix} |S_{HH}|e^{j\varnothing HH} & |S_{HV}|e^{j\varnothing HV} \\ |S_{VH}|e^{j\varnothing VH} & |S_{VV}|e^{j\varnothing VV} \end{bmatrix}, \quad (1)$$

where $|S_{XX}|$ and $\varnothing_{XX}$ represent the amplitudes and phases of the complex scattering coefficients, each complex element donates a polarization component. The two crossed-polarized terms are identical in Radarsat-2, i.e., $S_{HV} = S_{VH}$.

Polarization covariance matrix C and coherency matrix T contain abundant physical information of polarization characteristics of ocean objects. Cloude and Pottier outlined a scheme for parameterizing polarimetric scattering problems based on matrix T in 1997, the covariance matrix can be derived by

$$C_3 = \begin{bmatrix} \langle S_{HH}^2 \rangle & \langle \sqrt{2} S_{HH} S_{HV}^* \rangle & \langle S_{HH} S_{VV}^* \rangle \\ \langle \sqrt{2} S_{HV} S_{HH}^* \rangle & \langle 2S_{HV}^2 \rangle & \langle \sqrt{2} S_{HV} S_{VV}^* \rangle \\ \langle S_{VV} S_{HH}^* \rangle & \langle \sqrt{2} S_{VV} S_{HV}^* \rangle & \langle S_{VV}^2 \rangle \end{bmatrix}, \quad (2)$$

where $*$ represents conjugate, $\langle\ \rangle$ stands for multilook with an average window (we set the window size to 3, the same is true in later equations). The expression of matrix T is listed as follow:

$$T_3 = \frac{1}{\sqrt{2}} \begin{bmatrix} \langle |S_{HH} + S_{VV}|^2 \rangle & \langle (S_{HH} + S_{VV})(S_{HH} + S_{VV})^* \rangle & \langle (S_{HH} + S_{VV})(2S_{HV})^* \rangle \\ \langle (S_{HH} + S_{VV})^* (S_{HH} + S_{VV}) \rangle & \langle |S_{HH} + S_{VV}|^2 \rangle & \langle (S_{HH} - S_{VV})(2S_{HV})^* \rangle \\ \langle (S_{HH} + S_{VV})^* (2S_{HV}) \rangle & \langle (S_{HH} + S_{VV})^* (2S_{HV}) \rangle & \langle |(2S_{VV})|^2 \rangle \end{bmatrix}, \quad (3)$$

T_3 can be transformed into C_3 according to formula $C = A^T T A$, $A = \frac{1}{\sqrt{2}} \begin{bmatrix} 1 & 0 & 1 \\ 1 & 0 & -1 \\ 0 & \sqrt{2} & 0 \end{bmatrix}$, it can also be expressed in another form

$$T_3 = U_3 \begin{bmatrix} \lambda_1 & & \\ & \lambda_2 & \\ & & \lambda_3 \end{bmatrix} U_3^H, \quad (4)$$

where H donates transpose conjugate, and the formula of U_3 is

$$U_3 = \begin{bmatrix} \cos(\alpha_1)e^{j\varnothing_1} & \cos(\alpha_2)e^{j\varnothing_2} & \cos(\alpha_3)e^{j\varnothing_3} \\ \cos(\alpha_1)\cos(\beta_1)e^{j\delta_1} & \sin(\alpha_2)\cos(\beta_2)e^{j\delta_2} & \sin(\alpha_3)\cos(\beta_3)e^{j\delta_3} \\ \sin(\alpha_1)\sin(\beta_1)e^{j\gamma_1} & \sin(\alpha_2)\sin(\beta_2)e^{j\gamma_2} & \sin(\alpha_3)\cos(\beta_3)e^{j\gamma_3} \end{bmatrix}. \tag{5}$$

The column vectors $\vec{u}_1$, $\vec{u}_2$ and $\vec{u}_3$ of U_3 are the eigenvectors of matrix T_3, corresponding to eigenvectors λ_1, λ_2 and λ_3. Cloude decomposition regards the scattering behaviors of targets as the superposition of three independent scattering behaviors, and the probability of three eigenvectors, which represents the weights of each basic scattering can be calculated by

$$P_i = \frac{\lambda_i}{\sum_{j=1}^{3} \lambda_j}. \tag{6}$$

The polarimetric entropy describes the randomness of the scattering mechanisms and is defined by

$$H = -\sum_{i=1}^{3} P_i \log_3(P_i). \tag{7}$$

The formula of anisotropy is

$$A = \frac{(P_2 - P_3)}{(P_2 + P_3)}, \tag{8}$$

and the mean scattering angle is

$$\alpha = \alpha_1 P_1 + \alpha_2 P_2 + \alpha_3 P_3, \tag{9}$$

where $\alpha_i = \arccos(v_j)$, v_j donates the eigenvalue of T_3.

2.2.2. Single-Bounce Eigenvalue Relative Difference

Allain et al. proposed Single-Bounce Eigenvalue Relative Difference (SERD) based on Cloude decomposition in 2004. The correlation between co-polarized and cross-polarized channels is almost equal to 0 for sea surface microwave scattering, so the matrix T_3 can be simplified as

$$T_3 = K \cdot K^{*T} = \frac{1}{2}\begin{bmatrix} \langle |S_{HH} + S_{VV}|^2 \rangle & \langle (S_{HH} + S_{VV})(S_{HH} - S_{VV})^* \rangle & 0 \\ \langle (S_{HH} - S_{VV})(S_{HH} + S_{VV})^* \rangle & \langle |S_{HH} - S_{VV}|^2 \rangle & 0 \\ 0 & 0 & \langle \left|(2S_{HV})\right|^2 \rangle \end{bmatrix}, \tag{10}$$

and the eigenvalue of matrix T_3 can be calculated by

$$\lambda_{1nos} = \frac{1}{2}\left\{ \langle |S_{HH}^2| \rangle + \langle |S_{VV}^2| \rangle + \sqrt{\left(\langle |S_{HH}^2| \rangle - \langle |S_{VV}^2| \rangle \right) + 4|\langle S_{HH} S_{VV} \rangle|^2} \right\}, \tag{11a}$$

$$\lambda_{2nos} = \frac{1}{2}\left\{ \langle |S_{HH}^2| \rangle + \langle |S_{VV}^2| \rangle - \sqrt{\left(\langle |S_{HH}^2| \rangle - \langle |S_{VV}^2| \rangle \right) + 4|\langle S_{HH} S_{VV} \rangle|^2} \right\}, \tag{11b}$$

$$\lambda_{3nos} = 2\langle \left|(S_{HV})\right|^2 \rangle. \tag{11c}$$

The first two eigenvalues are related to the co-polarized backscatter coefficient, and the third one is related to the cross-polarized channel and multiple scattering. Calculate the value of scattering angle α_i according to the eigenvectors corresponding to eigenvalues λ_{1nos} and λ_{2nos} to distinguish the type of scattering mechanisms: the eigenvalue corresponds to a single scattering when $\alpha_i \leq \frac{\pi}{4}$, and it is a double scattering when $\alpha_i \geq \frac{\pi}{4}$. The SERD is defined as

$$SERD = \frac{\lambda_s - \lambda_{3nos}}{\lambda_s + \lambda_{3nos}}, \tag{12}$$

$\lambda_s = \lambda_{1nos}$ when $\alpha_1 \leq \frac{\pi}{4}$ or $\alpha_2 \geq \frac{\pi}{4}$, and $\lambda_s = \lambda_{2nos}$ when $\alpha_1 \geq \frac{\pi}{4}$ or $\alpha_2 \leq \frac{\pi}{4}$.

SERD is very sensitive to the surface roughness. A large value of SERD indicates a strong single scattering in the scattering process of the target, while the small SERD value indicates weak single scattering. For the high entropy scattering area of oil spill surface, the scattering is composed of many kinds of scattering mechanisms. Single scattering is not dominant, that is, the SERD value at oil film is relatively small, and then it can be used for oil spill detection.

2.2.3. Co- and Cross- Polarized Decomposition

This section will introduce two parameters based on the scattering matrix: co-polarized correlation coefficients and conformity coefficients. Correlation coefficients can be expressed as

$$\rho_{HH/VV} = \left| \frac{\langle S_{HH} S_{VV}^* \rangle}{\langle S_{HH}^2 \rangle \langle S_{VV}^2 \rangle} \right|. \tag{13}$$

The conformity coefficients were firstly introduced into compact polarimetric SAR to estimate soil moisture by Freeman et al. [18]. Extending the conformity coefficients to quad-polarimetric SAR, it can be expressed as

$$\mu = \frac{2(\mathrm{Re}(S_{HH} S_{VV}^*))}{|S_{HH}|^2 + 2\left|(S_{HV})\right|^2 + |S_{VV}|^2}. \tag{14}$$

2.2.4. Freeman 3-Component Decomposition

Freeman and Durden [11] proposed a three-component scattering model for polarimetric SAR data in 1998, and it includes three simple scattering mechanisms: volume (or canopy) scattering, double-bounce scattering and rough surface scattering. Assuming those three scatter components are uncorrelated, the scattering process of radar wave on the sea surface can be regarded as the composition of these three mechanisms, so the model for total backscatter is

$$\langle |S_{HH}|^2 \rangle = f_s \left|\beta\right|^2 + f_d |\alpha|^2 + f_v, \tag{15a}$$

$$\langle |S_{VV}|^2 \rangle = f_s + f_d + f_v, \tag{15b}$$

$$\langle S_{HH} S_{VV}^* \rangle = f_s \beta + f_d \alpha + f_v/3, \tag{15c}$$

$$\langle |S_{HV}|^2 \rangle = f_v/3, \tag{15d}$$

$$\langle S_{HH} S_{HV}^* \rangle = \langle S_{HV} S_{VV}^* \rangle = 0, \tag{15e}$$

where f_s, f_d and f_v are the contribution of surface, double-bounce and volume scattering to the VV cross section. Once f_s, f_d and f_v are estimated, we can also get contributions of three scatter to HH, HV and VH channels. α in the formula is defined by

$$\alpha = e^{j2(\gamma_h - \gamma_v)} \left(\frac{R_{gh} R_{th}}{R_{gv} R_{tv}} \right), \tag{16}$$

R_{th} and R_{tv} donate the reflection coefficients of vertical surface for H and V polarizations, while R_{gh} and R_{gv} are the Fresnel reflection coefficients of horizontal surface. The propagation factors $e^{j2\gamma_h}$ and $e^{j2\gamma_v}$ are used to make the model more general, γ represents any attenuation and phase change of the V

and H polarized waves when they propagate from the radar to the ground and back again. $\alpha = -1$ when $Re(S_{HH}S_{VV}{}^{*})$ is positive, if $Re(S_{HH}S_{VV}{}^{*})$ is negative, $\beta = 1$.

The volume scattering contribution can be calculated directly from Equation (15d). We can estimate the contribution of each scattering mechanism to the span P

$$P = P_s + P_d + P_v \equiv (|S_{HH}|^2 + 2|S_{HV}|^2 + |S_{VV}|^2), \tag{17}$$

with

$$P_s = f_s(1 + |\beta|^2), \tag{18a}$$

$$P_d = f_d(1 + |\alpha|^2), \tag{18b}$$

$$P_v = 8f_v/3. \tag{18c}$$

Then we can use Equations (15)–(18) to calculate the scattering power of three mechanisms: P_s, P_d and P_v. They are the result of Freeman decomposition.

2.2.5. Yamaguchi 4-Component Decomposition

In 2005, Yamaguchi et al. [12] proposed a four component decomposition method based on Freeman decomposition, included the helix scattering power as the fourth term for a more general model, which is essentially caused by the scattering matrix of helices and is mainly used in urban areas. What is more, Yamaguchi decomposition modify the volume scattering matrix according to the relative backscattering magnitudes of $\langle|S_{HH}|^2\rangle$ versus $\langle|S_{VV}|^2\rangle$.

Assume the magnitude of the helix scattering power f_c, the corresponding magnitude of $\langle S_{HV}S_{VV}{}^{*}\rangle$ becomes $f_c/4$, and the power relation becomes

$$\frac{f_c}{4} = \frac{1}{2}\left|\mathrm{Im}(\langle S_{HH}S_{HV}{}^{*}\rangle + \langle S_{HV}S_{VV}{}^{*}\rangle)\right|, \tag{19}$$

and we can get the following five equations $\alpha, \beta, f_s, f_d, f_v, f_c$ by comparing the covariance matrix element:

$$\langle|S_{HH}|^2\rangle = f_s|\beta|^2 + f_d|\alpha|^2 + \frac{8}{15}f_v + \frac{f_c}{4}, \tag{20a}$$

$$\langle|S_{HV}|^2\rangle = \frac{2}{15}f_v + \frac{f_c}{4}, \tag{20b}$$

$$\langle|S_{VV}|^2\rangle = f_s + f_d + \frac{3}{15}f_v + \frac{f_c}{4}, \tag{20c}$$

$$\langle S_{HH}S_{VV}{}^{*}\rangle = f_s\beta + f_d\alpha + \frac{2}{15}f_v - \frac{f_c}{4}, \tag{20d}$$

$$\frac{1}{2}\mathrm{Im}\{\langle S_{HH}S_{HV}{}^{*}\rangle + \langle S_{HV}S_{VV}{}^{*}\rangle\} = \frac{f_c}{4}. \tag{20e}$$

f_c can be measured directly. The volume scattering coefficient f_v is calculated by

$$f_v = \frac{15}{2}\left(\langle|S_{HV}|^2\rangle - \frac{f_c}{4}\right), \tag{21}$$

α and β is calculated as the same way as Freeman decomposition, so we can get contribution of four mechanisms: f_s, f_d, f_v and f_c. The scattering powers P_s, P_d, P_v and P_c corresponding to surface, double bounce, volume and helix scattering contributions can be obtained by

$$P_s = f_s(1 + |\beta|^2), \tag{22a}$$

$$P_d = f_d(1 + |\alpha|^2), \tag{22b}$$

$$P_v = f_v, \tag{22c}$$

$$P_c = f_c, \tag{22d}$$

$$P = P_s + P_d + P_v + P_c \equiv (|S_{HH}|^2 + 2|S_{HV}|^2 + |S_{VV}|^2), \tag{23}$$

P_s, P_d, P_v and P_c are the results of Yamaguchi decomposition.

2.3. SLIC Superpixel

The superpixel algorithm was first proposed in 2003 by Xiaofeng Ren et al. [32]. Adjacent pixels with the same attribute are divided in one region (one superpixel) and then the whole image can be indicated by a certain number of superpixels, which allows better performance for subsequent image processing. SLIC adopted k-means algorithm to generate superpixels. The algorithm limited the search space to a region proportional to the size of superpixels, and reduced the number of distance calculation in optimization and the linear complexity of pixels.

The SLIC segmentation result relies only on the number of superpixels k. Each superpixel has approximately the same size. These k initial cluster centers are sampled on a regular grid with S pixel intervals. S can be calculated by $S = \sqrt{\frac{N}{k}}$, where N is the number of pixels.

Each pixel i is assigned with the nearest clustering center if their search area could overlap its position, then SLIC allows faster clustering than traditional k-means does. The distance measurement D' determines the closest clustering center C_k for each pixel i. The expected spatial range of the superpixel is an area of approximate size S × S. A similar pixel search is performed in the area 2S × 2S around the center of the superpixel.

SLIC realizes the above steps based on *labxy* color image plane space. The value of the pixel is expressed as $[l \ a \ b]^T$ in CIELab color space. However, the position $[x \ y]^T$ of the pixel changes with the size of the image. In order to combine them into a single measurement, we need to standardize the color proximity and spatial proximity by their maximum distances N_c and N_s in the cluster. Then D' can be calculated by

$$d_c = \sqrt{\left(l_j - l_i\right)^2 + \left(a_j - a_i\right)^2 + \left(b_j - b_i\right)^2}, \tag{24a}$$

$$d_s = \sqrt{\left(x_j - x_i\right)^2 + \left(y_j - y_i\right)^2}, \tag{24b}$$

$$D' = \sqrt{(d_c/N_c)^2 + (d_s/N_s)^2}, \tag{24c}$$

where $N_s = S = \sqrt{\frac{N}{k}}$. When N_c is fixed as a constant m, the Equation (24c) can be listed as the following:

$$D' = \sqrt{(d_c/m)^2 + (d_s/S)^2}, \tag{25}$$

where $d_c = \sqrt{\left(l_j - l_i\right)^2}$ in gray-scale. m allows us to balance the relative importance between N_c and N_s. When m increases, the superpixel result depends more on the degree of spatial proximity.

Once each pixel has been associated with the nearest cluster center, the update step adjusts the cluster center to the average vector of all pixel. L_2 norm is used to calculate the residual error E between the new and the previous cluster center position. The allocation and iterative process ends when E is less than the setting threshold. In our experiments, we transform the HH, HV and VV channel SAR data to *labxy* color space for superpixel calculation.

2.4. Semantic Segmentation Algorithm

We constructed a refined segmentation method based on CNN to perform oil spill detection. The structure used in our network will be described in details in the followings.

2.4.1. Convolutional Layer and Dilated Convolution

CNN has been widely used in image classification and object detection for its good generation ability. Compared with traditional neural networks, CNN imitates the human visual nerve and allows automatic feature extraction. Two main processes in the training of CNN are forward propagation and backward propagation. Forward propagation expresses the transmission of characteristic information, while backward propagation mainly uses error information to correct model parameters.

Convolutional layer is the core component of CNN. In forward propagation, it sets up a filter kernel to slide on the input tensor and obtain image features, the number of layers of these kernels equals to the input tensor. Convolution operations can be expressed as

$$y_{ij} = f\left(\sum\nolimits_{i=1}^{m} \sum\nolimits_{j=1}^{n} x_{ij}\theta_k + b_k\right), \tag{26}$$

θ_k and b_k are weights and biases that need to be trained in neural networks, while $f(*)$ represents the activation function, herein Rectified Linear Units(ReLU) function and tanh() function were used, and their equations are

$$f(x) = \max(0, x), \tag{27}$$

$$\tanh(x) = \frac{\sinh(x)}{\cosh(x)} = \frac{e^x - e^{-x}}{e^x + e^{-x}}. \tag{28}$$

The backward propagation process of neural networks depends on the backward derivation of the output layer (loss function) and the calculation of errors. The parameter adjustment is further optimized by the error function, and Adam optimizer was adopted in this paper. It adjusts the value of weights and biases iteratively, which allows little error of the output of neural networks. The gradient of output layer transfer between convolutional layers can be expressed as

$$\delta^l = \frac{\partial J(W, b)}{\partial z^l} = \frac{\partial J(W, b)}{\partial a^l} \odot \sigma'\left(z^l\right), \tag{29}$$

where a^l donates the output tensor of layer l and $z^l = W^l a^{l-1} + b^l$, and $\sigma\left(z^l\right)$ is the formula of convolution. $\odot$ donates the Hadamard product. If matrix $A = [a_1, a_2, \cdots, a_n]^T$ and $B = [b_1, b_2, \cdots, b_n]^T$, $A \odot B = [a_1 b_1, a_2 b_2, \cdots, a_n b_n]^T$. $J(W, b)$ is the loss function between output tensor and ground truth. In our case we used the cross entropy, which is described by

$$H(p, q) = -\sum_{i=1}^{n} p(x_i) \log(q(x_i)), \tag{30}$$

$p(x_i)$ and $q(x_i)$ donate the probability of x_i classification of the output and ground truth.

The recurrence relation between layer l and layer $l-1$ is

$$z^l = a^{l-1} * W^l + b^l = \sigma\left(z^{l-1}\right) * W^l + b^l. \tag{31}$$

Then the gradient of layer $l-1$ is

$$\delta^{l-1} = \left(\frac{\partial z^l}{\partial z^{l-1}}\right)^T \delta^l = \delta^l * rot180\left(W^l\right) \odot \sigma'\left(z^{l-1}\right), \tag{32}$$

where $rot180()$ means that the convolution kernel is rotated 180 degrees when the derivative is calculated, and the gradients of all layers can be calculated. Assuming that the gradient after t iterations is $g_t = \delta^l(t)$, the exponential moving average of the gradient is calculated by

$$m_t = \beta_1 m_{t-1} + (1 - \beta_1) g_t, \tag{33a}$$

where β_1 is the exponential decay rate. The exponential moving average of gradient square is

$$v_t = \beta_2 v_{t-1} + (1 - \beta_2) g_t^2. \tag{33b}$$

Revised m_t and v_t as the formula

$$\hat{m} = \frac{m_t}{1 - \beta_1{}^t}, \tag{33c}$$

$$\hat{v} = \frac{v_t}{1 - \beta_2{}^t}. \tag{33d}$$

Then the formula for updating parameters is

$$\theta_t = \theta_{t-1} - \alpha * \frac{\hat{m}}{\sqrt{\hat{v}} + \varepsilon}, \tag{33e}$$

where α represents the learning rate.

In our paper, dilated convolution is applied to extract features from input layer, which adopts inject holes into traditional convolutional kernels and it can increase the reception field. The difference between standard kernel and dilated kernel is represented in Figure 4. The kernel will slide from left to right, top to bottom on the image. As shown in Figure 4a, the red points are standard kernel. For dilated kernel (see Figure 4b), several inject holes highlighted as blue or dark blue points were added. The values at these points are set as 0. Only values at red points are calculated. Suppose $k: \Omega_r \rightarrow R$, $\Omega_r = [-r, r]^2$ is a discrete filter with the size of $(2r+1)^2$, the discrete convolution operator can be defined as

$$(F * k)(\mathrm{p}) = \sum\nolimits_{\mathrm{s+t=p}} F(s)k(t). \tag{34}$$

When l is a dilation factor, $*_l$ should be defined as

$$(F *_l k)(\mathrm{p}) = \sum_{\mathrm{s}+l\mathrm{t=p}} F(s)k(t). \tag{35}$$

That is the calculation formula of dilated convolution.

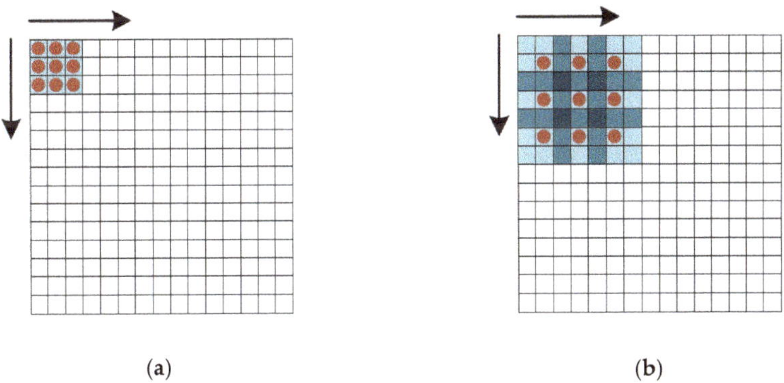

(a) (b)

Figure 4. Convolution kernels for (**a**) standard kernel, which has a receptive filed of 3 × 3, and (**b**) dilated kernel with dilation rate = 2, and its receptive field is 7 × 7.

2.4.2. Depthwise Separable Convolution with Dilated Kernel

Suppose the size of input tensor is N × H × W × C and there is a $h \times w \times k$ convolution kernel, the output of this layer would be an N × H × W × k tensor when $pad = 1$ and $stride = 1$. The whole process needs $h \times w \times k \times$ C parameters and $h \times w \times k \times$ C × H × W times multiplication.

Depthwise separable convolution decomposes traditional convolution layer into a depthwise convolution and a pointwise convolution. Depthwise process divides the N × H × W × C size input tensor into C groups. A convolution operation with a $h \times w$ kernel is carried out on each group. This process collects the spatial feature of each channel, i.e., depthwise features. The output N × H × W × C size output tensor is operated by a traditional $1 \times 1 \times k$ convolution kernel, which extract the pointwise feature from each channel. Its output is also a N × H × W × k. size tensor. Depthwise and pointwise can be regarded as a convolution layer with much lower amount of computation. The two processes need $(\mathrm{H} \times \mathrm{W} \times \mathrm{C}) \times (k + h \times w)$ times multiplication in total.

In order to combine the reception field of dilated convolution with the calculated performance of depthwise separable convolution, we adopt the strategy of adding holes into depthwise convolution kernel in several bottom layers of neural network.

2.4.3. Transposed Convolution

Transposed convolution, also known as deconvolution, is often used as decoder in neural networks. In the semantic segmentation task, transposed convolution upsample the feature map extracted by convolution layer. The final output is a fine classification map with the same size as the original image. In fact, it transposes the convolution kernel in the ordinary convolution we used in the encoder section and inverts the input and output. For example, Figure 5 shows a highly condensed feature map extracted by multilayer network and how it is decoded by a transposed convolution layer. For example, the 2 × 2 feature map padded with 2 × 2 border of zeros using 3 × 3 strides is convolved by a 3 × 3 kernel. Its output is a 6 × 6 tensor when there is no padding in convolution process.

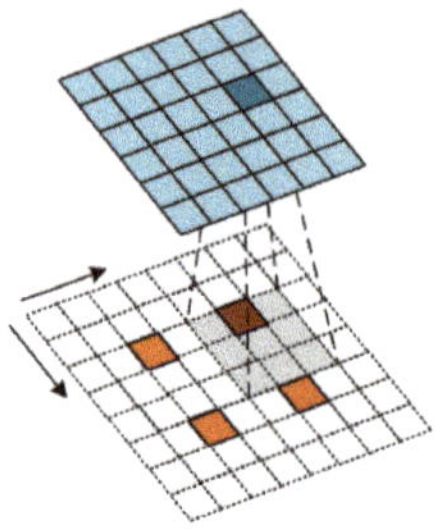

Figure 5. Convolution process of transposed convolution layer.

The detailed parameters of each layer are listed in Table 1. The encoder section contains 10 convolution layers and 2 residual blocks. Convolution layers 1 and 3 adopted depthwise separable convolution, and layer 5 was dilated convolution layer. The decoder section consisted of five deconvolution (transposed convolution) layers, in which layer 1 and layer 2 are connected with convolution layer 7 and layer 3, respectively.

Table 1. The detailed parameters of segmentation networks.

Layer [1]	Filter	Kernel Size	Strides
Conv 1	depthwise	5 × 5	1
Conv 2	Conv	6 × 6	2
Conv 3	depthwise	4 × 4	1
Conv 4	Conv	4 × 4	2
Conv 5	dilated	2 × 2	1
Conv 6	Conv	3 × 3	2
Conv 7	Conv	2 × 2	2
Conv 8	Conv	1 × 1	1
Conv 9	Conv	3 × 3	3
Conv 10	Conv	1 × 1	1

Table 1. *Cont.*

Layer [1]	Filter	Kernel Size	Strides
Res 1	Conv	6 × 6	4
Res 2	Conv	3 × 3	4
Deconv 1	transposed	3 × 3	1
Deconv 2	transposed	3 × 3	8
Deconv 2_1	transposed	6 × 6	4
Deconv 3	transposed	6 × 6	2
Deconv 3_1	transposed	6 × 6	2

[1] Conv, Res and Deconv represent the blue, green and purple-red blocks, respectively.

2.4.4. Evaluation Method

MIoU is usually used as an index to measure the accuracy in semantic segmentation task, it is to calculate the intersection between prediction and ground truth. MIoU can be expressed as

$$\text{MIoU} = \frac{1}{k+1} \sum\nolimits_{i=0}^{k} \frac{p_{ii}}{\sum\nolimits_{j=0}^{k} p_{ij} + \sum\nolimits_{j=0}^{k} p_{ji} - p_{ii}}, \tag{36}$$

which is equivalent to

$$\text{MIoU} = \frac{1}{k+1} \sum\nolimits_{i=0}^{k} \frac{TP}{FN + FP + TP}. \tag{37}$$

where TP is the abbreviation for true positive, which means the number of samples when real value and model prediction are both positive. FN represents false negative, which means real value is positive while model prediction is negative. FP represents false positive. k is the number of classifications.

3. Experiments and Results

3.1. SAR Data and Preprocessing

There are three images used in our experiments. Image 1 is a quad-pol oil spill image obtained by C-band Radarsat-2 satellite over the North Sea of England in 2011 during the oil-on-water exercise conducted by the Norwegian Clean Seas Association for Operating Companies (NOFO). The whole image contains five parts in total: clean sea, ships, biogenic look-alike film, emulsion and crude oil spill. The biogenic look-alike film was simulated by Radiagreen plant oil, while emulsion area was composed of Oseberg blend crude oil mixed with 5% IFO380 (Intermediate Fuel Oil). The oil spill area was the Balder crude oil. It was released 9h before SAR acquisition [16]. Emulsions are classified as an independent class in this paper since they have different composition and polarimetric scattering characteristics in SAR images. Image 2 and Image 3 are acquired by C-band SIR-C/X-SAR in 1994, the dark spots contained in images are biogenic look-alike and oil spill, respectively. The biogenic look-alike was composed of Oleyl Alcohol in the experiment [37]. The detailed information of SAR acquisition is listed in Table 2.

Table 2. Details of Synthetic Aperture Radar (SAR) image acquisition.

Image ID	137348	PR11588	PR44327
Radar Sensor	Radarsat-2	SIR-C/X-SAR	SIR-C/X-SAR
SAR Band	C	C	C
Pixel Spacing (m)	4.70 × 4.80	12.50 × 12.50	12.50 × 12.50
Radar Center Frequency (Hz)	5.405×10^9	5.298×10^9	5.304×10^9
Centre Incidence Angle (deg)	35.287144	23.600	45.878

The Single Look Complex (SLC) radar images experienced multi-look process and was filtered by Refined Lee Filter. Figure 6 shows the image extract from coherence matrix T before and after filtering.

It helped suppress speckle noise and enhance the edge of dark spots, and some early experiments have proved that Refined Lee Filter could help increase oil spill detection accuracy.

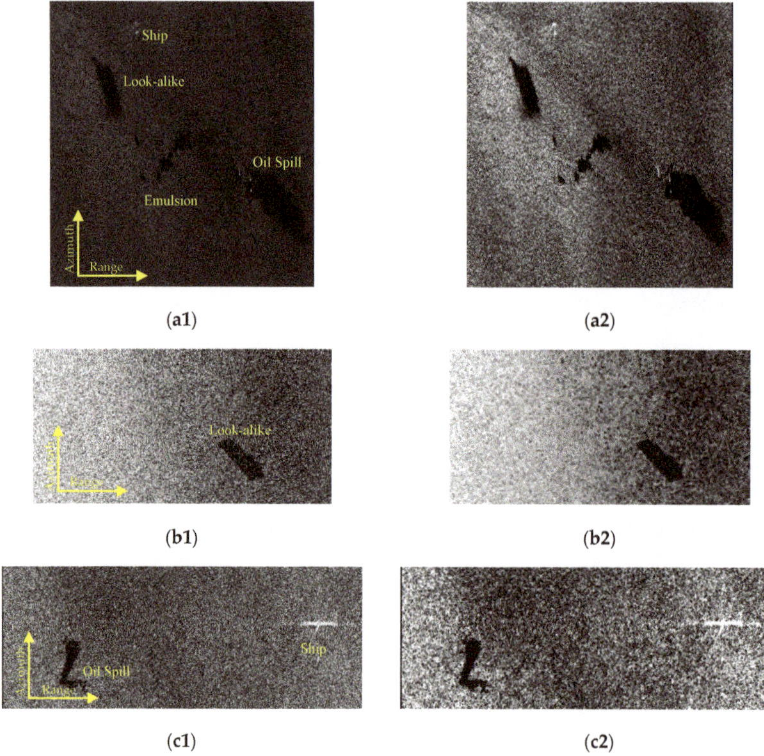

Figure 6. Three oil spill data used in the experiments. Left side shows the original image, and right side are images processed by Refined Lee Filter. (**a1**,**a2**) Image 1 acquired by Radarsat-2, (**b1**,**b2**) Image 2 (PR11588) acquired by SIR-C/X-SAR and (**c1**,**c2**) Image 3 (PR44327) acquired by Spaceborne Imaging Radar-C/X-Band Synthetic Aperture Radar (SIR-C/X-SAR).

The filtered images were processed by different polarized decomposition methods according to the steps listed in Section 2.2. Figure 7 lists the five groups of polarized parameters extracted from Image 1 as an example: H/A/Alpha, H/A/Alpha/SERD, correlation/conformity coefficients, Freeman decomposition and Yamaguchi decomposition, and characteristics of all these parameters are listed in Table 3.

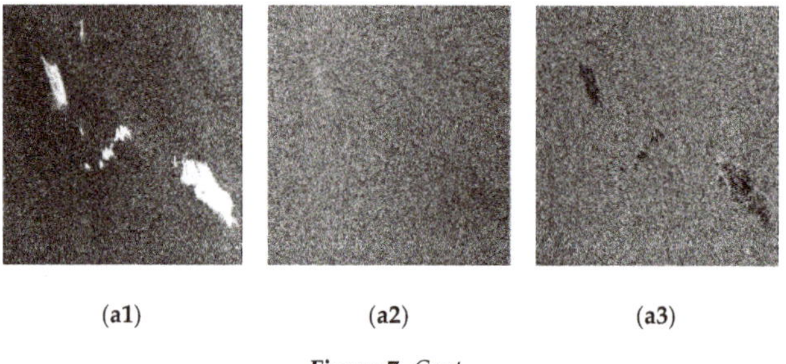

Figure 7. *Cont.*

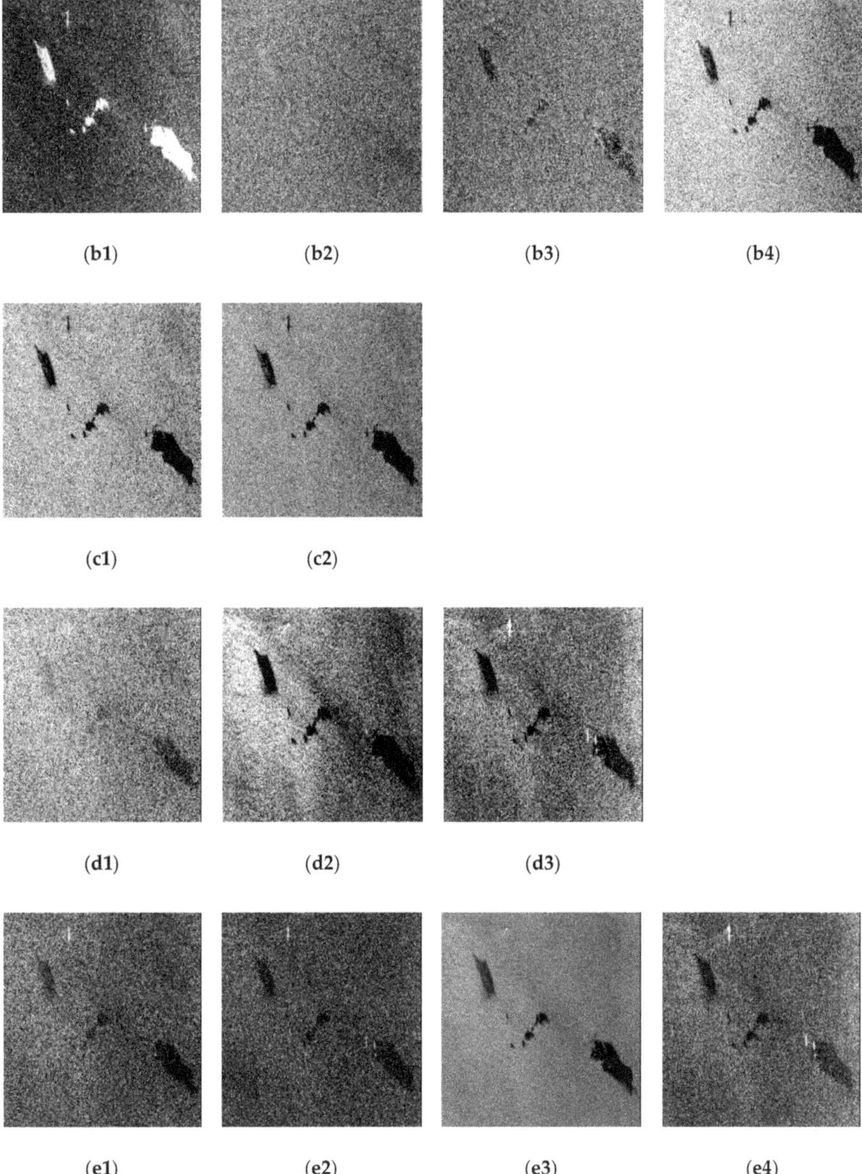

Figure 7. All the polarized features extracted from Radarsat2 data. (**a1–a3**) H/A/Alpha decomposition, a1 for entropy, a2 for anisotropy, a3 for alpha. (**b1–b4**) H/A/Alpha decomposition and Single-Bounce Eigenvalue Relative Difference (SERD), b1 for entropy, b2 for anisotropy, b3 for alpha, b4 for SERD. (**c1,c2**) Scattering coefficients calculated from scattering matrix, c1 for co-polarized correlation coefficients, c2 for conformity coefficients. (**d1–d3**) Freeman 3-component decomposition, d1 for double-bounce scattering, d2 for rough surface scattering, d3 for volume scattering. (**e1–e4**) Yamaguchi 4-component decomposition, e1 for double-bounce scattering, e2 for helix scattering, e3 for rough surface scattering, e4 for volume scattering.

Table 3. Characteristics of polarized parameters in experiments.

Parameter	Clean Sea	Look alike	Emulsion	Oil Spill	Ship
Entropy	low	high	higher	higher	higher
Anisotropy	low	low	low	low	low
Alpha	low	lower	low	low	low
SERD	high	low	lower	lower	lower
Correlation Coefficient	high	low	lower	lower	lower
Conformity Coefficient	high	low	lower	lower	lower
Freeman Double-Bounce	high	low	low	lower	higher
Freeman Rough-Surface	high	low	low	low	higher
Freeman Volume	high	lower	lower	low	higher
Yamaguchi Double-Bounce	high	low	low	lower	higher
Yamaguchi Helix	low	lower	lower	lower	high
Yamaguchi Rough-Surface	high	low	lower	lower	higher
Yamaguchi Volume	high	lower	lower	low	higher

3.2. SLIC Superpixel Segmentation

The HH, HV and VV data was taken as input data to perform SLIC superpixel segmentation. We used these three channels of SAR data to generate a new image, it was converted into CIElab color spaces. Following the steps of SCIC superpixel method described in Section 2.3, the superpixel segmentation results of SAR data are shown in Figure 8. The superpixel number of three images was set to 250, 40, 40, respectively. They are another type of input besides polarized parameters for CNN training. It can be seen from Figure 8 that SLIC superpixel divides the image into several independent areas, and can initially locate dark spots, especially in Image 2 and Image 3.

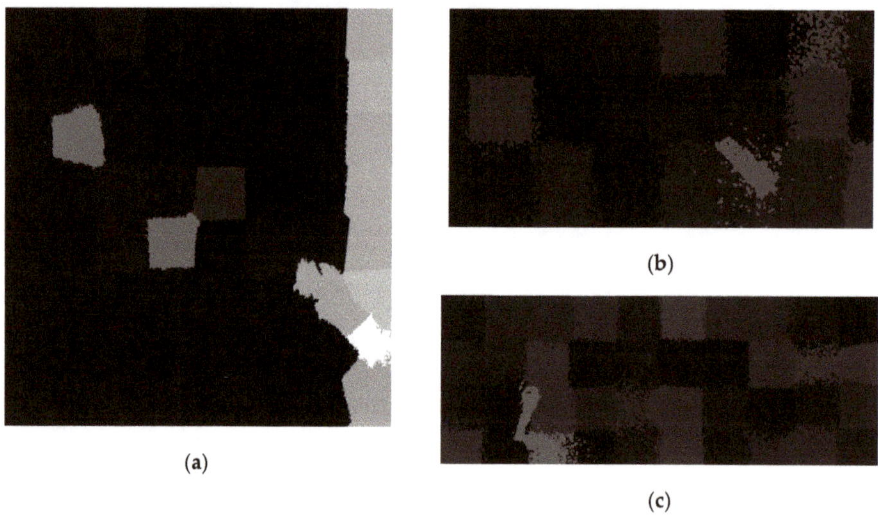

Figure 8. Simple Linear Iterative Clustering (SLIC) superpixel segmentation results. (**a**) Image 1, (**b**) Image 2 and (**c**) Image 3.

Polarimetric decomposition and superpixel images are divided into five groups as listed in Figure 2. The three SAR images are divided into five categories pixel by pixel: clean sea background (CS), emulsion (EM), biogenic look-alike (LA), oil spill (OS) and ships (SH). All the images are divided into 48×48 small pictures in the experiment. When multiple parameters are input into CNN, they are stacked along the third axis of images to form a three-dimensional array. The original SAR images only contained 5 ships, in order to increase the number of samples, especially ships, we sampled the same target area for multiple times. We extracted image of target areas from different positions, and these images are divided into 48*48. Thus, we can get several sampling images on the same area. We randomly selected training set and test set from sample images, the number of samples are listed in Table 4. The MIoU was calculated on the test set. They are trained with the proposed network described in Section 2.4 and the output segmentation results are verified with ground truth.

Table 4. Number of samples of each category.

Areas	CS	EM	LA	OS	SH
Training set	80	75	82	88	31
Test set	27	25	28	30	12

3.3. Oil Spill Classification

In order to evaluate the influence of SLIC superpixel on segmentation results, we carried out comparative experiments based on each group of polarized parameters with and without superpixel segmentation. Figure 9 presents the segmentation results of five groups of polarized parameters on five dark spots areas of three images. The oil spill area is marked with the dark spots and the light grey means the biogenic look-alikes. The medium grey represents emulsion.

As shown in Figure 9, the dark spots area can be extracted effectively and classified accurately in each group. The classification result of oil spill area in Image 3 showed the best. Among all the polarized decomposition parameters, the performance of Yamaguchi 4-component parameters was the best, followed by Freeman 3-component parameters and H/A/SERD/Alpha. H/A/Alpha could also distinguish each category in the images except ships. The parameter SERD effectively increased the classification accuracy on the basis of H/A/Alpha decomposition. The segmentation result of co-polarized correlation coefficients and conformity coefficients does not perform well nearly in all categories, indicating they are not optimal polarized parameters for detecting oil spill areas.

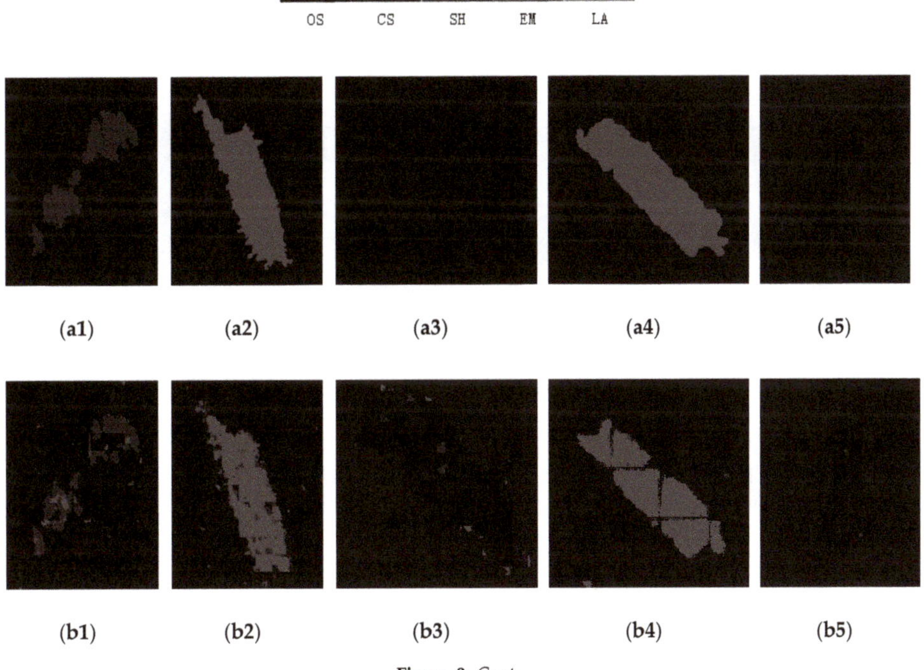

Figure 9. *Cont.*

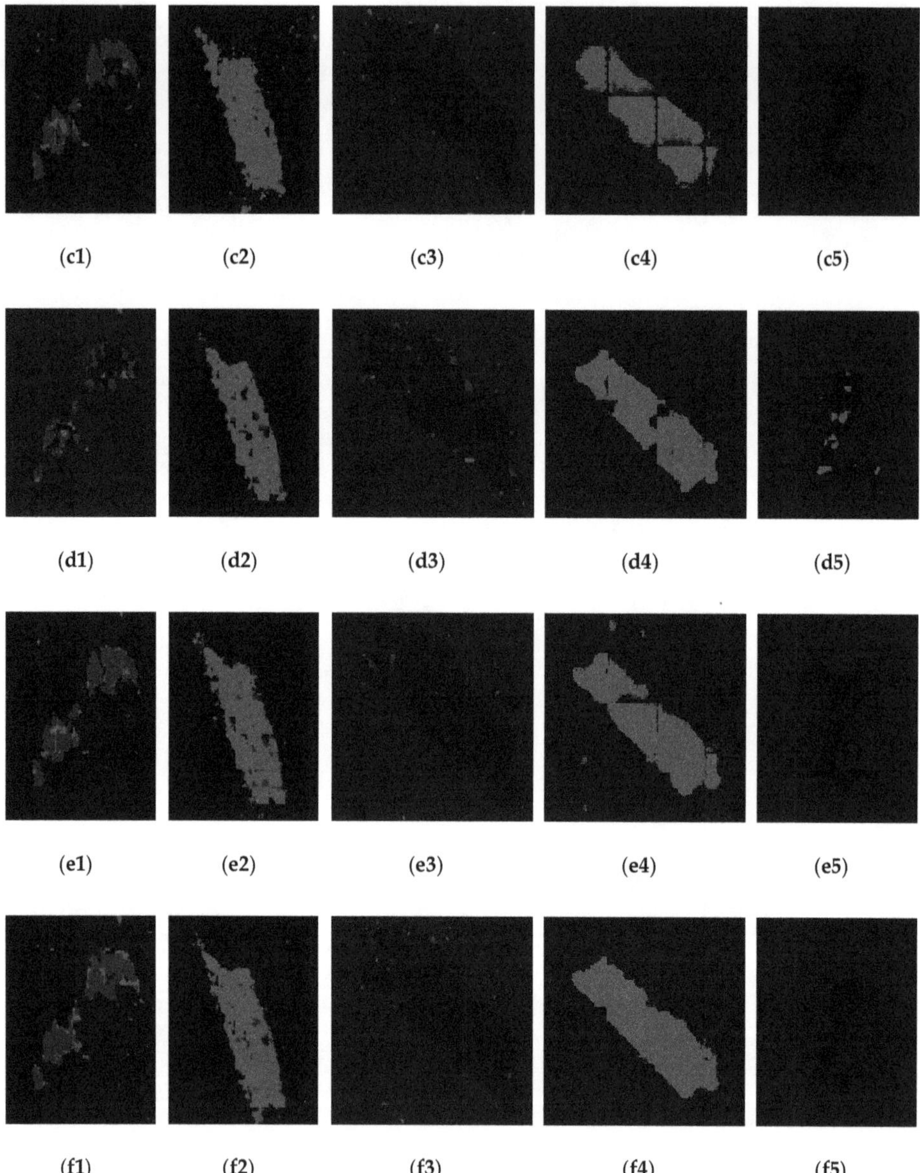

Figure 9. The results of dark spots area verified by polarized parameters, 1-3 in each group represents emulsion, 2 for biogenic look-alike, 3 for oil-spill area of Image 1, 4 and 5 represent biogenic look alike and oil spill area of Image 2 and Image 3. (**a1–a5**) Ground truth, (**b1–b5**) H/A/Alpha, (**c1–c5**) H/A/SERD/Alpha, (**d1–d5**) Scattering Coefficients, (**e1–e5**) Freeman 3-Component Decomposition, (**f1–f5**) Yamaguchi 4-Component Decomposition.

Considering all categories, the classification results of clean sea (CS) is the best. Then it is followed by oil spill (OS) areas, which is slightly better than look-alikes (LA). The classification accuracy of the categories of emulsions (EM) and ship (SH) are the lowest. The false detection mostly occurred in emulsions. A number of emulsion areas were misclassified into oil spill or look-alikes, especially in the experiments of H/A/Alpha, H/A/SERD/Alpha and co-polarized CC/conformity coefficients. Compared with those results, Freeman 3-component and Yamaguchi 4-component decomposition could distinguish most of these categories successfully. Moreover, the experiment results by applying these two groups of polarized parameters could also detect ships with high reliability, which are almost all misclassified as oil spill areas in other groups' experimental results.

In the followings, we added the SLIC segmentation result from SAR data as another input besides polarized parameters and inputted them together into neural network and repeated the above experiments. The output results are represented in Figure 10. The classification results of each category has been improved significantly, especially for emulsion areas. Compared with the segmentation results without applying superpixel model results, the edge of different classes become more distinct.

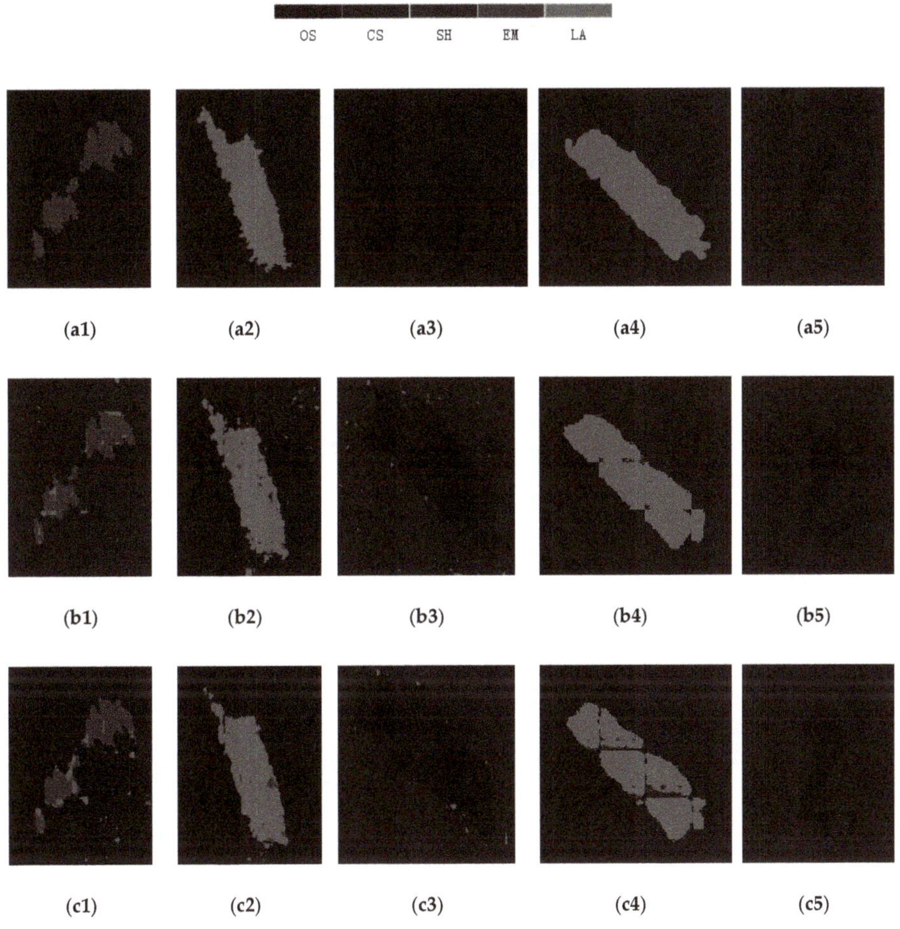

Figure 10. *Cont.*

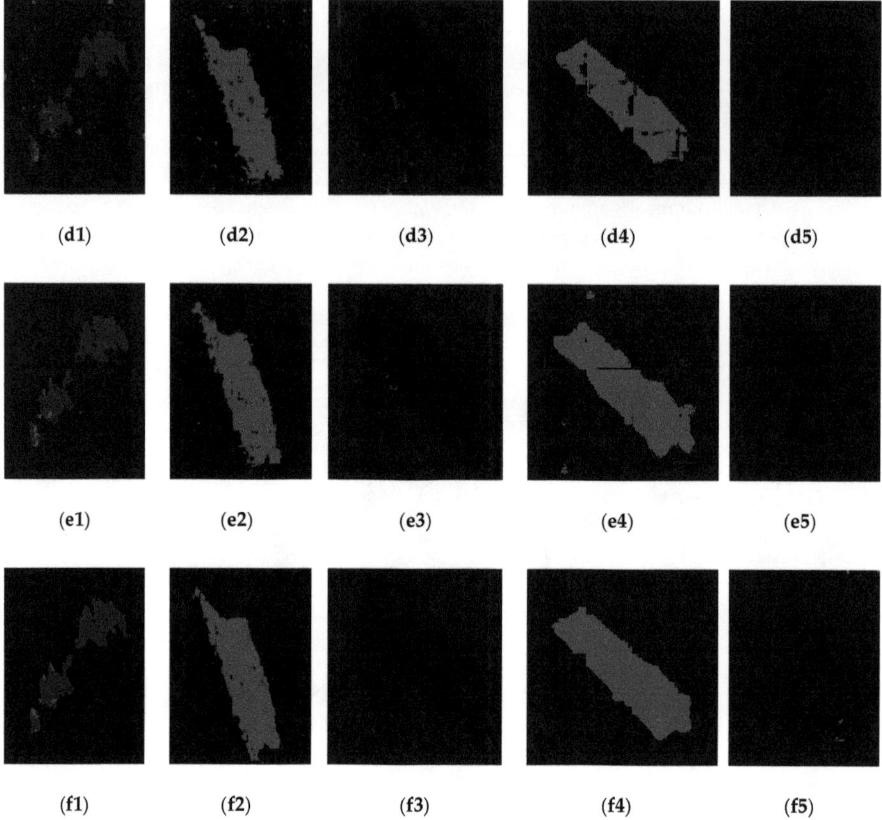

Figure 10. The results of dark spots area verified by polarized parameters combined with SLIC superpixel segmentation, 1-3 in each group represents emulsion, 2 for biogenic look-alike, 3 for oil-spill area of Image 1. Images 4 and 5 represent biogenic look-alike and oil spill area of Image 2 and Image 3. (**a1–a5**) Ground truth, (**b1–b5**) H/A/Alpha, (**c1–c5**) H/A/SERD/Alpha, (**d1–d5**) Scattering Coefficients, (**e1–e5**) Freeman 3-Component Decomposition, (**f1–f5**) Yamaguchi 4-Component Decomposition.

The numerical comparison was carried out by calculating the MIoU of each polarized parameter group on the test set. The compared results with and without SLIC superpixel are listed in Table 5. The accuracy of Yamaguchi and Freeman decomposition is significantly higher than other groups of polarized parameters, and that of each classification category has been also improved by SLIC superpixel to varying degree.

Table 5. Mean Intersection over Union (MIoU) result on each classification of polarized parameters experiments.

	Without SLIC Superpixel					With SLIC Superpixel				
Classification	**CS**	**EM**	**LA**	**OS**	**SH**	**CS**	**EM**	**LA**	**OS**	**SH**
H/A/Alpha	94.0%	70.8%	84.2%	85.7%	10.5%	95.6%	88.3%	89.7%	93.4%	39.7%
H/A/SERD/Alpha	94.5%	80.7%	84.9%	88.3%	11.7%	95.8%	91.0%	91.7%	95.1%	43.1%
Scattering coefficients [1]	94.1%	27.3%	82.3%	80.2%	6.8%	94.7%	85.8%	84.2%	90.1%	41.5%
Freeman	95.8%	80.7%	82.4%	90.6%	60.7%	96.3%	91.1%	91.5%	95.1%	48.3%
Yamaguchi	96.1%	81.8%	85.4%	94.0%	75.0%	96.9%	94.1%	94.6%	96.8%	70.2%

[1] scattering coefficient means the combination of correlation coefficients and conformity coefficients.

For further analysis, Table 6 shows the total MIoU of different polarized parameters decomposition methods. The average MIoU of each classification in all experiments is shown in Table 7. Both Tables 6 and 7 are calculated from the average value of Table 5. The overall accuracy of different polarimetric parameters after combined with SLIC superpixel segmentation maintained the same trend in previous analysis as illustrated in Tables 5 and 6. Yamaguchi 4-component decomposition achieved the highest MIoU by 90.5%, followed by Freeman parameters and H/A/SERD/Alpha. Although SLIC superpixel just provide a rough classification of dark spots area, it could also improve MIoU values of each polarimetric parameters, increased by 12.3%, 11.3%, 21.2%, 2.5%, 4.0% relatively. Take Yamaguchi parameters as example, the MIoU of OS area increased from 94.0% to 96.8%, and increased by 0.8%, 12.3% and 9.2% in CS, EM and LA area relatively. What's more, the largest increase of MIoU occurred in EM area, which increased by 21.9% in average in five groups of polarimetric parameters, as shown in Table 8. CS and OS areas achieved the highest MIoU by 95.9% and 94.1% in all experiments with and without SLIC superpixel, and SH was significantly lower than other parts.

Table 6. Total MIoU results of each group of polarimetric parameters.

Parameters	H/A/Alpha	H/A/SERD/Alpha	Scattering coefficients	Freeman	Yamaguchi
Without SLIC Superpixel	69.0%	72.0%	58.1%	82.0%	86.5%
With SLIC Superpixel	81.3%	83.3%	79.3%	84.5%	90.5%

Table 7. Average MIoU of each classification.

Areas	CS	EM	LA	OS	SH
Without SLIC Superpixel	94.9%	68.2%	83.8%	87.8%	32.9%
With SLIC Superpixel	95.9%	90.1%	90.3%	94.1%	48.6%

Table 8. Total MIoU results of Yamaguchi parameters combined with different SLIC parameters.

Superpixels	150	200	250	300	350	400
Total MIoU	87.1%	88.5%	91.0%	90.3%	90.5%	86.6%

It is worth noting that the number of superpixels in SLIC superpixel segmentation will also affect the final segmentation accuracy. We tested the number of superpixels from 150 to 400 with the step of 50 on Image 1 alone. Figure 11 shows the SLIC segmentation results of different numbers of superpixels. We carried out the comparison experiments with the use of the polarized parameter group of Yamaguchi 4-component decomposition, since it achieved the highest MIoU in the previous experiments. Table 10 lists the MIoU for oil spill segmentation accuracy under different superpixel numbers. The highest accuracy is 91.0% when superpixel number was set to 250.

Finally, the classification results of the whole image without and with SLIC superpixel by applying Yamaguchi parameters are represented in Figure 12. Each category on the sea surface can be distinguished with high accuracy. SLIC superpixel helped further improve the accuracy of each category, especially for emulsions. Biogenic look-alikes were also better classified with less misclassification pieces inside. Emulsions can be well detected from oil spill and biogenic look-alike areas, and the segmentation results of other categories also perform better. The improvement effect in Image 1 was the most obvious, while SLIC superpixel mainly helped improve the accuracy of CS area in Image 2 and Image 3.

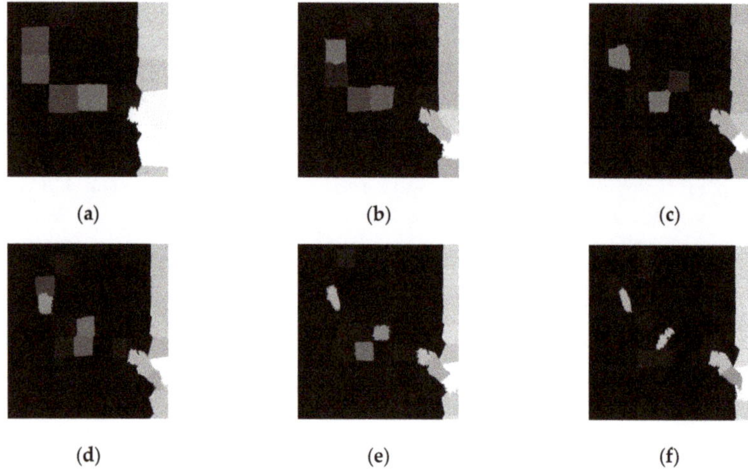

(a) (b) (c)

(d) (e) (f)

Figure 11. SLIC superpixel segmentation results with different superpixel numbers. (**a**) 150, (**b**) 200, (**c**) 250, (**d**) 300, (**e**) 350, (**f**) 400.

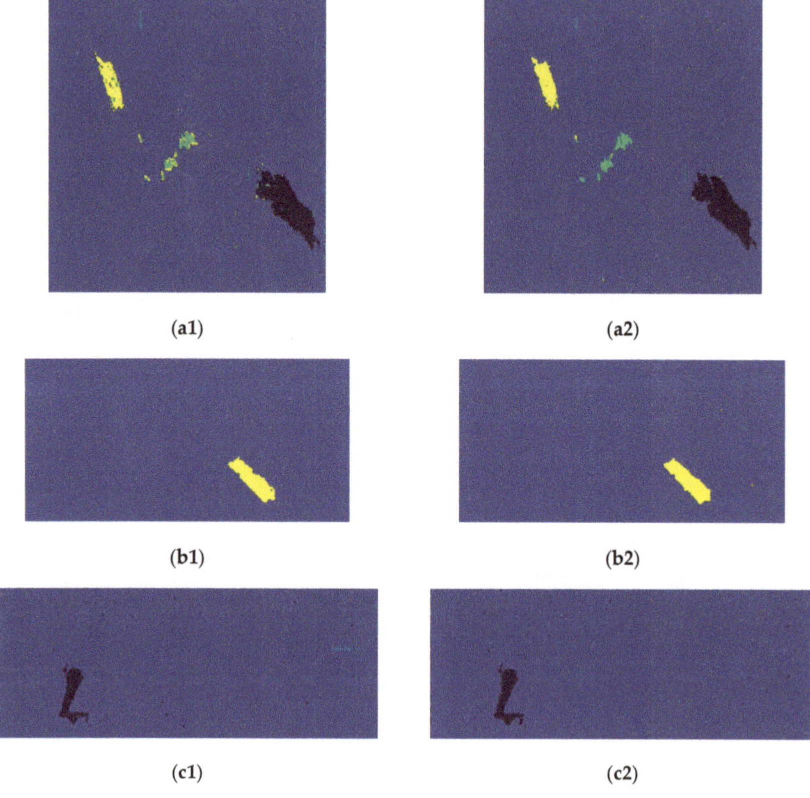

(a1) (a2)

(b1) (b2)

(c1) (c2)

Figure 12. The whole classification result of Yamaguchi 4-component parameters. Left: the results without SLIC superpixel; right: the results with SLIC superpixel. (**a1**,**a2**) Image 1, (**b1**,**b2**) Image 2, (**c1**,**c2**) Image 3.

In order to evaluate the algorithm complexity, we calculate the calculation time of the superpixel segmentation and CNN classification with different polarized parameters, the results are listed in Table 9. Table 10 shows the memory usage of different neural network models. Due to the limitation of experimental conditions, our experiments are carried out on a device without independence GPU. It should be noted that the processing speed will be more than several tens of times faster on a device with independence GPU, hence it will be no problem to achieve near-real time monitoring.

Table 9. Calculation time of each process (seconds).

		Image 1	Image 2	Image 3
Pixel Number in Experiments		2350 × 2450	1000 × 480	1550 × 600
Computer Configuration			i5-8250u CPU 8GB	
Superpixel Segmentation(s)		1742	576	482
CNN Classification with SLIC(s)	H/A/Alpha	1594	135	243
	H/A/Alpha/SERD	2107	186	359
	Scattering Coefficients	1023	102	159
	Freeman	1602	133	245
	Yamaguchi	2115	192	361

Table 10. Memory Usage Condition of CNN.

	Without SLIC	With SLIC
H/A/Alpha	59.2 M	91.3 M
H/A/Alpha/SERD	91.3 M	130 M
Scattering Coefficients	33.9 M	59.2 M
Freeman	59.2 M	91.3 M
Yamaguchi	91.3 M	130 M

4. Discussion

In order to improve the reception field and reduce parameters, we used depthwise convolution and dilated convolution in several bottom layers of our semantic segmentation model instead of traditional convolution kernel. These strategies achieved high accuracy with small amount of parameters in the experiments.

The emulsion marked in Image 1 we used is mixture of Oseberg oil, IFO380 and water, with water content of approximately 69%. The emulsion has different polarimetric characteristics from oil spill. Actually, they behave in between crude and clean sea surface on polarimetric SAR features. In actual cases, it should be also recognized as a type of oil leakage that will cause damage to ocean environment. The emulsions, biogenic look-alikes and oil spill area are independent of each other in actual SAR images, but dark area in the same test image may be classified into two or more categories. For example, many EM areas are classified into OS or LA as shown in Figure 9.

It was discovered that that parameters calculated from covariance matrix or correlation/conformity coefficients could mistakenly detect the ship as oil spill area. Experiments proved that Yamaguchi and Freeman decomposition parameters performed better in oil spill classification. Both of them are scattering model-based decomposition method, while Yamaguchi decomposition could better deal with the large cross-polarized component caused by complex ground target, which break the reflection symmetry. Hence Yamaguchi decomposition could distinguish each area with relatively high accuracy, especially on ship targets.

Moreover, SLIC can combine neighboring pixels together with special significance and associate adjacent pixels, thus forming connected blocks and greatly improves the classification accuracy, the MIoU of each group polarized parameters has been greatly improved in the demonstrated experiment. Further experiments on Image 1 showed that when superpixel number was set to 250, the recognition accuracy achieved the highest. In fact, the number of superpixels set strongly relies on the type and size of objects. That means that the SLIC superpixel numbers should be adjusted depending on real conditions.

5. Conclusions

In this paper, we proposed an oil spill detection method combining SLIC superpixel model and semantic segmentation algorithm based on CNN. The dilated convolution kernel and depthwise separable convolution kernel was adopted for better computing performance and larger sensing area. SLIC superpixel segmentation is set as an input for the CNN model for auxiliary classification.

The experiments were carried on a C-band fully polarized SAR data of Radarsat-2. We extracted several polarized parameters according to different methods, and tested their performance in oil spill classification based on the proposed method. The results showed that in each group of experiments, this network structure can effectively distinguish the oil spill area and other areas. The highest MIoU value appeared in Yamaguchi decomposition parameters experiment, followed by H/A/SERD/Alpha and Freeman decomposition.

The introduction of SLIC superpixel greatly improved the recognition accuracy. The MIoU values of each group are improved, and their numerical order of the polarimetric feature sets is almost the same as in experiments without SLIC superpixel. Hence, it is suggested that Yamaguchi parameters combined with superpixel segmentation is the most suitable method for oil spill detection.

Author Contributions: Funding acquisition, Q.L., J.W. and Y.L.; methodology, Q.L.; supervision, Q.L., H.F. and J.L.; writing—original draft, J.Z.; writing—review and editing, Q.L., Y.L. and J.W. All authors have read and agreed to the published version of the manuscript.

Acknowledgments: The RADARSAT-2 data in this paper is provided by Canadian Space Agency and MacDonald, Dettwiler and Associates. The authors are very thankful to the partially support of the National Natural Science Foundation of China (grant No. 41601446, 41706201 and 41801284).

Conflicts of Interest: The authors declare no conflict of interest.

References

1. Migliaccio, M.; Nunziata, F.; Montuori, A. A Multifrequency Polarimetric SAR Processing Chain to Observe Oil Fields in the Gulf of Mexico. *IEEE Trans. Geosci. Remote Sens.* **2011**, *49*, 4729–4737. [CrossRef]
2. Alpers, W.; Holt, B.; Zeng, K. Oil spill detection by imaging radars: Challenges and pitfalls. *Remote Sens. Environ.* **2017**, *201*, 133–147. [CrossRef]
3. Solberg, S.; Storvik, G.; Solberg, R.; Volden, E. Automatic Detection of Oil Spills in ERS SAR Images. *IEEE Trans. Geosci. Remote Sens.* **1999**, *37*, 1916–1924. [CrossRef]
4. Garcia-Pineda, O.; MacDonald, R.; Li, X.; Jackson, C.R.; Pichel, W.G. Oil Spill Mapping and Measurement in the Gulf of Mexico with Textual Classifier Neural Network Algorithm. *IEEE J. Sel. Top. Appl. Earth Obs. Remote Sens.* **2013**, *6*, 2517–2525. [CrossRef]
5. Solberg, S.; Brekke, C.; Husoy, O. Oil Spill Detection in Radarsat and Envisat SAR Images. *IEEE Trans. Geosci. Remote Sens.* **2007**, *45*, 746–755. [CrossRef]
6. Yang, C.-S.; Kim, Y.-S.; Ouchi, K.; Na, J.-H. Comparison with L-, C-, and X-band real SAR images and simulation SAR images of spilled oil on sea surface. In Proceedings of the 2009 IEEE International Geoscience and Remote Sensing Symposium, IGARSS 2009, Cape Town, South Africa, 12–17 July 2009; pp. 673–676.
7. Wismann, V. Radar Signatures of Mineral Oil Spills Measured by an Airborne Multi-Frequency Radar and the ERS-1 SAR. In Proceedings of the 1993 IEEE International Geoscience and Remote Sensing Symposium, IGARSS 1993, Tokyo, Japan, 18–21 August 1993; pp. 940–942.
8. Cheng, Y.; Liu, B.Q.; Li, X.; Nunziata, F.; Xu, Q.; Ding, X.; Migliaccio, M.; Pichel, W.G. Monitoring of Oil Spill Trajectories with COSMO-SkyMed X-Band SAR Images and Model Simulation. *IEEE J. Sel. Top. Appl. Earth Obs. Remote Sens.* **2014**, *7*, 2895–2901. [CrossRef]
9. Fingas, M.; Brown, E. A Review of Oil Spill Remote Sensing. *Sensors* **2018**, *18*, 91. [CrossRef]
10. Collins, J.; Denbina, M.; Minchew, B.; Jones, C.E.; Holt, B. On the Use of Simulated Airborne Compact Polarimetric SAR for Characterizing Oil-Water Mixing of the Deepwater Horizon Oil Spill. *IEEE J. Sel. Top. Appl. Earth Obs. Remote Sens.* **2015**, *8*, 1062–1077. [CrossRef]

11. Li, H.; Perrie, W.; He, Y.; Wu, J.; Luo, X. Analysis of the Polarimetric SAR Scattering Properties of Oil-Covered Waters. *IEEE J. Sel. Top. Appl. Earth Obs. Remote Sens.* **2015**, *8*, 3751–3759. [CrossRef]
12. Freeman, A.; Durden, L. A Three-Component Scattering Model for Polarimetric SAR Data. *IEEE Trans. Geosci. Remote Sens.* **1998**, *36*, 963–973. [CrossRef]
13. Yamaguchi, Y.; Moriyama, T.; Ishido, M.; Yamada, H. Four-Component Scattering Model for Polarimetric SAR Image Decomposition. *IEEE Trans. Geosci. Remote Sens.* **2005**, *43*, 1699–1706. [CrossRef]
14. Cloude, R.; Pottier, E. An Entropy Based Classification Scheme for Land Applications of Polarimetric SAR. *IEEE Trans. Geosci. Remote Sens.* **1997**, *35*, 68–78. [CrossRef]
15. Allain, S.; Lopez-Martinez, C.; Ferro-Famil, L.; Pottier, E. New Eigenvalue-based Parameters for Natural Media Characterization. In Proceedings of the 2005 IEEE International Geoscience and Remote Sensing Symposium, IGARSS 2005, Seoul, Korea, 29–29 July 2005; pp. 40–43.
16. Skrunes, S.; Brekke, C.; Eltoft, T. Characterization of Marine Surface Slicks by Radarsat-2 Multipolarization Features. *IEEE Trans. Geosci. Remote Sens.* **2014**, *52*, 5302–5319. [CrossRef]
17. Chen, G.; Li, Y.; Sun, G.; Zhang, Y. Application of Deep Networks to Oil Spill Detection Using Polarimetric Synthetic Aperture Radar Images. *Appl. Sci.* **2017**, *7*, 968. [CrossRef]
18. Zhang, Y.; Li, Y.; Liang, X.S.; Tsou, J. Comparison of Oil Spill Classifications Using Fully and Compact Polarimetric SAR Images. *Appl. Sci.* **2017**, *7*, 193. [CrossRef]
19. He, K.; Zhang, X.; Ren, S.; Sun, J. Deep Residual Learning for Image Recognition. Available online: https://arxiv.org/abs/1512.03385v1 (accessed on 10 December 2015).
20. Hadji, I.; Wildes, P. What Do We Understand About Convolutional Networks? Available online: https://arxiv.org/abs/1803.08834v1 (accessed on 23 May 2018).
21. Yang, Y.; Zhong, Z.; Shen, T.; Lin, Z. Convolutional Neural Networks with Alternatively Updated Clique. In Proceedings of the 2018 IEEE/CVF Conference on Computer Vision and Pattern Recognition, Salt Lake City, UT, USA, 18–22 June 2018.
22. Lin, M.; Chen, Q.; Yan, S. Network in Network. In Proceedings of the ICLR2014 International Conference on Learning Representations, Banff, AB, Canada, 14–16 April 2014.
23. Howard, G.; Zhu, M.; Chen, B.; Kalenichenko, D.; Wang, W. MobileNets: Efficient Convolutional Neural Networks for Mobile Vision Applicatoins. Available online: https://arxiv.org/abs/1704.04861 (accessed on 17 April 2017).
24. Howard, A.; Sandler, M.; Chu, G.; Chen, L.; Chen, B. Searching for MobileNetV3. Available online: https://arxiv.org/abs/1905.02244 (accessed on 6 May 2019).
25. Long, J.; Shelhamer, E.; Darrell, T. Fully Convolutional Networks for Semantic Segmentation. In Proceedings of the 2015 IEEE/CVF Conference on Computer Vision and Pattern Recognition, Boston, MA, USA, 7–12 June 2015.
26. Yu, F.; Koltun, V. Multi-Scale Context Aggregation by Dialted Convolutions. In Proceedings of the International Conference on Learning Representations 2016, San Juan, Puerto Rico, 2–4 May 2016.
27. Garcia-Garcia, A.; Orts-Escolano, S.; Opera, S.O.; Villena-Martinez, V.; Garcia-Rodriguez, J. A Review on Deep Learning Techniques Applied to Semantic Segmentation. Available online: https://arxiv.org/abs/1704.06857 (accessed on 22 April 2018).
28. Chen, L.-C.; Zhu, Y.; Papandreou, G.; Schroff, F.; Adam, H. Encoder-Decoder with Atrous Separable Convolution for Semantic Image Segmentation. Available online: https://arxiv.org/abs/1802.02611v3 (accessed on 22 August 2018).
29. Krestenitis, M.; Orfanids, G.; Ioannidis, K.; Avgerinakis, K.; Vrochidis, S.; Kompatsiaris, I. Oil Spill Identification from Satellite Images Using Deep Neural Networks. *Remote Sens.* **2019**, *11*, 1762. [CrossRef]
30. Yu, X.; Zhang, H.; Luo, C.; Qi, H.; Ren, P. Oil Spill Segmentation via Adversarial f-Divergence Learning. *IEEE Trans. Geosci. Remote Sens.* **2018**, *56*, 4973–4988. [CrossRef]
31. Kim, D.; Jung, H.-S. Mapping Oil Spills from Dual-Polarized SAR Images Using an Artificial Neural Network: Application to Oil Spill in the Kerch Strait in November 2007. *Sensors* **2018**, *18*, 2237. [CrossRef]
32. Ren, X.; Malik, J. Learning a Classification Model for Segmentation. In Proceedings of the ICCV 2003 IEEE International Conference on Computer Vision, Nice, France, 13–16 October 2003; Volume 1, pp. 10–17.
33. Achanta, R.; Shaji, A.; Smith, K.; Lucchi, A.; Fua, P.; Süsstrunk, S. SLIC Superpixels Compared to State-of-the-art Superpixel Methods. *IEEE Trans. Pattern Anal. Mach. Intell.* **2012**, *34*, 2274–2282. [CrossRef]

34. Bovenga, F.; Derauw, D.; Rana, F.M.; Barbier, C.; Refice, A.; Veneziani, N.; Vitulli, R. Multi-Chromatic Analysis of SAR Images for Coherent Target Detection. *Remote Sens.* **2014**, *6*, 8822–8843. [CrossRef]
35. Bovenga, F.; Derauw, D.; Barbier, C.; Rana, F.M.; Refice, A.; Veneziani, N.; Vitulli, R. Multi-Chromatic Analysis of SAR Images for Target Analysis. In Proceedings of the Conference on SAR Image Analysis, Modeling, and Techniques, Dresden, Germany, 24–26 September 2013.
36. Salberg, A.; Larsen, S. Classification of Ocean Surface Slicks in Simulated Hybrid-Polarimetric SAR Data. *IEEE Trans. Geosci. Remote Sens.* **2018**, *56*, 7062–7073. [CrossRef]
37. Nunziata, M. Single- and Multi-Polarization Electromagnetic Models for SAR Sea Oil Slick Observation. Ph.D. Thesis, Parthenope University of Naples, Naples, Italy, December 2008.

Article

Oil Spill Discrimination by Using General Compact Polarimetric SAR Features

Junjun Yin [1], Jian Yang [2,*], Liangjiang Zhou [3] and Liying Xu [4]

[1] School of Computer and Communication Engineering, University of Science and Technology Beijing, Beijing 100083, China; junjun_yin@ustb.edu.cn
[2] Department of Electronic Engineering, Tsinghua University, Beijing 100084, China
[3] National Key Lab of Microwave Imaging Technology, Institute of Electronics, Chinese Academy of Sciences, Beijing 100190, China; ljzhou@mail.ie.ac.cn
[4] Shanghai Institute of Satellite Engineering, Shanghai 200240, China; xuliying@509.sast.casc
* Correspondence: yangjian_ee@tsinghua.edu.cn

Received: 31 December 2019; Accepted: 27 January 2020; Published: 3 February 2020

Abstract: Ocean surveillance is one of the important applications of synthetic aperture radar (SAR). Polarimetric SAR provides multi-channel information and shows great potential for monitoring ocean dynamic environments. Oil spills are a form of pollution that can seriously affect the marine ecosystem. Dual-polarimetric SAR systems are usually used for routine ocean surface monitoring. The hybrid dual-pol SAR imaging mode, known as compact polarimetry, can provide more information than the conventional dual-pol imaging modes. However, backscatter measurements of the hybrid dual-pol mode depend on the transmit wave polarization, which results in lacking consistent interpretation for various compact polarimetric (CP) images. In this study, we will explore the capability of different CP modes for oil spill detection and discrimination. Firstly, we introduce the general CP formalism method to formulate an arbitrary CP backscattered wave, such that the target scattering vector is characterized in the same framework for all CP modes. Then, a recently proposed CP decomposition method is investigated to reveal the backscattering properties of oil spills and their look-alikes. Both intensity and polarimetric features are studied to analyze the optimal CP mode for oil spill observation. Spaceborne polarimetric SAR data sets collected over natural oil slicks and experimental biogenic slicks are used to demonstrate the capability of the general CP mode for ocean surface surveillance.

Keywords: general compact polarimetry; hybrid dual-polarization; oil spill discrimination; target decomposition; ocean environment

1. Introduction

Marine oil spills have been of tremendous concern due to the adverse impact on ocean economic and ecological systems. It results in serious effects on coastal fisheries, sea creatures, seabirds, and eco-environment regeneration. Oil spills are, regrettably, common around the world; e.g., the 2010 Deepwater Horizon oil spill in the Gulf of Mexico, oil leakage from the Penglai 19-3 oil rig platform in 2011 in Bohai Bay, and the Rena oil spill that occurred off a coast in New Zealand in 2011. The rapid increase in oil spill pollution is primarily due to increased ocean activities by humans. The spatial distribution of the spills showed that the most frequent occurrence of oil spills takes place along the main tanker routes, near offshore oil platform positions, as well as in the large ports. Optical and microwave remote sensing techniques are mostly used to monitor marine oil spills, with microwave sensing having significant capability for observing ocean ecosystems [1,2].

Synthetic aperture radar (SAR) has all-day and all-weather imaging capabilities, where the satellite systems can provide periodical observations of high-risk areas. Polarimetric SAR (PolSAR) offers multi-channel polarimetric information, and the fully or quad-polarimetric (quad-pol) SAR system allows the complete backscattering characterization. It has been widely demonstrated that polarimetric information greatly improves the performance of SAR systems [3]. The quad-pol system, alternatively transmitting two orthogonal polarizations and receiving in both polarizations simultaneously, has many advantages but suffers from system complexity, data volume, and limited imaging swath compared to the SAR systems, which transmit only a single polarization. The dual-pol system is a compromise for the trade-off between imaging spatial coverages and observation dimensionality. The hybrid dual-pol or the named compact polarimetric (CP) SAR refers to a unique polarization in transmission and coherent orthogonal polarizations in reception. At present, the Indian RISAT-1 (2012), Japan JAXA (Japan Aerospace Exploration Agency) ALOS/PALSAR-2 (2014), Argentine SAOCOM-1A (2018), and the Canadian RADARSAT Constellation Mission (RCM, 2019) have CP imaging modes. In the future, CP modes have also been planned for SAOCOM-1B.

The techniques of processing CP images are categorized into two groups. One is to reconstruct the pseudo quad-pol data from compact polarimetry [4–10], and then quad-pol methods can be applied to the reconstructed data for various applications. The other is to extract target scattering parameters directly from the backscattered waves [11–16]. In this study, we focus on the detection of oil spills by using polarimetric features measured by CP modes. Radar backscatter is sensitive to the ocean capillary–gravity waves [17–19]. Under low to moderate sea conditions with intermediate radar incidence angles ranging from 20° to 60°, the scattering mechanism of the sea surface is often predominated by Bragg resonant scattering [3,20]. Any process that affects the ocean surface roughness can be imaged with SAR. Oil slicks not only damp the ocean capillary and gravity waves, but also reduce the surface tension and friction between the wind and liquid surface [21]. Therefore, oil slicks have a low backscattering signature [17–19,22]; i.e., oil slicks appear as distinguishable dark patches compared to the ambient areas. However, low backscatter features could also be created by other ocean phenomena, known as look-alikes, such as biogenic films, low wind regions, rain affects, sea ice, and upwellings, etc. It is crucial to distinguish between oil slicks and their look-alikes, because false alarms could initiate the costly manual activities and more seriously delay the cleaning activities of the spills. The fully polarimetric features, such as the polarimetric signature and pedestal height [23], the Mueller matrix-based filter [24–26], and the co-polarized phase difference (CPD) [27], have been investigated for oil spill observation and discrimination. In compact polarimetry, the performance of the degree of polarization (m) were studied for both oil spill and ship detections under several typical dual-pol modes [16], but analysis related to oil look-alikes was not included. CP feature extraction methods mainly have the $m-\delta$ decomposition [11,13], the $m-\chi$ decomposition [12], and the $m-\alpha_s$ decomposition [14]. These methods were proposed based on the circular CP mode, not applicable to other CP modes without any modification. In [10], we extended the $m-\alpha_s$ decomposition to the linear $\pi/4$ mode.

In fact, there are numerous possibilities of transmit wave ellipses on the polarization plane, and thus theoretically we have numerous hybrid dual-pol imaging modes. However, the hybrid dual-pol features were only studied under the conventional HH/VH and HV/VV polarizations, as well as the circular and linear $\pi/4$ CP modes. In the open literature, there are no studies that investigate the general CP mode for ocean target characterization. In [28], we demonstrated that scattering characterization under compact polarimetry should be described in the same framework for the purpose that unified algorithms applicable for all CP modes can be developed. A formalism method was first proposed for the CP backscattered vector, and then a polarization ratio-based target decomposition method was developed to represent the scattering mechanism and the scattering randomness of targets for an arbitrary hybrid dual-pol mode [28]. In this paper, the performances of the general CP features for oil spill detection and discrimination are analyzed and the optimal CP mode for ocean environment monitoring is studied. The organization is given as follows. In Section 2, the formalism of the general CP descriptors and the CP decomposition method are introduced. In Section 3, data sets of RADARSAT-2, SIR-C/X-SAR, and ALOS/PALSAR-1 are analyzed to show the ability of different CP modes to distinguish between oil spills and biogenic look-alikes. Finally, conclusions are given in Section 4.

2. The General Compact Polarimetric Features

2.1. Formalism of the General CP Descriptors

For an arbitrary transmitting electromagnetic (EM) wave, the CP measurements are a function of both the target and the transmit wave polarization. The backscattered wave is represented by a Jones vector [29], which is a 2-dimensional complex vector. It can be formulated by an absolute coefficient and a complex channel ratio, which represents the vector nature (or the polarimetric property) of the backscattered wave to characterize target scattering mechanisms. Suppose the transmit transverse EM wave is

$$\vec{E}_i(\theta, \chi) = \begin{bmatrix} a \\ b \end{bmatrix} = \begin{bmatrix} \cos\theta & -\sin\theta \\ \sin\theta & \cos\theta \end{bmatrix} \begin{bmatrix} \cos\chi \\ j\sin\chi \end{bmatrix} = \begin{bmatrix} \cos\theta\cos\chi - j\sin\theta\sin\chi \\ \sin\theta\cos\chi + j\cos\theta\sin\chi \end{bmatrix} \tag{1}$$

where θ and χ are the ellipse orientation and ellipticity angles, and a and b are the complex transmitting wave elements with $|a|^2 + |b|^2 = 1$. For a given target $\mathbf{S}$, the received CP (or hybrid dual-pol) signal is totally dependent on a and b (or θ and χ), as follows:

$$\vec{E}_r(\theta, \chi) = \mathbf{S}\vec{E}_i(\theta, \chi) = \begin{bmatrix} S_{\rm HH} & S_{\rm HV} \\ S_{\rm VH} & S_{\rm VV} \end{bmatrix} \begin{bmatrix} a \\ b \end{bmatrix} = \begin{bmatrix} aS_{\rm HH} + bS_{\rm HV} \\ bS_{\rm VV} + aS_{\rm VH} \end{bmatrix}. \tag{2}$$

This formula is represented in the linear H/V polarization basis. It should note that the CP measurements are independent of the receiving polarization coordinates. Equation (2) shows that the complex vector direction of the scattering wave is highly affected by the transmit wave $\vec{E}_i = \begin{bmatrix} a & b \end{bmatrix}^{\rm T}$. When $a = 0$ or $b = 0$, $\vec{E}_r$ corresponds to the conventional HV/VV or HH/VH dual-pol case. We only consider the general CP mode. When $a \neq 0$ as well as $b \neq 0$, the backscattered wave $\vec{E}_r$ can be projected to another space by a scaling transformation as

$$\vec{k}_1 = \begin{bmatrix} E_1 \\ E_2 \end{bmatrix} = \begin{bmatrix} a^{-1} & 0 \\ 0 & b^{-1} \end{bmatrix} \vec{E}_r(\theta, \chi) = \begin{bmatrix} S_{\rm HH} + \frac{b}{a}S_{\rm HV} \\ S_{\rm VV} + \frac{a}{b}S_{\rm VH} \end{bmatrix}, \tag{3}$$

where $\vec{k}_1$ is the formalized CP vector, which is the sum of the co-polarized and cross-polarized components. Compared to $\vec{E}_r$, in which both the co-polarized and cross-polarized terms are affected by the transmitting wave's polarization, $\vec{k}_1$ has a fixed term $\begin{bmatrix} S_{\rm HH} & S_{\rm VV} \end{bmatrix}^{\rm T}$ to characterize the scattering properties of a target under all CP modes. Another vector can thus be obtained from (3) by a unitary transform:

$$\vec{k}_2 = \frac{1}{\sqrt{2}} \begin{bmatrix} 1 & 1 \\ 1 & -1 \end{bmatrix} \begin{bmatrix} E_1 \\ E_2 \end{bmatrix} = \frac{1}{\sqrt{2}} \begin{bmatrix} E_1 + E_2 \\ E_1 - E_2 \end{bmatrix}. \tag{4}$$

Then, the second-order products, named as the formalized CP covariance and coherency matrices, are accordingly obtained to describe the stochastic backscattering process.

$$C_2 = \vec{k}_1 \vec{k}_1^{\rm H} = \begin{bmatrix} \langle |E_1|^2 \rangle & \langle E_1 E_2^* \rangle \\ \langle E_2 E_1^* \rangle & \langle |E_2|^2 \rangle \end{bmatrix} \tag{5}$$

$$T_2 = \langle \vec{k}_2 \vec{k}_2^{\rm H} \rangle = \begin{bmatrix} \frac{\langle |E_1 + E_2|^2 \rangle}{2} & \frac{\langle (E_1 + E_2)(E_1 - E_2)^* \rangle}{2} \\ \frac{\langle (E_1 - E_2)(E_1 + E_2)^* \rangle}{2} & \frac{\langle |E_1 - E_2|^2 \rangle}{2} \end{bmatrix} \tag{6}$$

where ^{H} denotes the matrix conjugate transpose and $\langle\cdot\rangle$ denotes the ensemble average. In [28], we discussed the sensitivity of different CP modes to the target geometrical parameters. It showed that for the surface and trihedral scatterers, the polarization ratios of the formalized vector are always distributed around (1, 0) (see Figure 1a,b in [28]). The difference between the two vectors, i.e., $\vec{E}_r$ and $\vec{k}_1$, is schematically shown in Figure 1. For backscatter from natural areas, the cross-pol term is relatively small as compared to the co-pol terms. Suppose the term Δ is negligible for both vectors. When (a,b) varies, the direction of $\vec{E}_r$, determined by both $\begin{bmatrix} S_{\mathrm{HH}} & S_{\mathrm{VV}} \end{bmatrix}^{\mathrm{T}}$ and (a,b), can be dramatically affected by the transmitting wave's phase δ ($\delta = angle(b/a)$). The direction of $\vec{k}_1$ is only determined by $\begin{bmatrix} S_{\mathrm{HH}} & S_{\mathrm{VV}} \end{bmatrix}^{\mathrm{T}}$. In the real scattering case, the end point of $\vec{k}_1$ varies around $\begin{bmatrix} S_{\mathrm{HH}} & S_{\mathrm{VV}} \end{bmatrix}^{\mathrm{T}}$, which is taken as a reference point to characterize the scattering mechanism in the formalized vector. We use real measurements for intuitive illustration. Scattering matrices from the ocean surface and oil slicks are as follows:

$$\begin{gathered} \mathbf{S}_{\mathrm{sea}} = e^{j\varphi_1}\begin{bmatrix} 0.1555 & -0.0064-0.0051i \\ -0.0064-0.0051i & 0.1571-0.0500i \end{bmatrix} \\ \mathbf{S}_{\mathrm{oil-slick}} = e^{j\varphi_2}\begin{bmatrix} 0.025 & -0.0031-0.0063i \\ -0.0031-0.0063i & 0.0595-0.0114i \end{bmatrix}, \end{gathered} \tag{7}$$

which are randomly selected from the test data used in the experiments. Figure 2 shows variation in the polarization ratios of $\vec{E}_r$ and $\vec{k}_1$ with the varying CP modes ($\theta = \pi/4$, $\chi \in \begin{bmatrix} -\pi/4 & \pi/4 \end{bmatrix}$) for the scattering types in (7). It shows that with the formalized scattering vector, the effect of the transmitted polarization on the backscattered wave is greatly reduced, especially for the ocean surface. To distinguish between oil slicks and the sea surface in different CP modes, multiple thresholds or a nonlinear curve are needed when $\vec{E}_r$ is used to represent the target features, while only one threshold is needed when $\vec{k}_1$ is used. Polarimetric properties of targets can be explained consistently for all CP modes with $\vec{k}_1$, which facilitate developing unified explanation algorithms for target characterization.

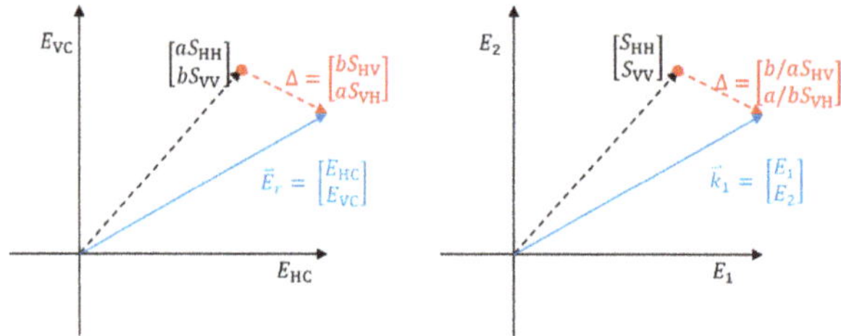

Figure 1. Schematic representation of $\vec{E}_r$ and $\vec{k}_1$.

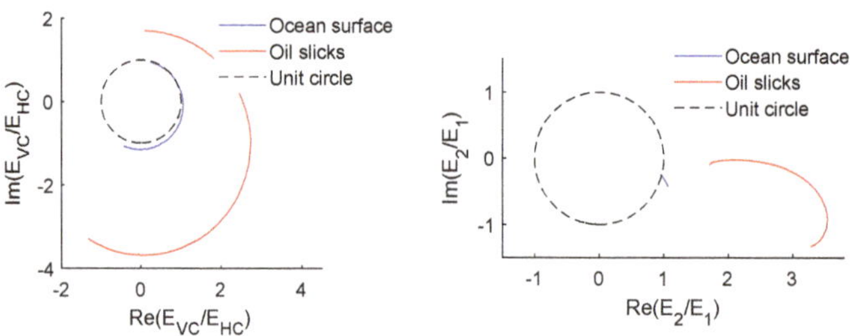

Figure 2. Variation of the polarization ratios of $\vec{E}_r$ and $\vec{k}_1$ with the varying compact polarimetric (CP) modes ($\theta = \pi/4, \chi \in [-\pi/4, \pi/4]$) for scatterers from the ocean surface and oil slicks.

2.2. Polarization Ratio-Based Decomposition for the General CP Images

By using the CP descriptors presented in (3)–(6), we proposed a polarization ratio-based decomposition method [28]. Polarization ratio is a fundamental parameter in revealing target scattering mechanisms. This idea was also employed for the $\Delta\alpha_B/\alpha_B$ method [30,31] in fully polarimetric imagery. The fully polarimetric and the general CP $\Delta\alpha_B/\alpha_B$ methods are mathematically equal, but there is difference in the physical interpretations. From matrix $\boldsymbol{T}_2$, we can define parameter α_{BCP} as

$$\alpha_{BCP} = \mathrm{atan}\left(\frac{\langle|E_1 - E_2|^2\rangle}{\langle|E_1 + E_2|^2\rangle}\right) \tag{8}$$

where $\alpha_{BCP} \in \begin{bmatrix} 0^\circ & 90^\circ \end{bmatrix}$ is used to describe the average scattering mechanism. For deterministic scatterers without obvious rotation, the cross-polarized term S_{HV} is usually small compared with the co-polarized terms. Then, by (3) and (8) it is easily known that when the transmitted wave is balanced in the channel amplitudes, i.e., $|a| \approx |b|$ (equivalent to $\theta \approx \pm\pi/4$ or $\chi \approx \pm\pi/4$), for surface scattering dominated areas, α_{BCP} is close to 0°; for double-bounce scattering dominated areas, α_{BCP} is close to 90°; and for random volume scattering, α_{BCP} is close to 45°. When the wave channel amplitudes are imbalanced, i.e., $|a| \gg |b|$ or $|a| \ll |b|$, the cross-polarized term will gradually play a leading role in determining the scattering mechanism with the imbalance increasing.

For the single-look data, α_{BCP} is equivalent to E_2/E_1. For the multi-look data, α_{BCP} is a function of the multi-look polarization ratio ρ_{CP} and the channel correlation coefficient $|r_{CP}|$ [28]. We defined another parameter to measure the effect of $|r_{CP}|$ on α_{BCP}, as follows:

$$\Delta\alpha_{BCP} = \alpha_{BCP} - \alpha_{0CP} \tag{9}$$

where

$$\alpha_{0CP} = \mathrm{atan}\left(\frac{|1-\rho_{CP}|^2}{|1+\rho_{CP}|^2}\right) \quad \text{and} \quad \rho_{CP} = \sqrt{\frac{\langle|E_2|^2\rangle}{\langle|E_1|^2\rangle}}e^{j\mathrm{angle}(\langle E_2 E_1{}^*\rangle)}. \tag{10}$$

α_{0CP} is only determined by the averaged polarization ratio ρ_{CP}. We use the distance between α_{BCP} and α_{0CP} to measure the scattering randomness. For all CP modes, $\Delta\alpha_{BCP}$ was distributed in the interval $\begin{bmatrix} -45^\circ & 45^\circ \end{bmatrix}$ and a larger $|\Delta\alpha_{BCP}|$ indicates a more random scattering process. The properties of α_{BCP} and $\Delta\alpha_{BCP}$ has been discussed in [28]. A diagram can be constructed as shown in Figure 3. The pixel distribution depends on the polarization phase difference, i.e., $\phi = \mathrm{angle}(\langle E_2 E_1{}^*\rangle)$.

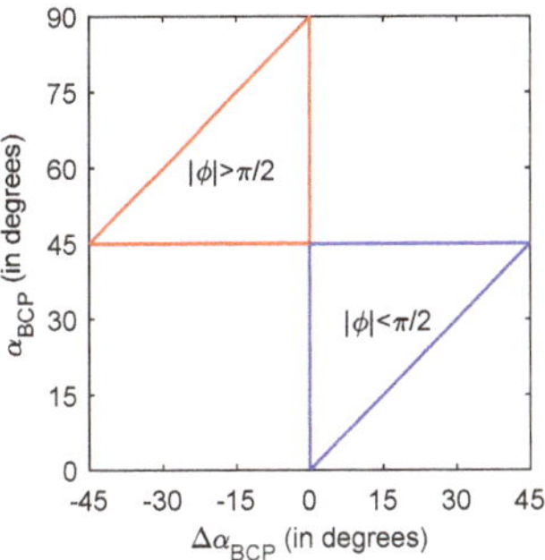

Figure 3. The $\Delta\alpha_{BCP}/\alpha_{BCP}$ plane. Pixels of all CP modes are distributed in the red and blue rectangles.

3. Experiments

3.1. Test Data Sets

To analyze the performances of various CP modes for oil spill detection and discrimination, data acquired by C-band RADARSAT-2, C-band SIR-C/X-SAR, as well as L-band ALOS/PALSAR-1 over oil slicks and biogenic slicks were used. The biogenic slicks were simulated using Oleyl Alcohol (OLA) by controlled experiments [26]. Data sets are specified in Table 1, in which the oil spills originated from oil platforms, such as the Penglai 19-3 oil slicks, as well as tanker accidents, such as the ALOS/PALSAR-1 data [25]. Figure 4 shows the Pauli-basis images. Before carrying out the analysis, polarimetric images were filtered by a sliding window for speckle reduction. Pixel spacing was in general taken into account for selection of the window size. However, it was found that when analyzing the scattering mechanisms, the filter window size does not affect the results too much if it varies within a small range, at least for the test data in this study. Thus, an appropriate and applicable window size of 5 was set for the experiments.

Table 1. Fully polarimetric synthetic aperture radar (SAR) images. The SIR-C/X-SAR data were measured in the C-band and p.n. is the processing number.

Sensor	Location or Site Identification	Pixel Spacing (in meters)	Incidence Angle (in degrees)	Acquisition Date	Object
ALOS/PALSAR-1	ALPSRP031440190	4.5*9.5	Center: 25.7°	2006-8-27	Oil slicks
RADARSAT-2	Penglai 19-3 oilfield, Bohai bay	4.7*5.5	36.5°–38.0°	2011-8-19	Oil slicks
SIR-C/X-SAR	p.n. 17041	12.5*12.5	35.4°–40.4°	1994-4-11	Oil slicks
SIR-C/X-SAR	p.n. 44327	12.5*12.5	44.1°–47.5°	1994-10-1	Oil slicks
SIR-C/X-SAR	p.n. 49939	12.5*12.5	47.2°–49.9°	1994-10-8	Oil slicks
SIR-C/X-SAR	p.n. 41467	12.5*12.5	25.8°–29.2°	1994-10-4	OLA
SIR-C/X-SAR	p.n. 11588	12.5*12.5	19.3°–24.4°	1994-4-15	OLA
SIR-C/X-SAR	p.n. 41370	12.5*12.5	26.2°–30.8°	1994-10-1	OLA

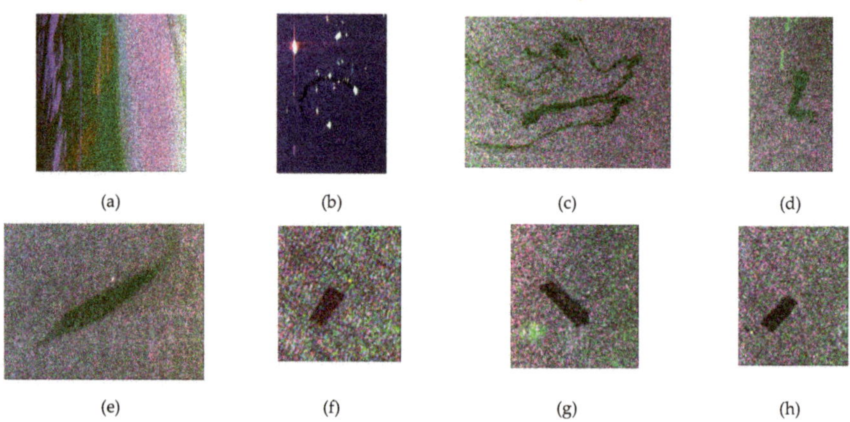

Figure 4. Pauli-basis images for oil slicks and Oleyl Alcohol (OLA). (**a**) ALOS/PALSAR-1 data, ALPSRP031440190; (**b**) RADARSAT-2 data acquired over the Penglai 19-3 oil field; (**c**–**h**) are SIR-C/X-SAR data sets with p.n. 17041, 44327, 49939, 41467, 11588, and 41370, respectively.

3.2. Oil Spill Detection

Radar signatures of natural slicks and oil spills are interpreted as dark patches in SAR images. The characteristic dark feature is a primary indicator for detection and mapping of potential oil spills. For various CP modes, the total backscattered energy varies with the transmit wave. The damping ratio has been widely used in SAR images for identification of surface slicks, including oil spills. For the hybrid dual-pol backscatter, we used the total intensity to define the contrast between the ocean surface and surface slicks as follows:

$$r_{ratio} = \frac{Tr(C_2)_{slick-free}}{Tr(C_2)_{observed\ area}} \tag{11}$$

where $Tr(\cdot)$ is the trace of a matrix. In damping ratio images, sea surface slicks will appear as bright spots against the dark background of water. We used SIR-C/X-SAR data with p.n. 49939 and 41370 for illustration. The damping ratio under the left circular mode in Figure 5a,b, i.e., $(\theta, \chi) = ([-\pi/2 \quad \pi/2], \pi/4)$, shows that the signature of the surface slicks is evident in the CP images. We used the areas outlined in Figure 5a,b to show the variations of the damping ratio with the transmitted wave polarizations, given in Figure 5c,d. The damping ratio in full polarimetry is also given, calculated based on the total backscattered energy, which was 2.63 ± 0.62 for the oil slicks and 2.59 ± 0.67 for OLA. It shows that when the transmitted wave approaches the conventional dual-pol modes, the contrast between the oil slicks and ocean surface is very small, which is not favorable to predict the surface slicks because it can generate missed detections. When the transmitting wave's polarization deviates from the H and V polarizations, the damping ratio increases and reaches its maximum at $\theta = \pm\pi/4$. Comparison of Figure 5c,d shows that the damping ratio is greatly affected by the ellipse orientation angle and that it varies only a little with the ellipticity angle. When the transmitting wave's channel amplitude is balanced, i.e., $\theta = \pm\pi/4$ or $\chi = \pm\pi/4$, the oil slick always has a larger contrast with the ambient water than the OLA slick. Damping ratios of oil slicks and OLA at the linear $\pi/4$ and the left circular modes are almost the same. When χ varies, the maximum difference between the damping ratios of oil slicks and OLA takes place at $(\theta, \chi) = (\pi/4, \pi/8)$. However, the overall difference is not significant.

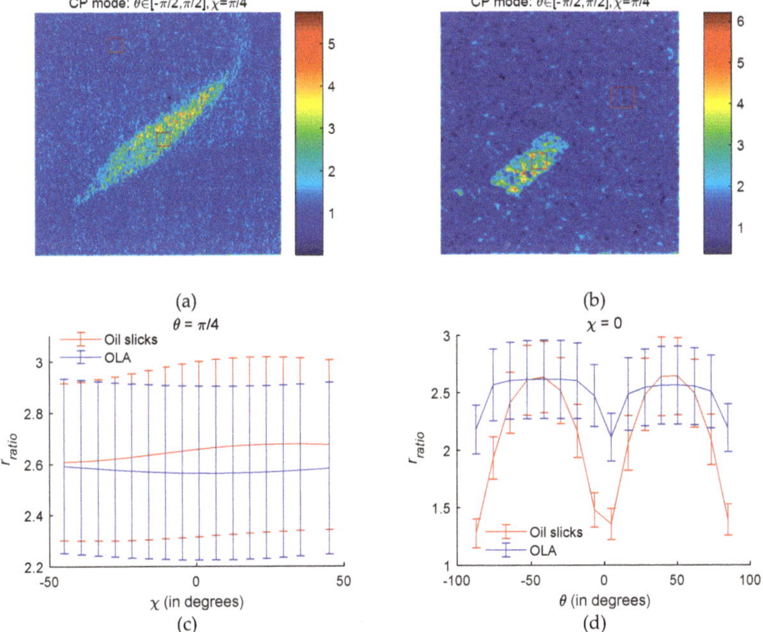

Figure 5. Damping ratios of oil slicks and OLA for the C-band SIR-C/X-SAR images with p.n. 49939 and p.n. 41370. (**a**) Damping ratio of oil slicks in the left circular mode; (**b**) damping ratio of OLA in the left circular mode; (**c**) variation of the damping ratio with χ changing from $-\pi/4$ to $\pi/4$; (**d**) variation of the damping ratio with θ changing from $-\pi/2 + eps$ to $\pi/2 - eps$.

The above analysis shows that low backscattered features are still visible in CP imagery. When the transmitted wave is balanced in the channel amplitude, i.e., $|a| = |b|$, the slicks that cause a dampening of the surface waves have a maximum contrast with the background surface. After the dark patches are detected, further analysis is needed to refine the results such that the possibility of false oil spill detection can be reduced. Next, the general CP features were analyzed to distinguish between oil slicks and OLA. α_{BCP} describes the physical scattering mechanism of targets in compact polarimetry, and $\Delta\alpha_{\mathrm{BCP}}$ relates to the scattering randomness. The left circular mode is used as an example. The general CP features for the test data sets are shown in Figure 6, where the $\Delta\alpha_{\mathrm{BCP}}/\alpha_{\mathrm{BCP}}$ scatter diagram is also given by using pixels of the outlined areas. The areas are only depicted in the α_{BCP} images for simplicity. It is observed that parameters $\Delta\alpha_{\mathrm{BCP}}$ and α_{BCP} can distinguish oil slicks from the background ocean surface for data of all the three sensors. The ocean surface always has small α_{BCP} and $\Delta\alpha_{\mathrm{BCP}}$ values, and oil slicks have relatively larger α_{BCP} and $|\Delta\alpha_{\mathrm{BCP}}|$ values. The oil slick images in Figure 6 indicate that in CP imagery, oil slicks and ocean surface can be discriminated by polarimetric features. We also observe that OLA and the ocean surface have very similar $\Delta\alpha_{\mathrm{BCP}}$ and α_{BCP} values. These two scattering types could not be separated in the $\Delta\alpha_{\mathrm{BCP}}/\alpha_{\mathrm{BCP}}$ diagram. The two-pixel groups overlap each other, indicating that OLA and the ocean surface have similar backscattering mechanisms in the circular CP mode.

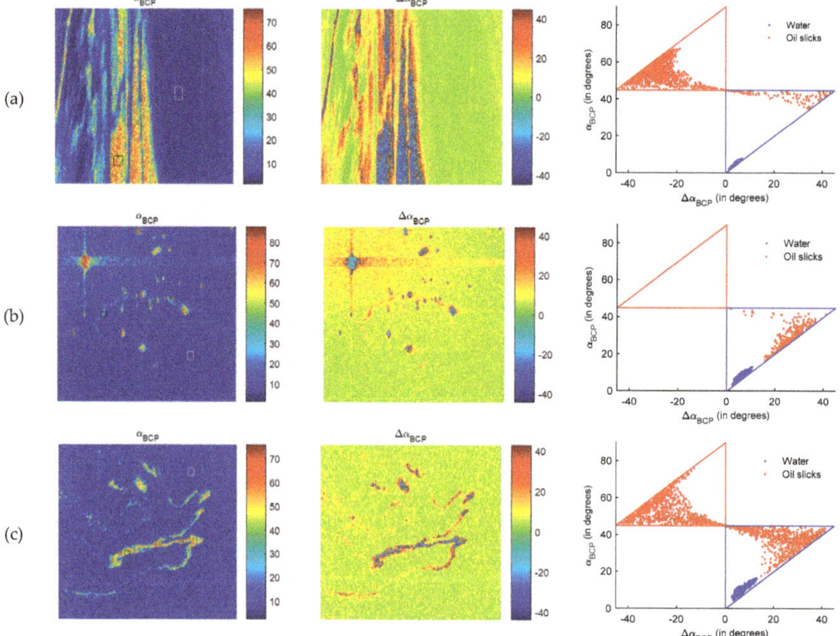

Figure 6. *Cont.*

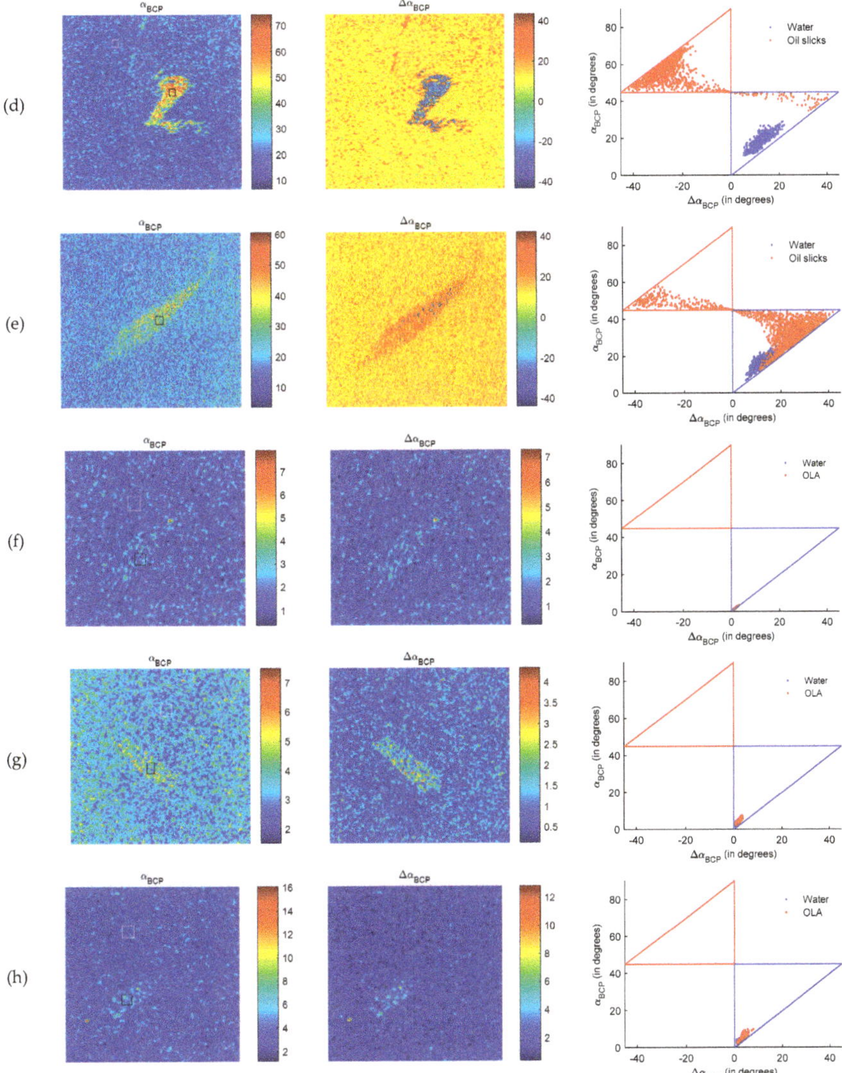

Figure 6. CP parameters in the left circular mode and the $\Delta\alpha_{\mathrm{BCP}}/\alpha_{\mathrm{BCP}}$ scatter plots. (**a**–**h**) used data from ALOS/PALSAR-1 with ALPSRP031440190, RADARSAT-2 (Penglai 19-3 oil field), and the SIR-C data sets with p.n. 17041, 44327, 49939, 41467, 11588, and 41370, respectively.

In [15], under the circular CP mode, we proposed a parameter ρ_{CTLR} (see (10) in [15]), which is a special case for α_{BCP}. It was shown that the performance of α_{BCP} in the circular mode is quite similar to that of α_{B} in the full-pol mode (see Figure 5a in [15]) for ocean surface characterization. For the test data sets in this study, experimental results show that the $\Delta\alpha_{\mathrm{BCP}}/\alpha_{\mathrm{BCP}}$ scatter plots in the circular CP modes have a good overall agreement with the FP $\Delta\alpha_{\mathrm{B}}/\alpha_{\mathrm{B}}$ distributions.

We further consider an arbitrary CP mode in the formalized vector to evaluate the effects of the transmitted wave's ellipticity and orientation angles on the surface slick representation. The data sets demonstrated in Figure 5 are used to keep the experimental analysis consistent. Averaged FP α_{B} and $\Delta\alpha_{\mathrm{B}}$ values of the outlined ocean surface, oil slick, and OLA areas in Figure 5, and their corresponding standard deviations, are given in Table 2. Variations of the CP α_{BCP} and $\Delta\alpha_{\mathrm{BCP}}$ values with the varying transmitted wave polarizations are shown in Figure 7. Table 2 shows the ocean surface and oil slicks have large differences in α_{B} and $\Delta\alpha_{\mathrm{B}}$. The averaged differences in both parameters for the two scatterers are about 15°. In contrast, the two parameters do not exhibit significant differences for ocean surface and OLA slicks. It is observed from Figure 7 that when the transmitted wave's amplitude is balanced in H and V polarization channels, variation trends of oil slicks, ocean surface, and OLA are very similar. When the transmitted wave is more circularly polarized, the difference between the oil slicks and ocean surface becomes larger with α_{BCP}. In the linear $\pi/4$ mode, the difference in α_{BCP} values is 6.5°, while in the circular mode, the difference increases to 15°. However, standard deviations of $\Delta\alpha_{\mathrm{BCP}}$ for oil slicks also increases with the increased polarization circularity. For all CP modes with the balanced channel amplitudes, the scattering mechanisms of ocean surface and OLA have little differences, with 1° on average in α_{BCP} and 1° on average in $\Delta\alpha_{\mathrm{BCP}}$. For the linearly polarized transmitted waves, the difference between the oil slicks and the ocean surface achieves its maximum when the wave orientation angle is at ±45°, indicating that when the transmit wave is with $|a| = |b|$ (equivalent to $\chi = \pm\pi/4$ or $\theta = \pm\pi/4$), oil slicks could be better detected from the ocean background. Differences between ocean surface and OLA slicks are very small with the varying wave orientation angles, indicating that ocean surface and OLA have a similar scattering mechanism. Figure 7 also shows that the scattering mechanisms of targets are greatly affected by transmitting wave orientations and less sensitive to wave ellipticity angles. When the wave orientation angle varies, a maximum change of 35° can be found in α_{BCP} and 15° in $\Delta\alpha_{\mathrm{BCP}}$ for ocean surface. While the maximum changes within both parameters for ocean surface are 5° with the varying ellipticity angles.

In Table 2 and Figure 7, target scattering mechanisms in polarimetric modes are analyzed based on 2 representative data sets. To further validate the performances of the CP parameters for surface slick characterization, results of the other six data sets in Table 1 are also given. Table 3 shows the averaged FP α_{B} and $\Delta\alpha_{\mathrm{B}}$ values for the ocean surface, oil-slick, and OLA areas outlined in Figure 6. Variations of the CP parameters with the varying transmit polarizations for these areas are shown in Figure 8. From Table 3 and Figure 8, similar analysis results can be observed as those from Table 2 and Figure 7, verifying that the CP $\Delta\alpha_{\mathrm{BCP}}$ and α_{BCP} is effective in differentiating target scattering mechanisms, and CP modes with balanced transmit channel amplitudes are better for detection of oil spills.

Table 2. α_{B} and $\Delta\alpha_{\mathrm{B}}$ values (in degrees) of the FP images for the outlined areas in Figure 5.

mean(·) ± std (·)	SIR-C Data with p.n. 49939		SIR-C Data with p.n. 41370	
	Water	Oil Slicks	Water	OLA
α_{B}	17 ± 3	32 ± 7	3.7 ± 0.7	4.5 ± 1
$\Delta\alpha_{\mathrm{B}}$	12 ± 3	27 ± 6	1.8 ± 0.5	2.9 ± 1

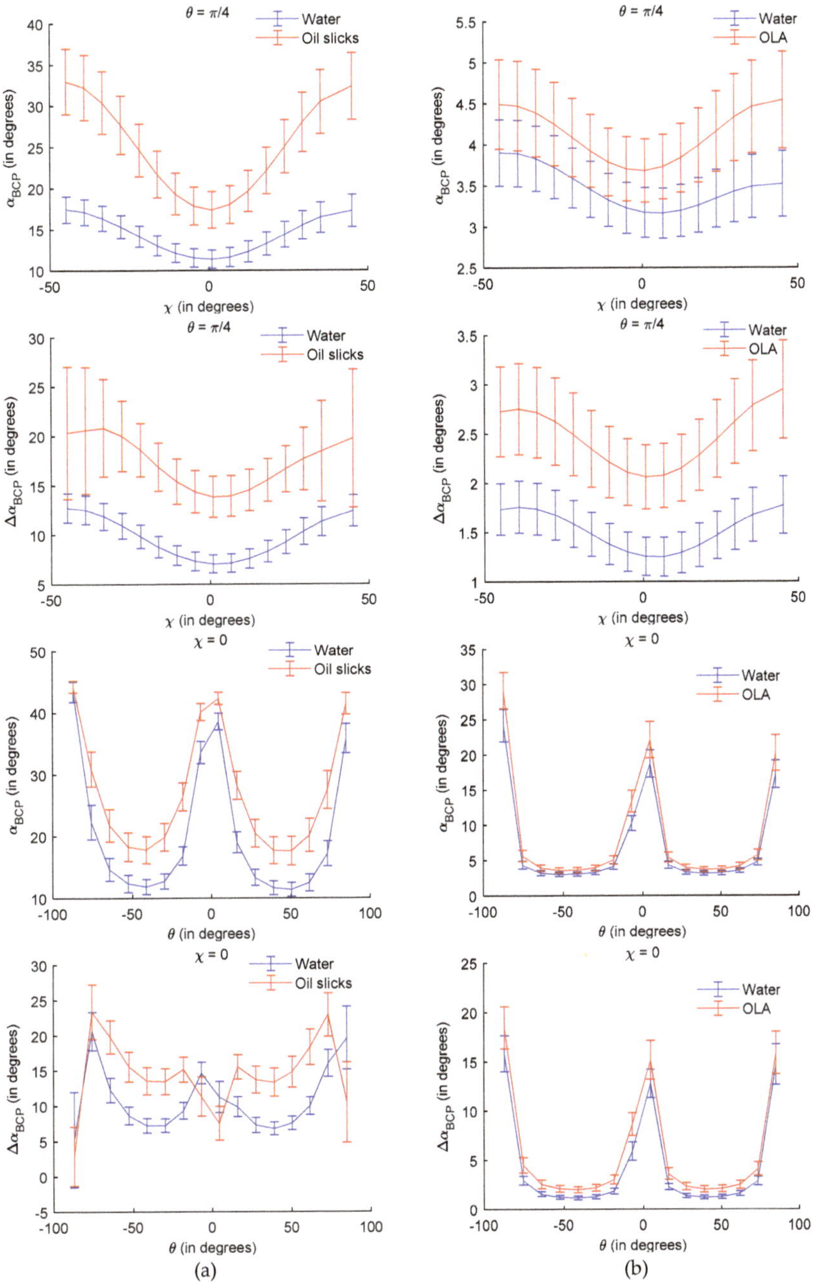

Figure 7. Variation of α_{BCP} and $\Delta\alpha_{BCP}$ values with varying transmitted wave polarizations for the water surface, oil slick, and OLA areas outlined in Figure 5. (**a**) SIR-C data with p.n. 49939. (**b**) SIR-C data with p.n. 41370.

Table 3. α_B and $\Delta\alpha_B$ values (in degrees) of the other six FP images. Areas outlined in Figure 6 were used for calculation.

Mean (·) ± Std (·)	ALPSRP031440190		RADARSAT-2		SIR-C (17041)		SIR-C (44327)		SIR-C (41467)		SIR-C (11588)	
	Water	**Oil slicks**	**Water**	**Oil slicks**	**Water**	**Oil slicks**	**Water**	**Oil slicks**	**Water**	**OLA**	**Water**	**OLA**
α_B	4.3 ± 0.8	51.8 ± 5	7.5 ± 1.8	30.2 ± 6	8.8 ± 1.8	44 ± 10	18 ± 4.5	53 ± 7.2	1.4 ± 0.3	1.8 ± 0.6	3 ± 0.5	4 ± 0.9
$\Delta\alpha_B$	3.4 ± 0.7	35 ± 19	5.7 ± 1.6	28.4 ± 6	5.2 ± 1.5	28 ± 17	12 ± 5	27 ± 23	1.1 ± 0.3	1.6 ± 0.6	1 ± 0.3	1.7 ± 0.6

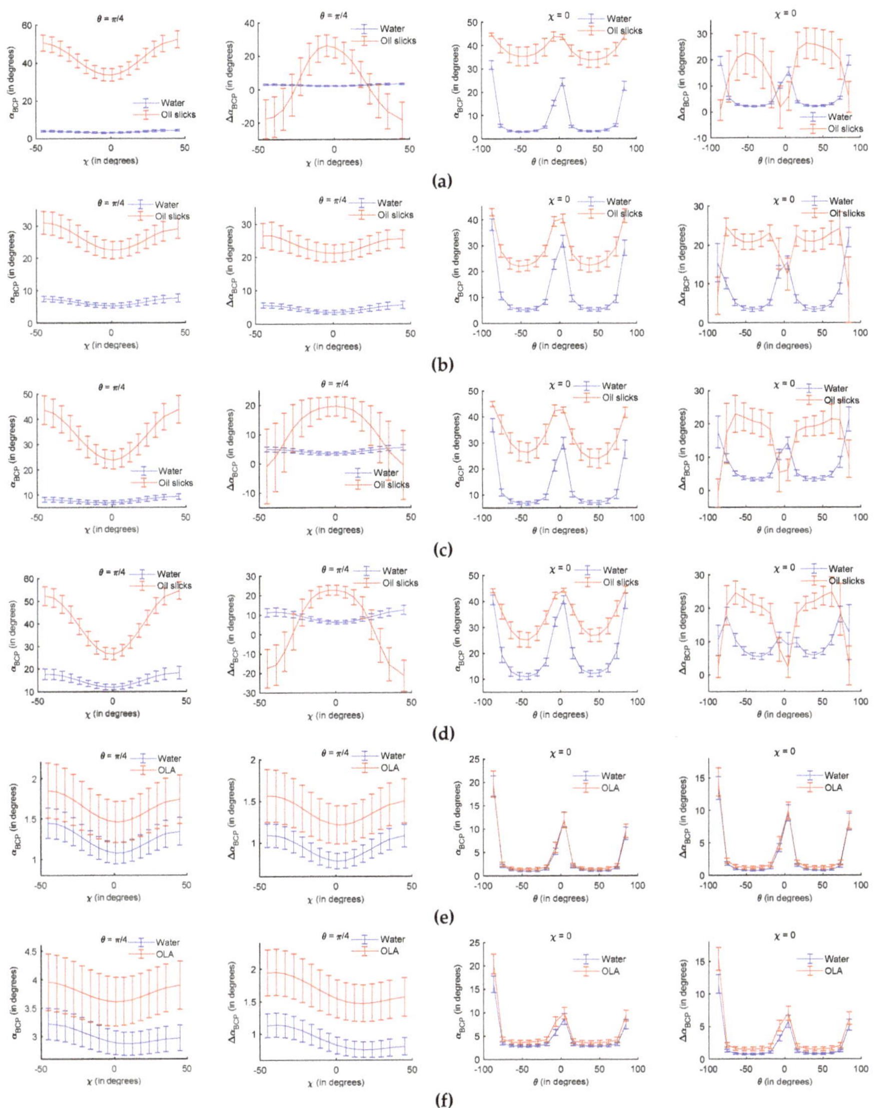

Figure 8. Variation of α_{BCP} and $\Delta\alpha_{\text{BCP}}$ values with transmitting polarizations for the water surface, oil slick, and OLA areas outlined in Figure 6. (**a**) ALOS/PALSAR-1 data (ALPSRP031440190); (**b**) RADARSAT-2 data (Penglai 19-3 oilfield); (**c**) SIR-C data with p.n. 17041; (**d**) SIR-C data with p.n. 44327; (**e**) SIR-C data with p.n. 41467; (**f**) SIR-C data with p.n. 11588.

Finally, an example carried out in the circular and linear $\pi/4$ CP modes is given for an intuitive visualization of the oil spill detection results. Both the characteristic low backscattered feature and polarimetric features are utilized. The experiment follows steps as follows. First, the damping ratio is used to detect surface slicks, where the Parzen window with a Gaussian Kernel is employed to model the ocean clutter. Low backscattered features are detected based on a given false alarm rate, which is set to $p_{fa} = 0.5\%$ in this example. Then, the α_{BCP} and $\Delta\alpha_{\text{BCP}}$ parameters are used to discriminate between oil slicks and OLA. We simply use the Euclidean distance to measure the dissimilarity of a detected slick and ocean surface. According to Tables 2 and 3 and Figure 7 and Figure 8, we set the threshold as 5°, which is the median value of the distances between oil slicks and ocean surface and the distances between OLA and ocean surface. Detection results are shown in Figure 9. The detected oil spills are shown in yellow and the detected low backscatter features are shown in green. Figure 9 shows that in compact polarimetry, oil slicks and OLA can be distinguished from the ocean surface by combining polarimetric features and the backscattered intensity. Detection results of the circular and linear $\pi/4$ CP modes are quite similar, but comparatively the circular CP mode can detect more areas of oil spills. Both modes can discriminate oil slicks from OLA.

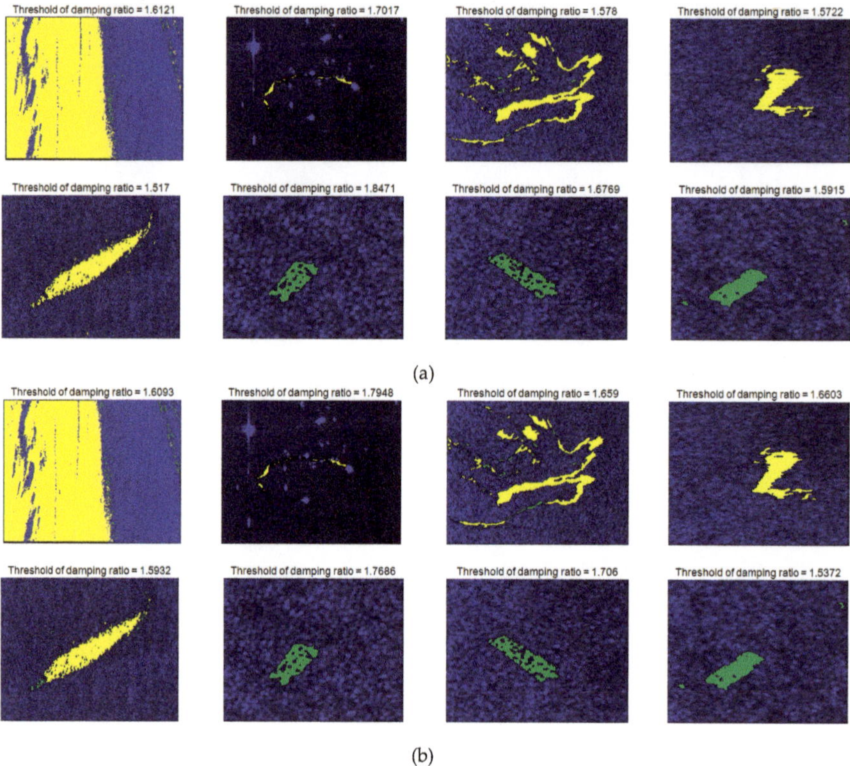

Figure 9. An example of an oil spill and the OLA detection results. The pseudo-color composite image is fused by R = Detection of oil slicks, G = Detection of dark features, and B = Span image. (**a**) The left circular mode; (**b**) the linear $\pi/4$ mode.

4. Conclusions

Oil spill detection is a very important step for ocean environment managing. In this paper, the general compact polarimetric (CP) mode, which refers to a coherent dual-pol system with an arbitrary transmitted elliptical polarized wave, was first analyzed to observe oil spills and biogenic slicks. A previously proposed formalism method is employed to describe the general CP measurement. We showed that the formalized vector is better in characterizing target scattering mechanism for ocean surface with and without slicks as compared with the original scattering vector. Both the backscattered intensity, which is favorable in detection of surface slicks due to their characteristic dark signatures in SAR images, and polarimetric parameters, which can discriminate between scattering mechanisms, are analyzed. Polarimetric SAR data from C-band SIR-C/X-SAR and RADARSAT-2, as well as L-band ALOS/PALSAR-1 were used in experiments. Results demonstrated the effectiveness of the general CP parameters, i.e., α_{BCP} and $\Delta\alpha_{BCP}$, for oil spill detection. Analysis of these two parameters showed that the scattering mechanism of the ocean surface is very similar to that of the OLA slicks but different from that of oil spills. The CP modes for which the transmit wave amplitude is balanced in the H and V coordinates, i.e., $\theta = \pm\pi/4$ or $\chi = \pm\pi/4$, enable better detection performances. Compared to the linear $\pi/4$ mode, the circular mode is better in preserving the integrity of the detected oil spill areas.

Author Contributions: Conceptualization and methodology, J.Y.; validation, J.Y.; investigation, L.Z. and L.X. All authors have read and agreed to the published version of the manuscript.

Funding: This work was funded in part by NSFC under Grant 61771043, NSFC under Grant 61490693, the National Key R&D Program of China under Grant 2017YFB0502703, the Fundamental Research Funds for the Central Universities under Grant FRF-IDRY-19-008 and FRF-TP-18-013A2, the USTB-NTUT Joint Research Program under Grant TW2019010, and the foundation SAST2018-037 of CASC.

Conflicts of Interest: The authors declare no conflict of interest.

References

1. Sandven, S.; Hamre, T.; Babiker, M.; Kloster, K.; Hansen, M.; Wåhlin, J.; Kudryavtsev, V.; Myasoedov, A.G.; Alexandrov, V.; Melentyev, V.V.; et al. *MONRUK SAR Data Analysis Report*; NERSC Technical Report no. 310; Nansen Environmental and Remote Sensing Center: Bergen, Norway, 2005–2007.
2. Alpers, W.; Espedal, H.A. Chapter 11 oils and surfactants. In *Synthetic Aperture Radar Marine User's Manual*; NOAA: Washington, DC, USA, 2004; pp. 263–277.
3. Crisp, D.J. *The State-Of-The-Art In Ship Detection In Synthetic Aperture Radar Imagery*; Defence Science and Technology Organization (DSTO), Information Sciences Laboratory: Edinburgh, SA, Australia, 2004.
4. Souyris, J.C.; Imbo, P.; Fjørtoft, R.; Mingot, S.; Lee, J.-S. Compact polarimetry based on symmetry properties of geophysical media: the π/4 mode. *IEEE Trans. Geosci. Remote Sens.* **2005**, *43*, 634–646. [CrossRef]
5. Nord, M.; Ainsworth, T.; Lee, J.-S.; Stacy, N. Comparison of compact polarimetric synthetic aperture radar modes. *IEEE Trans. Geosci. Remote Sens.* **2009**, *47*, 174–188. [CrossRef]
6. Collins, M.; Denbina, M.; Atteia, G. On the reconstruction of quad-pol SAR data from compact polarimetry data for ocean target detection. *IEEE Trans. Geosci. Remote Sens.* **2013**, *51*, 591–600. [CrossRef]
7. Li, Y.; Zhang, Y.; Chen, J.; Zhang, H. Improved compact polarimetric SAR quad-pol reconstruction algorithm for oil spill detection. *IEEE Geosci. Remote Sens. Lett.* **2014**, *11*, 1139–1142. [CrossRef]
8. Yin, J.; Yang, J. Multi-polarization reconstruction from compact polarimetry based on modified four-component scattering decomposition. *J. Syst. Eng. Electron.* **2014**, *25*, 399–410. [CrossRef]
9. Yin, J.; Moon, W.M.; Yang, J. Model-based pseudo-quad-pol reconstruction from compact polarimetry and its application to oil-spill observation. *J. Sens.* **2015**, *2015*, 1–8. [CrossRef]
10. Yin, J.; Papathanassiou, K.; Yang, J.; Chen, P. Least-squares estimation for pseudo quad-pol image reconstruction from linear compact polarimetric SAR. *IEEE J. Sel. Topics Appl. Earth Observ. Remote Sens.* **2019**, *12*, 3746–3758. [CrossRef]
11. Raney, R.K. Hybrid-polarity SAR architecture. *IEEE Trans. Geosci. Remote Sens.* **2007**, *45*, 3397–3404. [CrossRef]
12. Raney, R.K.; Cahil, J.T.S.; Patterson, G.W.; Bussey, D.B.J. The M-Chi decomposition of hybrid dual-polarimetric radar data. In Proceedings of the IEEE International Geoscience and Remote Sensing Symposium, Munich, Germany, 22–27 July 2012; pp. 5093–5096.
13. Charbonneau, F.J.; Brisco, B.; Raney, R.K.; McNairn, H.; Liu, C.; Vachon, P.W.; Shang, J.; DeAbreu, R.; Champagne, C.; Merzouki, A.; et al. Compact polarimetry overview and applications assessment. *Can. J. Remote Sens.* **2010**, *36*, S298–S315. [CrossRef]
14. Cloude, S.; Goodenough, D.; Chen, H. Compact decomposition theory. *IEEE Geosci. Remote Sens. Lett.* **2012**, *9*, 28–32. [CrossRef]
15. Yin, J.; Yang, J.; Zhou, Z.-S.; Song, J. The extended Bragg scattering model-based method for ship and oil-spill observation using compact polarimetric SAR. *IEEE J. Sel. Top. Appl. Earth Obs. Remote Sens.* **2015**, *8*, 3760–3772. [CrossRef]
16. Shirvany, R.; Chabert, M.; Tourneret, J.-Y. Ship and oil-spill detection using the degree of polarization in linear and hybrid/compact dual-pol SAR. *IEEE J. Sel. Top. Appl. Earth Observ. Remote Sens.* **2012**, *5*, 885–892. [CrossRef]

17. Pablo, C.-C.; Yan, X.-H. Low-backscatter ocean features in synthetic aperture radar imagery. *Johns Hopkins APL Tech. Dig.* **2000**, *21*, 116–121.
18. Gade, M.; Alpers, W.; Hühnerfuss, H.; Wismann, V.R.; Lange, P.A. On the reduction of the radar backscatter by oceanic surface films: scatterometer measurements and their theoretical interpretation. *Remote Sens. Environ.* **1998**, *66*, 52–70. [CrossRef]
19. Solberg, A.H.S. *Remote Sensing Of Ocean Oil-Spill Pollution*; Institute of Electrical and Electronics Engineers: Piscataway, NJ, USA, 2012; Volume 100, pp. 2931–2945.
20. Schuler, D.L.; Lee, J.-S. Mapping ocean surface features using biogenic slick-fields and SAR polarimetric decomposition techniques. *IEE Proc. -Radar Sonar Navig.* **2006**, *153*, 260–270. [CrossRef]
21. Nunziata, F.; Sobieski, P.; Migliaccio, M. The two-scale BPM scattering model for sea biogenic slicks contrast. *IEEE Trans. Geosci. Remote Sens.* **2009**, *47*, 1949–1956. [CrossRef]
22. Kim, D.-J.; Moon, W.M.; Kim, Y.-S. Application of TerraSAR-X data for emergent oil-spill monitoring. *IEEE Trans. Geosci. Remote Sens.* **2010**, *48*, 852–863.
23. Migliaccio, M.; Nunziata, F.; Gambardella, A. Polarimetric signature for oil spill observation. In Proceedings of the 2008 IEEE/OES US/EU-Baltic International Symposium, Tallinn, Estonia, 27–29 May 2008; pp. 1–5.
24. Nunziata, F.; Gambardella, A.; Migliaccio, M. On the Mueller scattering matrix for SAR sea oil slick observation. *IEEE Geosci. Remote Sens. Lett.* **2008**, *5*, 691–695. [CrossRef]
25. Migliaccio, M.; Gambardella, A.; Nunziata, F.; Shimada, M.; Isoguchi, O. The PALSAR polarimetric mode for sea oil slick observation. *IEEE Trans. Geosci. Remote Sens.* **2009**, *47*, 4032–4131. [CrossRef]
26. Nunziata, F. Single- and Multi-Polarization Electromagnetic Models for SAR Sea Oil Slick Observation. Ph.D. Thesis, Parthenope University of Naples, Naples, Italy, December 2008.
27. Velotto, D.; Migliaccio, M.; Nunziata, F.; Lehner, S. Dual-polarized TerraSAR-X data for oil-spill observation. *IEEE Trans. Geosci. Remote Sens.* **2011**, *49*, 4751–4762. [CrossRef]
28. Yin, J.; Papathanassiou, K.; Yang, J. Formalism of compact polarimetric descriptors and extension of the $\Delta\alpha B/\alpha B$ method for general compact-pol SAR. *IEEE Trans. Geosci. Remote Sens.* **2019**, *57*, 10322–10335. [CrossRef]
29. Lee, J.-S.; Pottier, E. *Polarimetric Radar Imaging from Basics to Applications*; CRC Press: Boca Raton, FL, USA, 2009; pp. 31–98.
30. Yin, J.; Moon, W.M.; Yang, J. Novel model-based method for identification of scattering mechanisms in polarimetric SAR data. *IEEE Trans. Geosci. Remote Sens.* **2016**, *54*, 520–532. [CrossRef]
31. Yin, J.; Yang, J.; Zhang, Q. Assessment of GF-3 polarimetric SAR data for physical scattering mechanism analysis and terrain classification. *Sensors* **2017**, *17*, 2785. [CrossRef] [PubMed]

Article

Combining Segmentation Network and Nonsubsampled Contourlet Transform for Automatic Marine Raft Aquaculture Area Extraction from Sentinel-1 Images

Yi Zhang [1,2], Chengyi Wang [1,*], Yuan Ji [3], Jingbo Chen [1], Yupeng Deng [1,2], Jing Chen [1,2] and Yongshi Jie [1,2]

1 Aerospace Information Research Institute, Chinese Academy of Sciences, Beijing 100101, China; zhangyi@radi.ac.cn (Y.Z.); chenjb@aircas.ac.cn (J.C.); dengyp@radi.ac.cn (Y.D.); chenjing185@mails.ucas.ac.cn (J.C.); jieys@radi.ac.cn (Y.J.)

2 School of Electronic, Electrical and Communication Engineering, University of Chinese Academy of Sciences, Beijing 100049, China

3 People's Liberation Army 91039 Troop, Beijing 102401, China; jiyuan13@163.com

* Correspondence: wangcy@radi.ac.cn; Tel.: +86-136-9327-5492

Received: 9 November 2020; Accepted: 17 December 2020; Published: 21 December 2020

Abstract: Marine raft aquaculture (MFA) plays an important role in the marine economy and ecosystem. With the characteristics of covering a large area and being sparsely distributed in sea area, MFA monitoring suffers from the low efficiency of field survey and poor data of optical satellite imagery. Synthetic aperture radar (SAR) satellite imagery is currently considered to be an effective data source, while the state-of-the-art methods require manual parameter tuning under the guidance of professional experience. To preclude the limitation, this paper proposes a segmentation network combined with nonsubsampled contourlet transform (NSCT) to extract MFA areas using Sentinel-1 images. The proposed method is highlighted by several improvements based on the feature analysis of MFA. First, the NSCT was applied to enhance the contour and orientation features. Second, multiscale and asymmetric convolutions were introduced to fit the multisize and strip-like features more effectively. Third, both channel and spatial attention modules were adopted in the network architecture to overcome the problems of boundary fuzziness and area incompleteness. Experiments showed that the method can effectively extract marine raft culture areas. Although further research is needed to overcome the problem of interference caused by excessive waves, this paper provides a promising approach for periodical monitoring MFA in a large area with high efficiency and acceptable accuracy.

Keywords: marine raft aquaculture; Sentinel-1; nonsubsampled contourlet transform; semantic segmentation; fully convolutional network

1. Introduction

According to the data from the Food and Agriculture Organization of the United Nations, aquaculture production worldwide has surpassed that of capture fisheries, and it has been steadily increasing year by year [1]. As the primary mean for coastal aquaculture, marine raft aquaculture plays an important role in the development of the global marine economy and has a considerable impact on the global marine ecosystem. For example, the 2008 Yellow Sea shade green tide outbreak induced by marine raft aquaculture had a negative impact on the Olympic sailing event and led to the loss of RMB 1.3 billion. This serious environmental catastrophe stemmed largely from the environmental pressure caused by the increasing number of marine raft aquaculture areas [2]. The distribution and number of marine raft aquaculture areas reflect the development status of the fishery as well as the quality of the water environment. The monitoring of marine raft aquaculture areas is of great significance for the protection of marine ecosystems and the sustainable use of marine fishery resources.

Marine raft aquaculture covers a wide area, is far from land, and is sparsely distributed. At present, the relevant government departments mainly rely on field surveys to monitor it. As a result of sea conditions, travel means, and weather, it is difficult to use manual inspection to detect illegal areas and keep track of the distribution and quantity of floating rafts in a timely manner. Remote sensing satellites have the capacity to periodically observe wide areas with few ground restrictions, which enables regular and quick monitoring of marine raft aquaculture areas. A combination of remote sensing monitoring technology and field surveys can provide comprehensive and efficient monitoring of marine aquaculture areas, which is conducive to the orderly development of aquaculture and the protection of natural ecology [3].

Raft culture is a form of aquaculture that uses floats and ropes to form rafts on the sea surface that are fixed to the seabed with cables, which have algae or shellfish suspended on slings [4]. As shown in Figure 1, the floating raft is divided into two parts, i.e., above and below water, and the water surface mainly contains floating balls. The structure of the floating raft makes it difficult for passive remote sensing to capture its reflected signal. Limited by imaging modality, it is difficult to accurately describe the marine raft culture area on optical satellite images [5,6]. Additionally, sea environmental elements such as wind and waves as well as fog render it difficult to extract a research target using optical remote sensing imagery. Synthetic aperture radar (SAR) actively emits electromagnetic waves (fixed frequency beams) and collects reflected and back-scattered signals. It is considered to be the best method for monitoring marine environments because it is not affected by the above elements [7]. Therefore, marine raft aquaculture monitoring through SAR images has practical research significance, especially in coastal cities where mariculture is the main economic activity. Timely and effective monitoring of marine raft aquaculture areas can effectively assist in the planning of marine aquaculture resources.

(**a**)

(**b**)

Figure 1. Raft culture. A floating raft has two parts: underwater (**a**) and water (**b**) [5].

In recent years, researchers have focused on SAR-based methods for aquaculture area extraction. Chu et al. extracted raft aquaculture areas using various filtering methods and human–computer interaction [8]. Fan et al. proposed a joint sparse representation classification method to construct meaningful texture features of raft aquaculture on the basis of wavelet decomposition and gray-level co-occurrence matrix (GLCM) statistical methods [9]. Hu et al. improved the statistical region merging algorithm for superpixel segmentation and used a fuzzy compactness and separation clustering algorithm to identify raft aquaculture areas from SAR images [10]. Geng et al. extracted raft aquaculture areas by means of weighted fusion classifiers and sparse encoders [11,12]. These methods are efficient in certain regions with the help of professional experience. However, knowledge-intensive feature engineering always leads to low robustness. The empirical parameter tuning causes the above methods to not work well with different data and in different regions.

The emergence of convolutional neural networks has provided a way to avoid intensive parameter tuning through deep learning and has led to the focus on object extraction based on semantic segmentation network. Long et al. proposed fully convolutional network (FCN) [13] as the pioneering work of deep learning semantic segmentation model based on full convolutional network, and subsequent algorithms have improved on this framework. Ronneberger et al. proposed U-net [14] to improve the FCN's loss of information in task practice with an encoder–decoder network structure. Then, this method was followed by several models such as DeepLab (v1/v2/v3) [15–17], multi-path refinement networks (RefineNet) [18], and pyramid scene parsing network (PSPNet) [19]. Advances in semantic segmentation network have made it possible to improve the accuracy and efficiency of marine raft aquaculture area extraction. Yueming et al. used richer convolutional features network (RCF) [20] to extract rafts through edge detection in a raft aquaculture area in Sanduao, China [21]. Shi et al. used dual-scale homogeneous convolutional neural network (DS-HCN) to extract rafts in a dual-scale full convolutional network, finding it had superior performance on marine raft aquaculture in Dalian, China [22]. Cui et al. proposed improved U-net with a pyramid upsampling and squeeze-excitation (PSE) structure (UPS-net), which captures both boundary and background information by adding PSE structures to the decoder part of U-net, with this method being effectively verified in marine raft aquaculture in eastern Lianyungang, China [23]. However, Yue's method suffers from partial edges [21]. Shi's method is mainly aimed at rafts, and the segmentation results are incomplete and suffer from the adhesion problem [22]. Cui's method has been experimentally demonstrated to be more accurate than other popular networks based on the FCN model framework. It was proposed to solve the adhesion problem of the DS-HCN method, and it is more suitable for marine aquaculture than the DS-HCN method, but it does not take advantage of the characteristics of the raft itself, and the edge of the raft is rough and incomplete [23].

It can be seen that state-of-the-art works, such as those by Chu, Fan, and Hu, mainly rely on artificial parameter adjustments and feature designs. The deep learning method for semantic segmentation avoids a large amount of manual work, but there are still poor integrity and boundary fuzzy flaws in the detection results.

This paper proposes a segmentation method, which combines a semantic segmentation network with the nonsubsampled contourlet transform (NSCT), to extract marine raft aquaculture areas and to overcome the phenomena of rough edges, adhesion, and incomplete results in the existing methods. To the best of our knowledge, this paper is the first to attempt to use a semantic segmentation network to extract marine raft aquaculture areas from SAR images.

The method is characterized by improvements in feature enhancement and model optimization on the basis of the feature analysis of marine raft aquaculture areas, as follows:

1. To address the low signal-to-noise ratio problem in SAR images, we enhanced Sentinel-1 images with the NSCT [24] to strengthen the subject contour features and directional features.
2. To capture better feature representations, we combined several modules in U-net. Multiscale convolution was used to fit the multisize characteristics of marine raft aquaculture areas, asymmetric convolution was selected to address the floating raft strip geometric features, and the attention module was adopted to focus on both spatial and channel interrelationships.

This paper is organized as follows. The first part introduces the background significance of marine raft aquaculture area extraction and the current research status, the second part analyzes the characteristics of marine raft aquaculture areas, the third part introduces the details of the method proposed in this paper, the fourth part shows the experimental results and analyzes the results, the fifth part provides the discussion, and the sixth part is the conclusion.

2. Feature Analysis of Marine Raft Aquaculture Areas

Feature analyses of the marine raft aquaculture areas provide the basis for the design of the method. A SAR image is the reflection of the target on radar beam, and the single-band echo information reflects more scattering characteristics and structural characteristics of the target. Hence, this section focuses on the scattering characteristics and structural characteristics of raft aquaculture areas.

2.1. Scattering Characteristics

Rafts are basically floating with floating balls on the surface of the water, and thus the scattering from the raft culture area consists mainly of surface scattering from the seawater and the balls, with two-sided angles and spirals scattering between them [25]. Therefore, an area where a raft exists has a different scattering intensity than areas with only seawater. Due to the presence of waves, surges, currents, and internal waves in various regions of the ocean, the backscatter characteristics of the ocean are very irregular. Furthermore, the backscatter characteristics of floating raft, influenced by the sea state, vary in different areas of the ocean. Therefore, enhancing the features of marine raft aquaculture areas in SAR images is necessary to enhance the commonalities among marine raft aquaculture areas in order to overcome the lack of floating raft features in SAR images and to mitigate background effects.

SAR images are visualized by the coherent processing of echoes from successive radar pulses, in which coherent speckle noise is unavoidable. Speckle noise exhibits a granular, black and white dotted texture on an image. Due to noise, some pixels in a homogeneous region are brighter than average, while the others are darker. Thus, the speckle effect makes a radar image of a floating raft look like a random matrix, and the magnitude values of the backscattering coefficient obey the Rayleigh distribution [26]. Raft culture increases the roughness of the sea surface, and the backscattering signal of seawater in a floating raft region is enhanced. Nevertheless, considering the influence of periodic ocean waves, backscattering coherence superposition is more prominent, resulting in more severe coherent speckle noise in SAR images [25]. As shown in Figure 2, the grayscale values of the pixels depict the amplitude values of the backscattering at each pixel, and the variability in grayscale values within the raft aquaculture area leads to the blurred edges of the area, as well as inconspicuous local features. The global features can better characterize the raft culture area.

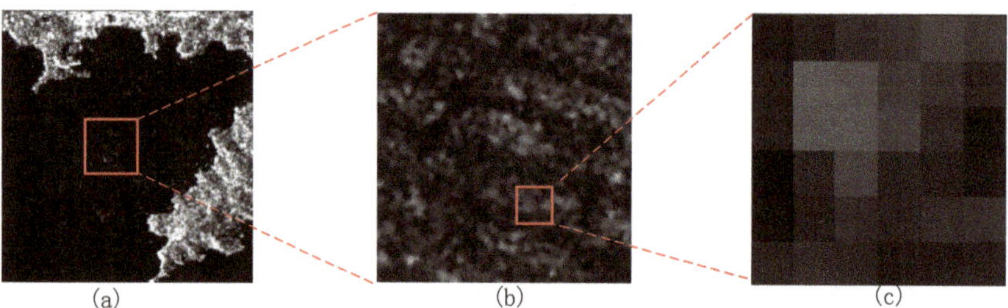

Figure 2. Difference in gray levels of a floating raft affected by noise. (**a**) The raft aquaculture area is globally distinguishable. (**b**) The local culture area is poorly characterized. (**c**) Differences in grayscale values exist at various pixels within the floating raft.

2.2. Structural Characteristics

Marine raft aquaculture areas tend to extend outward from offshore areas near islands and cover wide areas with distinct structural characteristics. The structural characteristics of raft culture areas help us to distinguish them from the seawater background, which includes but not limited to the following aspects:

- Multisize characteristics:

  The multisize nature of marine raft aquaculture areas is twofold. Overall, the aquaculture regions are scattered, with varying regional range sizes and inconsistent densities. Locally, the strips in the aquaculture areas are uniform in width, vary in length, and have narrow sea lanes, which vary in width between rafts. Thus, the method design needs to consider a method that can fit multisize features, and the use of a single feature sensibility field makes avoiding missing detailed information difficult.
- Strip-like geometric contour characteristics:

  Floating rafts are made of ropes in series with floating balls and have distinct strip geometric characteristics in an image. The non-centric symmetry of this type of rectangle needs to be noted when using convolution to extract targets.
- Outstanding directionality:

  The arrangement of floating rafts within an aquaculture area is directional, has explicit main directions, and is generally parallel to the shoreline.

Figure 3 shows the structure of the marine raft aquaculture areas. The floating rafts in areas A and B have the same strip-like geometric features and are aligned in the same direction within each zone. The size, density, and alignment direction between regions are different, i.e., zone A is more tightly packed than zone B, and the rafts are arranged horizontally in zone A and vertically in zone B.

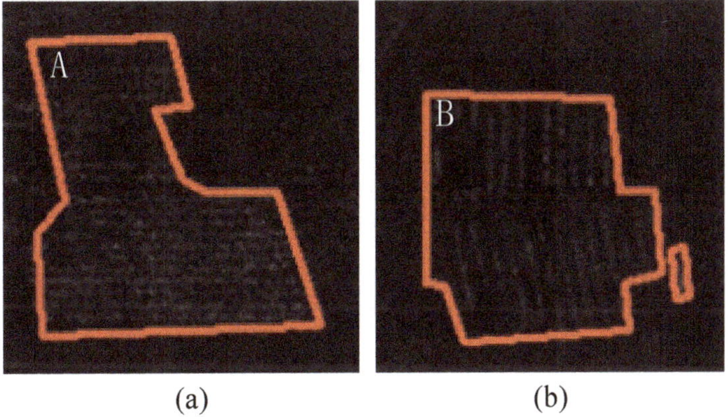

Figure 3. (**a**,**b**) show two independent examples of marine raft aquaculture areas. Raft aquaculture areas A in (**a**) and B in (**b**) are inconsistent in range size, density, and arrangement orientation.

In summary, the scattering features cause marine raft aquaculture areas to have weak local features on SAR images and detailed features can be missed, which make it difficult to distinguish raft aquaculture areas from seawater using only scattering features. The structural characteristics indicate that when designing an approach, attention needs to be paid to the multisize and internal uniform directional features, as well as the geometric features of the floating raft strip contour. On the basis of the features of marine raft aquaculture areas, this paper proposes a segmentation method involving feature enhancement and a semantic segmentation network similar to U-net, which will be introduced in Section 3.

3. Methods

There are four main steps in the semantic segmentation method for extracting the target area: dataset construction, model construction, model training, and final testing. In addition, the accuracy of target extraction is calculated on the basis of the results of the final testing. The method for extracting marine raft aquaculture areas in this paper adds a feature enhancement step between dataset construction and model construction. The overall process of the method in this paper is shown in Figure 4.

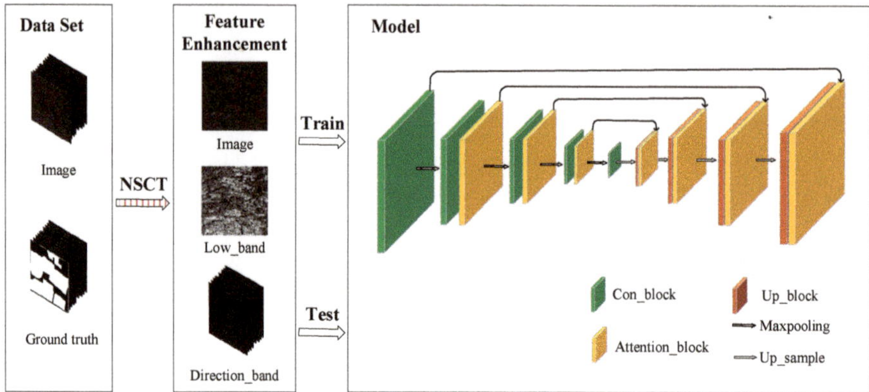

Figure 4. Overall flow chart.

In the dataset processing stage, this paper collected Sentinel-1 image data. After basic image processing of the data, we used the ArcGIS software to mark the image and generate binary maps called ground truth maps. Then, the images and ground truth maps were divided into training, validation, and test datasets at a 3:1:1 ratio, and data augmentation methods including mirroring, panning, and other operations on the training and validation samples were used to expand the dataset. The specific details are presented in Section 4.1.

After the construction of the dataset was completed, this paper used the NSCT to enhance the contour and orientation features of the image, and the obtained low-frequency sub-band and direction sub-bands were synthesized with the original image into a 26-channel image. This step is explained in detail in Section 3.1.

In the model construction phase of this paper, the task of detecting marine raft aquaculture areas was implemented by constructing a semantic segmentation algorithm model similar to the U-net model. The details are presented in Section 3.2.

During the model training phase, the training samples and the validation samples were input into the network model of this paper, and then the weight information was saved. The saved weights were applied to the test samples in the final testing stage to extract the floating raft region and calculate the final accuracy.

3.1. Feature Enhancement

During the feature analysis in Section 2, it was shown that the raft culture area was less distinguishable from seawater on the SAR image, but the raft culture area had significant contour and directional features. The NSCT method is well known for its capacity to highlight the main contour and directional features [27,28]. In this paper, the NSCT was used to enhance the main contour features of marine raft aquaculture areas, clarify the direction of the raft arrangement, and thus improve its distinguishability.

The NSCT is an improved method of the contour wave transform with anisotropy, multi-directionality, and translation invariance and consists of the nonsubsampled pyramid filter bank (NSPFB) and the nonsubsampled directional filter bank (NSDFB). As illustrated in Figure 5, the NSPFB acquires sub-band images with different frequencies through an iterative filter bank for multiscale decomposition of images, and the NSDFB acquires directional sub-band images with different directional divisions through a directional filter bank.

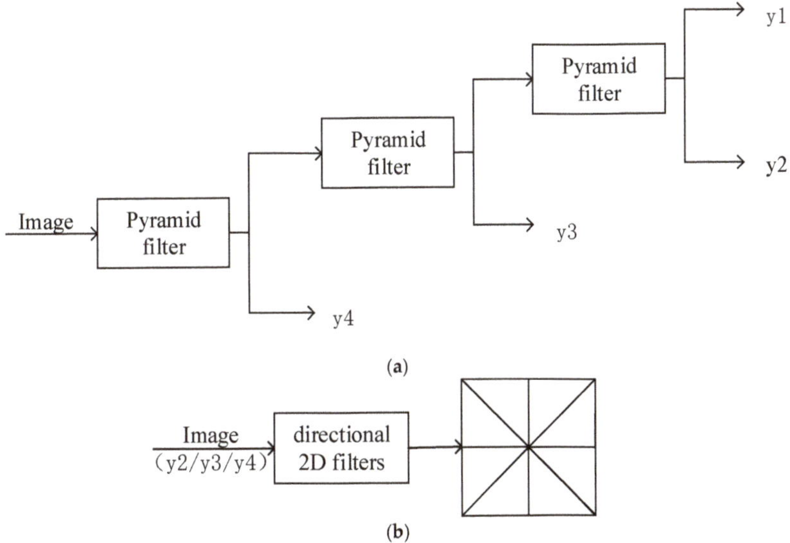

Figure 5. The two parts of the nonsubsampled contourlet transform (NSCT) for the decomposition diagram. (**a**) Nonsubsampled pyramid filter bank (NSPFB) decomposition diagram with the pyramid filter (y1 is the final obtained low-frequency sub-band). (**b**) Nonsubsampled directional filter bank (NSDFB) decomposition diagram with the directional 2D filters (decomposition of high-frequency images).

Figure 6 shows a sample image decomposed by the NSCT. Figure 6a is the original image, Figure 6b corresponds to y1 in Figure 5, and Figure 6c–i is the directional sub-band of the decomposed image. Figure 6b concentrates most of the energy of the original diagram and describes the main contour features well. Figure 6h is the main directional sub-band image of the original diagram, where the directionality of the floating raft arrangement can be clearly observed. It can be seen that the sub-band images obtained from the NSCT, which are decoupled from the SAR image, describe the main profile features of the floating raft and the directional features of the floating raft arrangement in the marine raft aquaculture area well and make full use of the information in the SAR images to enrich the data features. Therefore, this paper enhanced the data features with the NSCT before importing the data into the network for training. The scale parameter of the NSPFB was set to 2, and the direction parameter of the NSDFB was set to 8 according to the image size (512) of the whole scene image in the dataset.

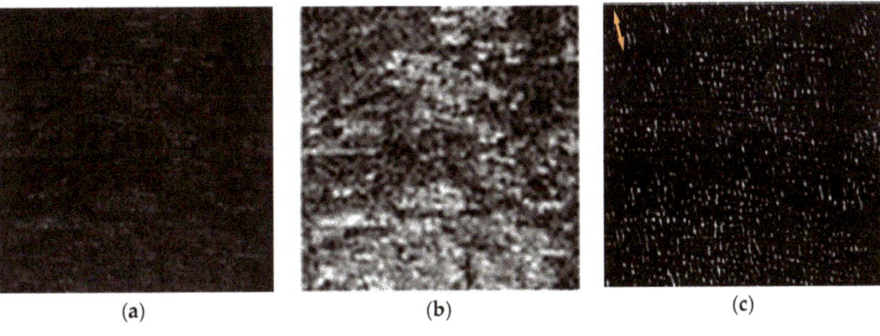

Figure 6. *Cont.*

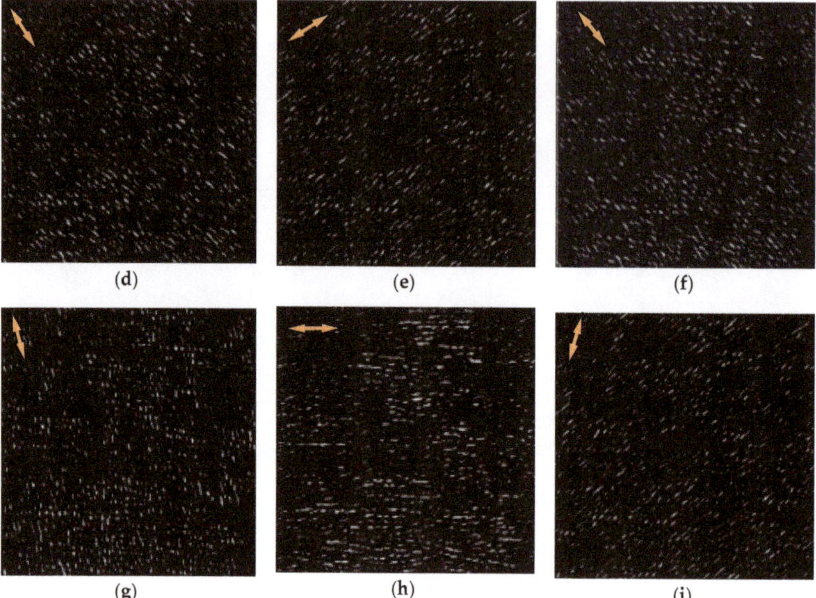

Figure 6. NSCT decomposition image results. (a) Original image. (b) Low-frequency sub-bands. (c–i) High-frequency directional sub-bands.

3.2. Fully Convolutional Networks

This section specifies the structure of the semantic segmentation network, which is similar to the U-net structure proposed in this paper for the extraction of marine raft aquaculture areas from Sentinel-1 images and includes the design of the convolution module and the integration of the attention mechanism shown in Figure 7.

The original U-net uses only a simple 3-by-3 convolution layer, which makes it difficult to fit the multisize features and regular geometric features of the raft culture area. Therefore, in the encoder stage, we introduced multiscale convolution to adapt for multisize features and asymmetric convolution in order to filter geometric features, which is shown in Figure 7b. To fully use the explicit global feature while discarding vague local features, we introduced a spatial attention mechanism at the encoder stage to calculate the spatial relationship among the pixels, and the global features were assigned to each pixel by weighting. A channel attention mechanism was adopted to direct more attention to the contour and directional features acquired through the NSCT decomposition. Figure 7c shows the attention mechanism that is used.

3.2.1. Convolution Block

As stated in Equation (1), the essence of convolution is a kind of weighted superposition. In the field of image processing, the size and weight of the convolution are designed to extract the required features from the image. The extracted multiple features constitute a multi-dimensional feature space, where inter-class variance is expected to be enhanced and intra-class difference is expected to be suppressed. In a fully convolutional network, the image is mapped to a high-dimensional feature space by means of convolutional modules, and the weights of the convolution are learned through data training to avoid the uncertainty of artificial design.

$$H = F * G$$

which can be written as

$$H(i,j) = \sum_{m}\sum_{n} F(m,n)G(i-m,j-n) \quad (1)$$

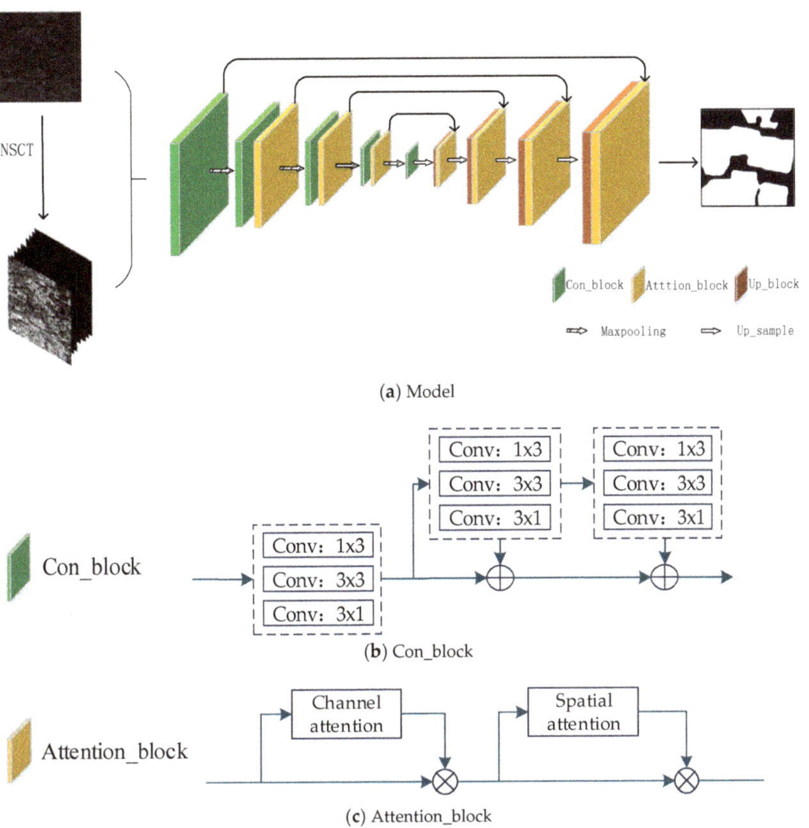

Figure 7. The model for the marine raft aquaculture area extraction task. (**a**) The overall model structure; (**b**) the Con_block with an asymmetric and multiscale structure; (**c**) the structure of the Attention_block with channel attention and spatial attention.

Therefore, in this paper, a full convolutional network was designed for the marine raft aquaculture area extraction task. It used multiscale convolution to extract multisize characteristics and asymmetric convolution to extract the strip-like geometric characteristics of marine raft aquaculture areas.

- Multiscale Convolution

Since the marine raft aquaculture areas vary in size and the spatial structures of the areas consist of large strip rafts and narrow sea lanes, we proposed the extraction of the features by a multiscale convolutional kernel, which is an appropriate choice. On the one hand, multiscale convolution extracts the information of the large-scale strip rafts and the detailed information of the narrow sea lanes. On the other hand, multiscale convolution can also capture features effectively, regardless of the size differences among the areas. When convolution kernels of different sizes, such as 3×3, 5×5, or 7×7, are applied simultaneously to extract feature maps, the computational complexity of the model increases. Inspired by the GoogLeNet architecture, we designed a multiscale convolutional kernel, as shown in Figure 8a. Due to the computational characteristics of convolution, the computational effects of two 3 × 3 convolution kernels are equivalent to that of a 5 × 5 convolution kernel, and the computational effects of three 3 × 3 convolution kernels are equivalent to that of a 7 × 7 convolution kernel [29–32]. Therefore, in this paper, the feature map fusion of multiscale convolutional kernels was achieved through series and parallel convolution kernels, which led to the extraction of features. Then, in the basic unit of each encoder, 3/5/7, three-scale receptive field information was obtained through three 3 × 3 convolution kernels.

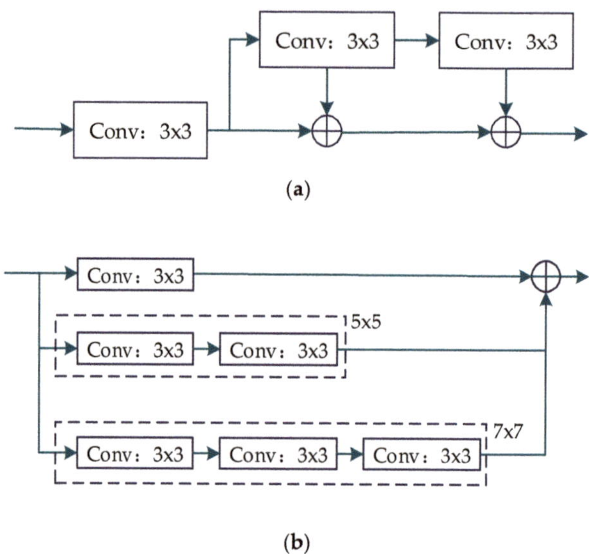

Figure 8. Multiscale convolution kernel was achieved through series and parallel convolution kernels. (**a**) A schematic diagram of the multiscale convolutional structure advanced in this paper; (**a**) is equivalent to (**b**).

- Asymmetric convolution

  The sensory field of common convolution is a rectangle with equal length and width, and thus it is difficult to capture the shape features of the non-centric symmetrical target. In consideration of the remarkable geometric structure of strip rafts, we selected asymmetric convolution kernels of sizes 1 × 3 and 3 × 1 for additive fusion with the results extracted from the 3 × 3 convolutional kernels.

3.2.2. Attention Block

Although the original SAR image was enhanced by the NSCT, there was still a need to better characterize the overall global features to overcome the interference of noise and the sea state. To address this problem, the proposed method combined the channel attention and spatial attention mechanisms in a series between the convolution modules. Channel attention is designed to direct attention to channels that contain the main and directional features of the raft culture after the NSCT. The spatial attention mechanism converts overall spatial relationship into weights assigned to each point of the raft culture area to better extract the global features. Convolutional block attention module (CBAM) [33] showed that channel attention and spatial attention can be used in chains, and inspired by efficient channel attention for deep convolutional neural networks (ECA-Net) [34], we simplified the calculations for channel attention weights to make them easier and faster.

- Channel Attention

The simple 2D convolution operation focuses only on the relationship among pixels within the sensory field and ignores the dependencies between channels. Channel attention links the features of each channel in order to focus on key information, such as the primary direction of the raft culture area, more effectively. As shown in Figure 9, the feature map was globally averaged to obtain a feature map of size [batch-size, channel, 1, 1]. Then, a 1 × 1 convolution was used to learn the correlation between each channel. Finally, the sigmoid function was used to obtain information about the weights assigned to each channel to adjust the feature information for the next level of inflow.

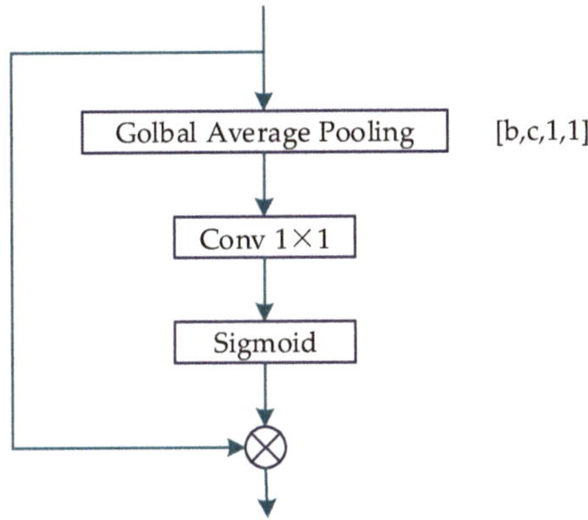

Figure 9. Channel attention.

- Spatial Attention

In addition to the dependency among channels, the overall spatial relationship also has a great influence on the extraction result. As shown in Figure 10, the spatial attention module first normalized the number of channels and then learnt the higher-dimensional features under a larger sensory field through convolution, thus reducing the flow of redundant information of low-dimensional features to the lower convolution and focusing on the overall information of the target.

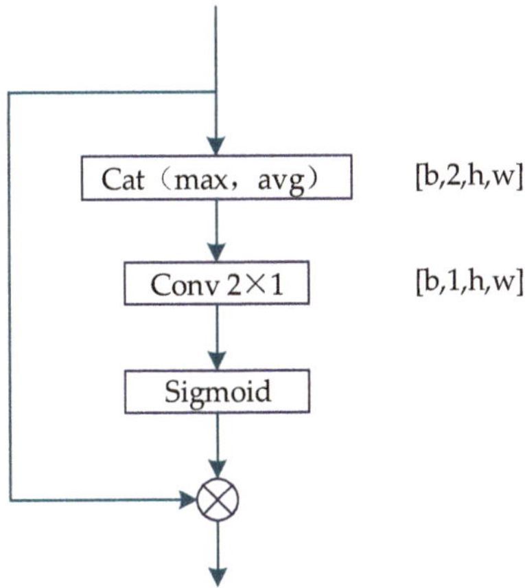

Figure 10. Spatial attention.

Although the semantic segmentation network proposed in this paper adopts a U-shaped structure similar to U-net, it is different from U-net. The key to the difference lies in the design of the encoder. The network proposed in this paper additively merges multi-scale convolution and asymmetric convolution at the encoder stage to form basic coding units, and connects channel attention and spatial attention in series between these basic units.

4. Experiment, Results, and Analysis

This section illustrates the experiments on the test data in the study area. Section 4.1 presents the study area and experimental data. Section 4.2 proves the interpretability and generality of the proposed method through validation experiments, and Section 4.3 verifies the superiority of the method through comparative experiments. The experiments were carried out under 64-bit Linux system, using GeForce RTX 1080.

To quantify the experimental results, we used the following metrics: intersection over union (IOU) and F1-Score (F1). IOU and F1, at present, are commonly used as accuracy indexes. IOU means the intersection ratio of the predicted image and ground truth image. The F1 is an accuracy index that considers precision and recall. The formulas to calculate the metrics are as follows:

$$\mathrm{IOU} = \frac{\mathrm{TP}}{\mathrm{TP}+\mathrm{FP}+\mathrm{FN}} \tag{2}$$

$$\mathrm{Precision} = \frac{\mathrm{TP}}{\mathrm{TP}+\mathrm{FP}} \tag{3}$$

$$\mathrm{Recall} = \frac{\mathrm{TP}}{\mathrm{TP}+\mathrm{FN}} \tag{4}$$

$$\mathrm{F1} = \frac{2 \times \mathrm{Precision} \times \mathrm{Recall}}{\mathrm{Precision}+\mathrm{Recall}} \tag{5}$$

4.1. Study Area and Experimental Data

Currently, there is no authoritative open dataset for marine raft aquaculture area extraction, and thus this paper collected Sentinel-1 images of the Changhai region to construct a dataset to use as a basis for research.

4.1.1. Study Area

Changhai County is located in the Yellow Sea on the east side of the Liaodong Peninsula at longitude 122°13′18″ E–123°17′38″ E and latitude 38°55′48″ N–39°18′26″ N, as shown in Figure 11. It is under the jurisdiction of Dalian City and has a land area of 142.04 square kilometers, sea area of 10,324 square kilometers, coastline of 358.9 km [35], and sea use area of 244.82 square kilometers for raft culture, which is a typical large-scale marine raft aquaculture area [36].

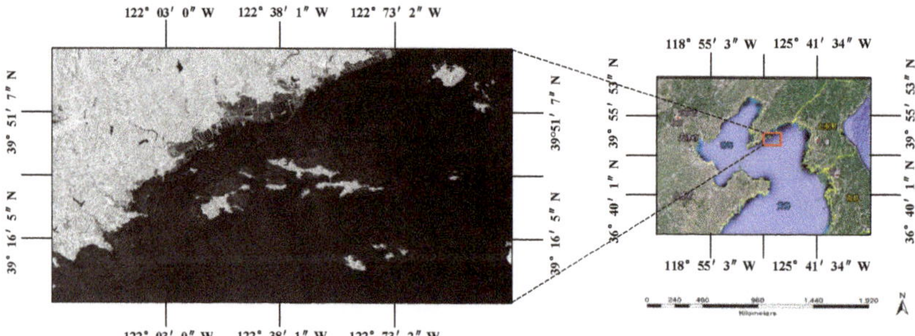

Figure 11. Study area.

4.1.2. Dataset

With the advantages of high coverage, free access, and stable updates, the Sentinel-1 interferometric wide swarth (IW) ground range detected (GRD) data from the European Space Agency (ESA)'s Copernicus project's dual polarized C-band SAR were chosen by this study as the data source (https://vertex.daac.asf.alaska.edu/).

The dataset contains four Sentinel-1 images from Changhai County (September 16, September 28, October 10, and October 22), each containing both vertical–horizontal (VH) and vertical–vertical (VV) polarization data. The cross-polarized VH data were less permeable than the isotropic polarized VV data, and it can be seen from the images in Figure 12 that the marine raft aquaculture area was difficult to observe with the cross-polarized data; thus the isotropic polarization (VV) image was used to construct the dataset.

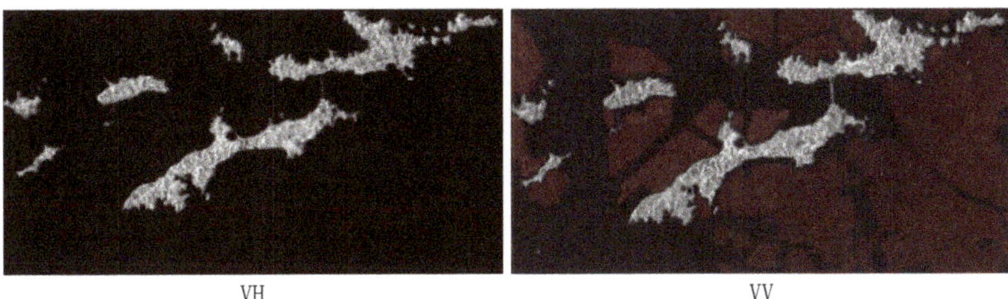

Figure 12. Sentinel-1 dual polarization imaging.

It is difficult for SAR images to avoid speckle noise, which leads to a jump in the digital number (DN) value in a homogeneous region. Although the existing SAR image noise suppression methods have significantly improved image grayscale resolution, the texture information is too smooth and loses its unique features and information after denoising [37]. In this paper, before the image data were annotated, we utilized the preprocessing operations are of histogram equalization and linear stretch. On the basis of this operation, we used ArcGIS to annotate data and generate a .shp file. Then, the vector files were converted to binary images, called ground truth maps. The size of a Sentinel-1 image is too large, and thus the original images and the labeled ground truth maps were clipped into 10038 patch pairs with sizes of 512 × 512. Figure 13 illustrates the main steps of dataset construction.

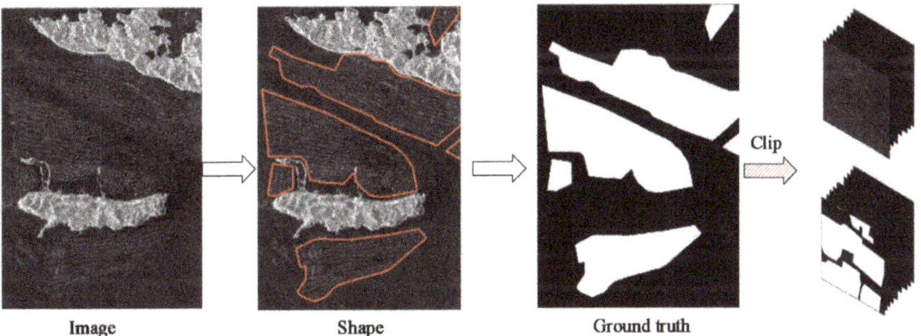

Figure 13. After selecting a data source, dataset construction requires three main steps: data labeling, generating a truth map, and cropping.

The complete dataset should include training data, validation data, and test data. To better verify the validity of the methods in this paper, we selected the image from Oct 16, which was not included in the dataset, as independent test data. The data used in experiments is shown in Table 1.

Table 1. Training, validation, and test data of the experiments.

Dataset	Time	Number of Images	Size	Number of Patches
Training	9.16/9.22/10.10	3	(512, 512)	7421
Validation	10.22	1	(512, 512)	2617
Testing	10.16	1	(6217, 3644)	/

4.2. Verification Experiment

4.2.1. Ablation Experiment

To prove that the strategy used in the presented method is effective, this section uses the results of the test image obtained by the original U-net (denoted by U-net), the network that introduced the attention layer (denoted by Attention_block + U-net), and the network that modified the convolutional structure (denoted by Attention_block + Con_block + U-net) and compares them with the result obtained by the proposed method (denoted by NSCT + Attention_block + Con_block + U-net).

Figure 14 shows the prediction maps for a typical region. Table 2 shows the evaluation of the whole test map segmentation results.

Table 2. Precision, recall, F1-Score (F1), and intersection over union (IOU) scores on the test data of the ablation experiment.

Method	IOU	F1	Precision	Recall
U-net	77.2%	87.1%	93.0%	81.9%
Attention_block + U-net	78.2%	87.8%	86.4%	89.2%
Attention + Con_block + U-net	81.4%	89.7%	87.4%	92.1%
NSCT + Attention + Con_block + U-net	83.0%	90.7%	89.1%	92.3%

From the results shown in Figure 14c, we can see that the segmentation result of U-net had poor integrity and relatively inward shrinking edges with burr. The result was greatly affected by speckle noise. U-net had difficulty capturing multisize information from the raft culture area, such as the narrower gaps within the floating rafts, as shown by the false positive (FP) pixels in the red box of Figure 14c; the smaller range and density of the raft culture area caused false negative (FN) pixels, as shown by the blue box in Figure 14c. There was considerable noise at the edges of the results, as shown in the green box in Figure 14c. U-net extracted features with only two tandem 3*3 convolutions, did not fully exploit the features of the floating raft, did not pay attention to the multisize information and the geometric features of the floating raft, and was highly influenced by speckle noise. There were a large number of omissions and neglected areas of interregional sea lanes in the U-net detection results, and thus the U-net segmentation result had low IOU and F1 but high precision.

As shown in Figure 14d, the addition of Attention_block effectively reduced the number of FN pixels and smoothed the edges, while the number of FP pixels was increased. In addition, the overflow and adhesion problems at the edges were serious, and the intervals between the raft culture areas were ignored. It can be seen in Table 2 that recall was improved by 7.3%, while precision decreased. Considering that the lack of information about the area inside the marine raft aquaculture area also contributed to this phenomenon, this paper overcomes this problem by designing a tailored convolutional structure.

The convolution block design with a multiscale convolution kernel and an asymmetric convolution kernel was more compatible in addressing the structures of marine raft aquaculture areas. As shown in Figure 14e, the number of FP pixels was reduced but FN pixels were displayed in areas where the outline of the subject was not visible in the yellow box. Asymmetric convolution was used to filter the striped geometric information of the raft. Multiscale convolution was used to adapt for the different sizes and densities of the raft culture area, as well as the different sizes of the raft and seaway. The IOU increased by 3.2% compared to Attention_block + U-net.

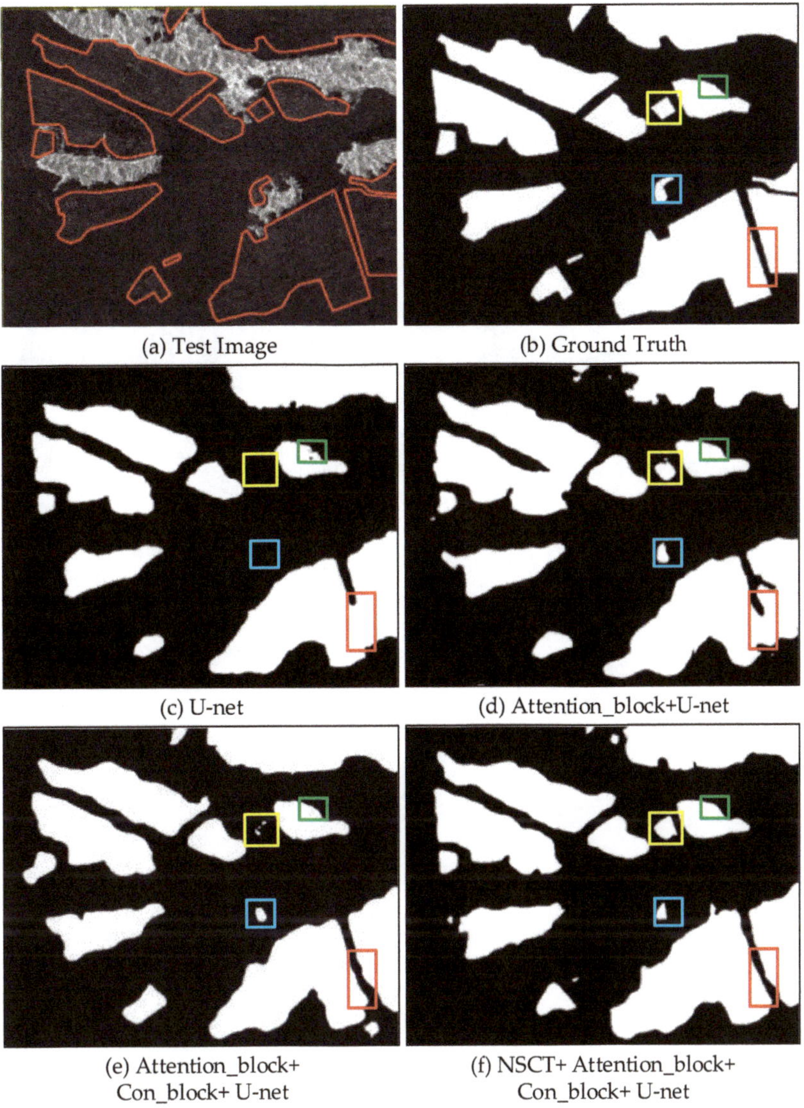

Figure 14. Results of the ablation experiment for a subset of the study area. (**a**) Test image. (**b**) Ground truth. (**c**) The results obtained by the original U-net. (**d**) The results obtained by the network that introduced the attention layer. (**e**) The results obtained by the network that modified the convolutional structure. (**f**) The results obtained by the proposed method.

As simply changing the network structure would not be sufficient to address the low signal-to-noise ratio due to the SAR imaging mechanism and because areas with obscure subject contours were still missed, we used the NSCT in the proposed method to enhance the features. Figure 14f shows the results obtained by the proposed approach. The result was similar to the actual distribution of the raft culture areas and had fewer FN pixels in areas where the main contour was not obvious, such as the area in the yellow box, and even single floating rafts could be extracted. The NSCT improved the information utilization and emphasized the main contour features of the floating raft, making the directional information clearer. Therefore, the feature enhancement operation increased the difference between the raft culture region and the background and helped to distinguish similarly structured but irregularly arranged waves. As shown in Table 2, the proposed method in this paper was optimal regardless of whether it used IOU or F1 as the evaluation index.

4.2.2. Applied Experiment

To verify the generality of the method, we in this section select demonstration areas in the coastal region of Shandong for experiments. Figure 15 shows the extraction results of the proposed method in subsets of the demonstration area. Table 3 shows the evaluation of the results for the whole demonstration area along the coast of Shandong.

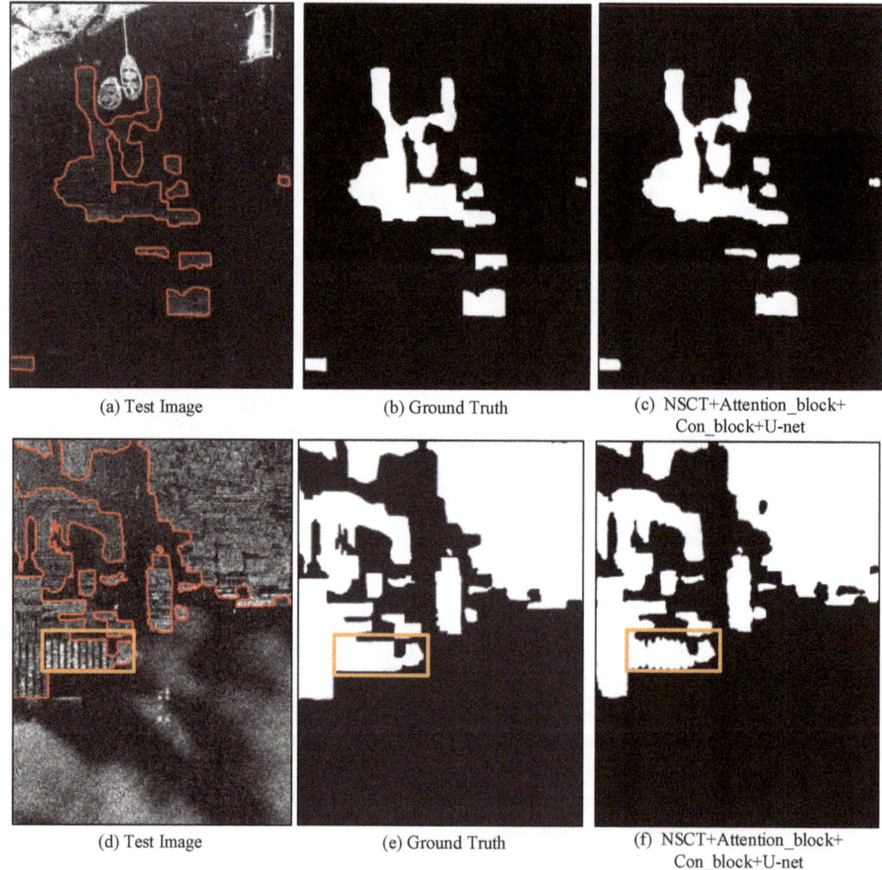

Figure 15. Results of the applied experiment for subsets of the study area. (**a**,**d**) Test image. (**b**,**e**) Ground truth. (**c**,**f**) The results obtained by the proposed method.

Table 3. Precision, recall, F1, and IOU scores on the demonstration areas of the applied experiment.

Method	IOU	F1	Precision	Recall
NSCT + Attention + Con_block + U-net	83.8%	91.2%	96.1%	86.8%

As shown in the orange box in Figure 15d, the raft culture area in Shandong coastal area was slightly different from Changhai. Some of the floating rafts in the inner part of the marine raft aquaculture area were wider while they were arranged more sparsely. This led to larger fluctuation and more FN pixels at the edge when extracting the whole area, as shown in the orange box in Figure 15f. However, scattering characteristics and structural features of the raft culture area as a whole did not change, and thus the method proposed in this paper was also effective in this region. As for the results, the proposed method was applicable in others areas with the same characteristic performance besides Changhai.

4.3. Comparative Experiment

In recent similar studies, UPS-net [23] was demonstrated to be more accurate than other popular networks based on the FCN and more applicable than DS-HCN [22] to extract marine raft aquaculture area. Therefore, UPS-net was chosen as the comparison method to verify the superiority of the proposed method.

Figure 16 shows the prediction map for a typical region. Table 4 shows the evaluation of the whole test map segmentation results.

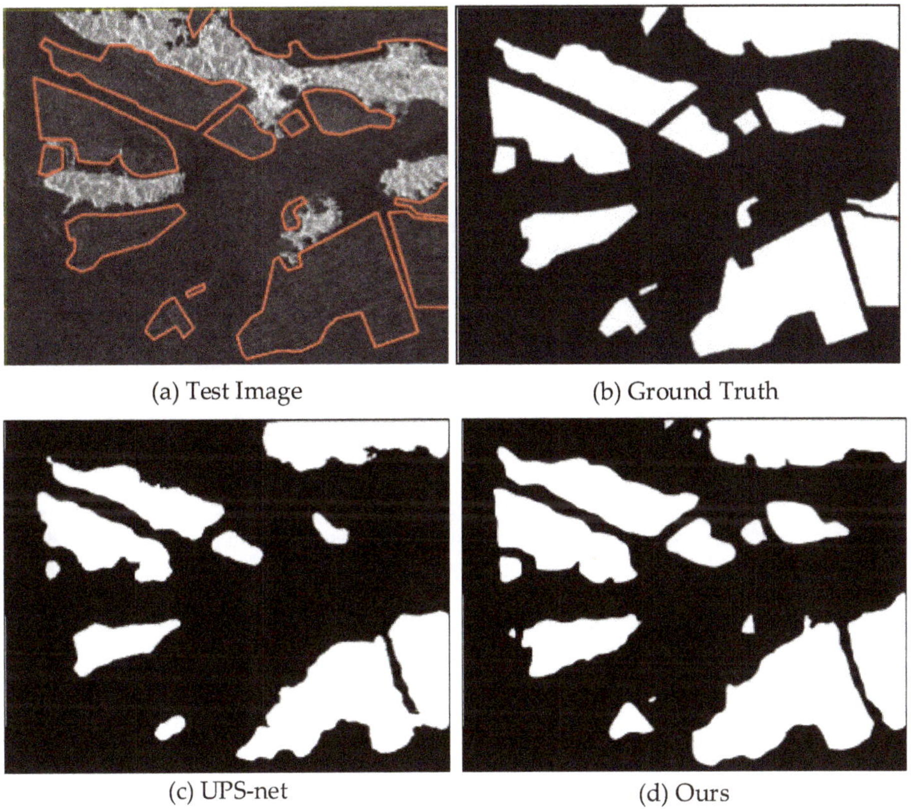

Figure 16. Results of the comparison experiment for a subset of the study area. (**a**) Test image. (**b**) Ground truth. (**c**) The results obtained by UPS-net. (**d**) The results obtained by the proposed method.

Table 4. Precision, recall, F1, and IOU scores on the test data of the comparative experiment.

Method	IOU	F1	Precision	Recall
UPS-net	74.8%	85.6%	91.7%	80.3%
Ours	83.0%	90.7%	89.1%	92.3%

As shown in Figure 16, the results of UPS-net showed FN pixels in the sparsely arranged or the small raft culture area and FP pixels in narrow sea lanes, with severe edge shrinkage. In contrast, the result of the proposed method was less affected by the background, had more complete edges, and provided better discrimination of sea lanes. Table 4 shows the evaluation results of the comparative experiments, and it can be seen that compared to UPS-net, the proposed method improved recall by 12%, IOU by 8.2%, and F1 by 5.1%. UPS-net adds PSE modules to U-net to obtain more contextual information and discards some redundant information at the decoder stage. However, simply adding multiscale information without considering the scattering features, geometry, and orientation of the marine raft aquaculture area makes it difficult to ensure the integrity of the extraction results. The method in this paper used asymmetric convolution to fit the geometric information of the raft culture area while adding multiscale information fusion, and used the attention mechanism and NSCT method to make better use of the scattering features and directionality of the raft culture area, which resulted in better outcomes.

5. Discussion

The state-of-the-art works for marine raft aquaculture areas are mostly dependent on professional experience. Although the deep learning method for semantic segmentation avoids a large amount of manual work, it does not work well when directly migrated to SAR images. Marine raft aquaculture areas in SAR images exhibit large differences in grayscale values (influenced by speckle noise) and distinct structural characteristics (striped contours and directionality). In consideration of these characteristics, this paper proposes a segmentation network combined with NSCT. This combination of frequency domain priors and semantic segmentation models provides a promising idea for future research in marine raft aquaculture areas extraction from SAR images.

The optimized model in this paper is more suitable for the task of marine raft aquaculture area extraction on SAR images. Although semantic segmentation models have been widely used in optical remote sensing image target extraction, these models do not yield good results by direct migration to SAR images. As shown in Figure 14c, the original U-net could not extract the raft culture area as a whole. The inward adhesion at the edges was caused by the inherent speckle noise in SAR images, which caused great loss to the detail and structure information, and this led us to focus more on global and structural information instead of local information [38]. Thus, attention module and multiscale asymmetric convolution were introduced to capture global and structural information, respectively, leading to a 4.2% increase in IOU and a 2.6% increase in F1.

Furthermore, the results show that the combination of NSCT and semantic segmentation model is useful for obtaining better results. Studies by Yin and Wang show that the naturally trained model focuses more on low-frequency information and is poorly robust to high-frequency information [39,40]. To enhance both the low and high frequency information at the same time, the proposed method decouples the original SAR image using NSCT. This improved IOU by 1.6% and F1 by 1%.

Overall, the proposed method obtained more satisfactory results than the state-of-the-art methods. However, there are some matters requiring attention. Firstly, post-processing techniques such as conditional random field (CRF) can be used to remove noise masks and obtain integration results [41], but this was not the focus of this article. Secondly, it is worth noting that the applicability of the method is related to the image resolution. Marine raft aquaculture areas extraction using the medium resolution imagery is validated in this paper, while imagery with higher resolution should be used for refined information extraction inside the area. Transfer learning will be a good way for the model transfer between imagery with different resolutions [42].

6. Conclusions

This paper proposes a segmentation algorithm for the marine raft aquaculture area extraction using Sentinel-1 images and is characterized by feature enhancement and an improved semantic segmentation network.

1. Feature enhancement: In response to the low signal-to-noise ratio problem in SAR images, the floating raft features were enhanced using the NSCT. The low-frequency sub-band obtained by decomposing the original SAR image with the NSCT was used to enhance the contour features, and the high-frequency sub-bands were used to supplement details with direction information.
2. Improved semantic segmentation network: Multiscale feature fusion was introduced to better recognize large rafts and small seaways with less edge adhesion. Asymmetric convolution was adopted to capture the characteristics of floating raft strip distribution by screening the geometric features. Attention module was added to improve the integrity and smoothness in view of the grayscale variance in the homogeneous region of the SAR image caused by speckle noise.

In summary, the segmentation method makes full use of the scattering and structural features, and is effective in marine raft aquaculture area extraction. It is worth mentioning that the data in this paper had not been denoised, which eliminates the tedious step of extracting targets from an SAR image and has good application prospects.

In regions with poor sea conditions, the present method still suffers from errors caused by coherent spot noise enhanced by wave cascades. Therefore, further research is needed to address the issue of wave interference.

Author Contributions: Conceptualization, Y.Z.; methodology, Y.Z.; validation, Y.Z., C.W., and Y.D.; investigation, Y.Z., J.C. (Jing Chen), and Y.J. (Yongshi Jie); writing—original draft preparation, Y.Z.; writing—review and editing, Y.Z., C.W., Y.J. (Yuan Ji), and J.C. (Jingbo Chen); funding acquisition, C.W. All authors have read and agreed to the published version of the manuscript.

Funding: This research was funded by "Nation Key R&D Program of China under grant [YFC0821900]" and "Big data on Earth in Support of Ocean Sustainable Development Goals Research [XDA19090123]".

Acknowledgments: We would like to thank the Copernicus program of the European Space Agency for making Sentinel-1 SAR data freely available.

Conflicts of Interest: The authors declare no conflict of interest.

Abbreviations

FCN	Fully convolutional network
RefineNet	Multi-path refinement networks
PSPNet	Pyramid scene parsing network
RCF	Richer convolutional features network
UPS-net	Improved U-Net with a PSE structure
DS-HCN	Dual-scale homogeneous convolutional neural network
CBAM	Convolutional block attention module
ECA-Net	Efficient channel attention for deep convolutional neural networks

References

1. FAO. *World Fisheries and Aquaculture Overview 2018*; FAO Fisheries Department: Rome, Italy, 2018.
2. Zhang, Y.; He, P.; Li, H.; Li, G.; Liu, J.; Jiao, F.; Zhang, J.; Huo, Y.; Shi, X.; Su, R.; et al. Ulva prolifera green-tide outbreaks and their environmental impact in the Yellow Sea, China. *Neurosurgery* **2019**, 6, 825–838. [CrossRef]
3. Chen, X.; Fan, W. *Theory and Technology of Fishery Remote Sensing Applications*; Science China Press: Beijing, China, 2014.

4. Deng, G.; Wu, H.Y.; Guo, P.P.; Li, M.Z. Evolution and development trend of marine raft cultivation model in China. *Chin. Fish. Econ.* **2013**, *31*, 164–169.
5. Fan, J.; Zhao, J.; An, W.; Hu, Y. Marine Floating Raft Aquaculture Detection of GF-3 PolSAR Images Based on Collective Multikernel Fuzzy Clustering. *IEEE J. Sel. Top. Appl. Earth Observ. Remote Sens.* **2019**, *99*, 1–14. [CrossRef]
6. Fan, J.; Zhang, F.; Zhao, D.; Wen, S.; Wei, B. Information extraction of marine raft aquaculture based on high-resolution satellite remote sensing SAR images. In Proceedings of the Second China Coastal Disaster Risk Analysis and Management Symposium, Hainan, China, 29 November 2014.
7. Zhu, H. Review of SAR Oceanic Remote Sensing Technology. *Mod. Radar* **2010**, *2*, 1–6.
8. Chu, J.L.; Zhao, D.Z.; Zhang, F.S.; Wei, B.Q.; Li, C.M.; Suo, A.N. Monitormethod of rafts cultivation by remote sense-A case of Changhai. *Mar. Environ. Sci.* **2008**, *27*, 35–40.
9. Fan, J.; Chu, J.; Geng, J.; Zhang, F. Floating raft aquaculture information automatic extraction based on high resolution SAR images. In Proceedings of the 2015 IEEE International Geoscience and Remote Sensing Symposium (IGARSS), Milan, Italy, 26–31 July 2015; pp. 3898–3901.
10. Hu, Y.; Fan, J.; Wang, J. Target recognition of floating raft aquaculture in SAR image based on statistical region merging. In Proceedings of the 2017 Seventh International Conference on Information Science and Technology (ICIST), Da Nang, Vietnam, 16–19 April 2017; pp. 429–432.
11. Geng, J.; Fan, J.; Wang, H. Weighted fusion-based representation classifiers for marine floating raft detection of SAR images. *IEEE Geosci. Remote Sens. Lett.* **2017**, *14*, 444–448. [CrossRef]
12. Geng, J.; Fan, J.C.; Chu, J.L.; Wang, H.Y. Research on marine floating raft aquaculture SAR image target recognition based on deep collaborative sparse coding network. *Acta Autom. Sin.* **2016**, *42*, 593–604.
13. Long, J.; Shelhamer, E.; Darrell, T. Fully convolutional networks for semantic segmentation. In Proceedings of the IEEE Conference on Computer Vision and Pattern Recognition (CVPR), Boston, MA, USA, 7–12 June 2015; pp. 3431–3440.
14. Ronneberger, O.; Fischer, P.; Brox, T. U-net: Convolutional networks for biomedical image segmentation. In Proceedings of the International Conference on Medical Image Computing and Computer-Assisted Intervention, Munich, Germany, 5–9 October 2015; Springer: Cham, Switzerland, 2015; pp. 234–241.
15. Chen, L.C.; Papandreou, G.; Kokkinos, I.; Murphy, K.; Yuille, A.L. Semantic image segmentation with deep convolutional nets and fully connected crfs. *arXiv* **2014**, arXiv:1412.7062.
16. Chen, L.C.; Papandreou, G.; Schroff, F.; Adam, H. Rethinking atrous convolution for semantic image segmentation. *arXiv* **2017**, arXiv:1706.05587.
17. Chen, L.C.; Papandreou, G.; Kokkinos, I.; Murphy, K.; Yuille, A.L. Deeplab: Semantic image segmentation with deep convolutional nets, atrous convolution, and fully connected crfs. *IEEE Trans. Pattern Anal. Mach. Intell.* **2017**, *40*, 834–848. [CrossRef] [PubMed]
18. Lin, G.; Milan, A.; Shen, C.; Reid, I. Refinenet: Multi-path refinement networks for high-resolution semantic segmentation. In Proceedings of the IEEE Conference on Computer Vision and Pattern Recognition (CVPR), Honolulu, HI, USA, 21–26 July 2017; pp. 1925–1934.
19. Zhao, H.; Shi, J.; Qi, X.; Wang, X.; Jia, J. Pyramid scene parsing network. In Proceedings of the IEEE Conference on Computer Vision and Pattern Recognition (CVPR), Honolulu, HI, USA, 21–26 July 2017; pp. 2881–2890.
20. Liu, Y.; Cheng, M.M.; Hu, X.; Wang, K.; Bai, X. Richer convolutional features for edge detection. In Proceedings of the IEEE Conference on Computer Vision and Pattern Recognition (CVPR), Honolulu, HI, USA, 21–26 July 2017; pp. 3000–3009.
21. Yueming, L.; Xiaomei, Y.; Zhihua, W.; Chen, L. Extracting raft aquaculture areas in Sanduao from high-resolution remote sensing images using RCF. *Acta Oceanol. Sin.* **2019**, *41*, 119–130.
22. Shi, T.; Xu, Q.; Zou, Z.; Shi, Z. Automatic raft labeling for remote sensing images via dual-scale homogeneous convolutional neural network. *Remote Sens.* **2018**, *10*, 1130. [CrossRef]
23. Cui, B.; Fei, D.; Shao, G.; Lu, Y.; Chu, J. Extracting raft aquaculture areas from remote sensing images via an improved U-net with a PSE structure. *Remote Sens.* **2019**, *11*, 2053. [CrossRef]
24. Da Cunha, A.L.; Zhou, J.; Do, M.N. The nonsubsampled contourlet transform: Theory, design, and applications. *IEEE Trans. Image Process.* **2006**, *15*, 3089–3101. [CrossRef]
25. Fan, J.; Wang, D.; Zhao, J.; Song, D.; Han, M.; Jiang, D. National sea area use dynamic monitoring based on GF-3 SAR imagery. *J. Radars* **2017**, *6*, 456–472.

26. Madsen, S.N. Spectral properties of homogeneous and nonhomogeneous radar images. *IEEE Trans. Aerosp. Electron. Syst.* **1987**, *4*, 583–588. [CrossRef]
27. Geng, J. Research on Deep Learning Based Classification Method for SAR Remote Sensing Images. Ph.D. Thesis, Dalian University of Technology, Dalian, China, 2018.
28. Jiao, J.; Zhao, B.; Zhang, H. A new space image segmentation method based on the non-subsampled contourlet transform. In Proceedings of the 2010 Symposium on Photonics and Optoelectronics, Chengdu, China, 19–21 June 2010; pp. 1–5.
29. Szegedy, C.; Liu, W.; Jia, Y.; Sermanet, P.; Reed, S.; Anguelov, D.; Erhan, D.; Vanhoucke, V.; Rabinovich, A. Going deeper with convolutions. In Proceedings of the IEEE Conference on Computer Vision and Pattern Recognition (CVPR), Boston, MA, USA, 7–12 June 2015; pp. 1–9.
30. Ioffe, S.; Szegedy, C. Batch normalization: Accelerating deep network training by reducing internal covariate shift. *arXiv* **2015**, arXiv:1502.03167.
31. Szegedy, C.; Vanhoucke, V.; Ioffe, S.; Shlens, J.; Wojna, Z. Rethinking the inception architecture for computer vision. In Proceedings of the IEEE Conference on Computer Vision and Pattern Recognition (CVPR), Las Vegas, NV, USA, 27–30 June 2016; pp. 2818–2826.
32. Szegedy, C.; Ioffe, S.; Vanhoucke, V.; Alemi, A. Inception-v4, inception-resnet and the impact of residual connections on learning. *arXiv* **2016**, arXiv:1602.07261.
33. Woo, S.; Park, J.; Lee, J.Y.; So Kweon, I. Cbam: Convolutional block attention module. In Proceedings of the European Conference on Computer Vision (ECCV), Munich, Germany, 8–14 September 2018; pp. 3–19.
34. Wang, Q.; Wu, B.; Zhu, P.; Li, P.; Zuo, W.; Hu, Q. ECA-net: Efficient channel attention for deep convolutional neural networks. In Proceedings of the IEEE/CVF Conference on Computer Vision and Pattern Recognition (CVPR), Seattle, WA, USA, 16–18 June 2020; pp. 11534–11542.
35. Party History Research Office of the CPC Changhai County Committee. *Changhai Yearbook (Archive) 2017*; Liaoning Nationality Publishing House: Dandong, China, 2018; pp. 29–31.
36. Yan, J.; Wang, P.; Lin, X.; Tang, C. Analysis of the Scale Structure of Mariculture Using Sea in Changhai County Based on GIS. *Ocean Dev. Manag.* **2016**, *33*, 99–101.
37. Liu, S.; Hu, Q.; Liu, T.; Zhao, J. Overview of Research on Synthetic Aperture Radar Image Denoising Algorithm. *J. Arms Equip. Eng.* **2018**, *39*, 106–112+252.
38. Yu, W. SAR Image Segmentation Based on the Adaptive Frequency Domain Information and Deep Learning. Master's Thesis, XIDIAN University, Xi'an, China, 2014.
39. Yin, D.; Gontijo Lopes, R.; Shlens, J.; Cubuk, E.D.; Gilmer, J. A Fourier Perspective on Model Robustness in Computer Vision. *Adv. Neural Inform. Process. Syst.* **2019**, *32*, 13276–13286.
40. Wang, H.; Wu, X.; Huang, Z.; Xing, E.P. High-frequency Component Helps Explain the Generalization of Convolutional Neural Networks. In Proceedings of the IEEE/CVF Conference on Computer Vision and Pattern Recognition (CVPR), Seattle, WA, USA, 16–18 June 2020; pp. 8684–8694.
41. Hoberg, T.; Rottensteiner, F.; Feitosa, R.Q.; Heipke, C. Conditional Random Fields for Multitemporal and Multiscale Classification of Optical Satellite Imagery. *IEEE Trans. Geoence Remote Sens.* **2015**, *53*, 659–673. [CrossRef]
42. Pan, S.J.; Yang, Q. A Survey on Transfer Learning. *IEEE Trans. Knowl. Data Eng.* **2009**, *22*, 1345–1359. [CrossRef]

Publisher's Note: MDPI stays neutral with regard to jurisdictional claims in published maps and institutional affiliations.

Technical Note

Refocusing High-Resolution SAR Images of Complex Moving Vessels Using Co-Evolutionary Particle Swarm Optimization

Lei Yu [1], Chunsheng Li [1], Jie Chen [1], Pengbo Wang [1,2,*] and Zhirong Men [1]

1 School of Electronics and Information Engineering, Beihang University, Beijing 100083, China; ylanlei@buaa.edu.cn (L.Y.); lics@buaa.edu.cn (C.L.); chenjie@buaa.edu.cn (J.C.); menzhirong@buaa.edu.cn (Z.M.)

2 Collaborative Innovation Center of Geospatial Technology, Wuhan 430079, China

* Correspondence: wangpb7966@buaa.edu.cn; Tel.: +86-135-2099-5334

Received: 30 August 2020; Accepted: 5 October 2020; Published: 11 October 2020

Abstract: To increase the global convergence and processing efficiency of particle swarm optimization (PSO) applied in the adaptive joint time-frequency, in this study an improved PSO is proposed to refocus the high-resolution SAR images of complex moving vessels in high sea states. According to the characteristics of the high-order multi-component polynomial phase signal, this algorithm provides parallel processing and co-evolution methods by setting the different permissions of the sub-population and sharing its search information. As a result, the multiple components can be extracted simultaneously. Experiments were conducted using the simulation data and Gaofen-3 (GF-3) SAR data. Results showed the processing speed increased by more than 40% and the global convergence was significantly improved. The imaging results verify the efficiency and robustness of this co-evolutionary PSO.

Keywords: synthetic aperture radar (SAR); moving vessel; multicomponent polynomial phase signal(mc-PPS); adaptive joint time-frequency (AJTF) decomposition; co-evolutionary particle swarm optimization

---

1. Introduction

Synthetic aperture radar (SAR) has the distinct ability to be able to observe vessels at all times, and is an important method in the detection and monitoring of marine moving targets [1–3]. Marine application research of SAR has been carried out around the world, and ship detection systems based on space-borne SAR have been developed and used in practical applications, e.g., the ocean monitoring workstation (OMW) system of Canada, the automated maritime surveillance tool (MaST) system of England, the Kongsberg satellite services (KSAT) system of Norway, the collect localization satellite (CLS) system of France, and the Ship Surveillance system of China. The GF-3 satellite is China's first C-band multi-polarization SAR, and has a maximum resolution of spotlight mode of 1 m. As an ocean surveillance and monitoring satellite, it has played an increasingly important role in the field of marine theory and applications [4]. Vessels can be detected effectively using constant false alarm rate (CFAR) processing and its derivatives using high-resolution GF-3 images [5]. Wang et al. used the analytic hierarchy process by calculating the vessel's kernel density estimation, aspect ratio, and pixel number to finally obtain optimized vessel detection results [6]. Similarly, the identification of vessels can also be achieved with GF-3 SAR images using artificial intelligence techniques, such as convolutional neural networks (CNNs) and Region-CNN algorithms [7,8]. However, SAR imaging uses the relative motion of the satellite and ground targets to obtain high azimuth resolution in principle, assuming that the ground is static during synthetic aperture acquisition (i.e., dwell time).

For a stationary vessel, the main structure of the hull surface is distinguishable in SAR images, and is easily detected and identified. On the contrary, vessels moving due to sea wind and waves appear as various distortions and are blurred (defocused) in SAR images. This can cause the shape of vessels to be distorted. As a result, the length, width, and scattering distribution cannot be accurately obtained, thus affecting the application of target detection and recognition. The motions of a marine vessel have multiple periodicities and a high degree of randomness. The different motion characteristics of each scattering point in every range cell cause different distance migration and initial phase, which result in frequency folding and wrapping in the azimuth dimension, and generate high-order non-cooperative phase errors in the SAR echo signal, ultimately reducing the precision of compensation [9]. In particular, the number of signal components in each cell increases and significant mutual interference between the components exists, which reduces the reliability of the component extraction. The SAR images of vessels are less clear, and vessels may even be unrecognizable, with worse sea conditions, longer dwell time, and higher SAR resolution. This phenomenon is frequently found in high-resolution images of space-borne SAR and airborne SAR.

Therefore, refocusing of the SAR images of vessels in complex motion has consistently been an important research subject for marine remote sensing applications. Generally, the refinement of image processing for moving SAR marine targets is divided into two main aspects: translation compensation and rotation compensation. The compensation method for translation is relatively mature, and usually use the inverse SAR (ISAR) principle and a self-focusing method, such as the phase gradient autofocus algorithm (PGA), to achieve high-quality SAR images [10]. Liu et al. [11] presented a model for numerical simulation and quantitative evaluation of the image distortions caused by each rotation of a ship. Martorella [12] applied ISAR processing to the Cosmo-SkyMed SAR system and refocused moving targets. These methods are suitable for vessels with relatively stable motion or medium-resolution SAR. However, due to the complex three-dimensional rotation in high sea states, there may be no effect of using the envelope alignment and phase correction method of ISAR processing steps. The remaining uncompensated translational and rotational terms after ISAR processing still have a large influence on the high-resolution SAR image in particular.

The time-frequency analysis method is an effective method for rotational compensation [13], and utilizes the instantaneous Doppler frequency of the target to avoid blurring. In the literature [14], the relatively stable interval of the target motion is selected for imaging using different time-frequency analysis methods. However, for high-resolution SAR, the signal of a moving vessel can be represented by a high-order multicomponent polynomial phase signal (mc-PPS), which includes complex envelope migration and Doppler wrapping. In these circumstances, the traditional time-frequency analysis methods, such as short-time Fourier transform, Wigner Ville distribution, and polynomial phase transformation [15,16], are seriously affected by cross-terms and cannot adapt to the practical applications of the mc-PPS. For an effective extraction of signal components, the adaptive joint time-frequency (AJTF) method, as an improved maximum likelihood method, is proposed to represent the mc-PPS in ISAR imaging, and offers better results without being affected by cross-terms. Li et al. [17], based on the AJFT method, estimated the phase of multiple scattering centers of ISAR data. According to the linearity relationship between the scattering point location and the Doppler change in the echo phases, refined ISAR images are achieved using the data of the best imaging period time. Searching for optimal parameter components with different extremal solutions in the solution space is a multidimensional optimization problem during AJTF processing. Therefore, optimization algorithms, such as genetic algorithms (GA) and particle swarm optimization (PSO) algorithms, are used to reduce the computing complexity of searching, and simulation results confirm the efficiency of these approaches during processing [18,19]. The frequency-domain extraction-based AJTF decomposition method has been proposed to deal with the high-resolution space-borne SAR. Although its feasibility and effectiveness have been verified, the imaging processing time is long and the SAR imaging results show limited improvement [1].

In this paper, a refocusing method is proposed to deal with the mc-PPS of the vessels' SAR data using a co-evolutionary PSO optimizing AJTF. Because the design of the method uses parallel processing, this algorithm improves the effectiveness and computing speed by extracting multiple components simultaneously. Furthermore, the convergence and global optimal ability are enhanced through the co-evolution of multiple sub-populations. The simulation high-resolution SAR data and GF-3 SAR data were processed and compared with ISAR and the classic PSO algorithm. Results show that the imaging technical indicators and image visualization were clearly improved. These results also verify the robustness and efficiency of the presented algorithm, particularly under high sea conditions.

2. AJTF Decomposition Method

AJTF is a global phase compensation method that has been proved to be an effective focused imaging algorithm to address the problem of non-cooperative targets [20]. This method performs parameter estimation and phase compensation, which reduces the influence of mutual interference between scattering centers and is particularly suitable for situations with different phase changes of each scattering center of the vessels in high-resolution SAR.

The echo of a moving vessel for a range cell in high-resolution SAR can be expressed in the form of the mc-PPS as follows [21]:

$$s(t) = \sum_{m=1}^{M} A_m \cdot rect\left[\frac{t}{T}\right] \exp\left\{ j2\pi \sum_{n=0}^{N_p} a_{m,n} t^n \right\}, \quad (1)$$

where M is the number of components; A_m represents the component intensity; $rect[\cdot]$ is the rectangular time window of width T; N_p is the polynomial order of the signal phase; $a_{m,0}$ is a time-independent constant phase; $a_{m,1}$ is the linear term of time t, which is related to the real position of the scatter point; and $a_{m,2}$ Skimmed higher-order parameters are the phase errors generated by the target motion and need to be compensated in the SAR imaging process.

For the m-th component signal s_m of the mc-PPS, the phase compensation function $s_h(t)$ is expressed as follows:

$$s_h(t) = rect\left[\frac{t}{T}\right] \exp\left\{ -j2\pi \sum_{n=2}^{N_p} a_n t^n \right\}, \quad (2)$$

By multiplying this component signal s_m and the phase compensation function s_h, the m-th motion compensation is achieved. The frequency spectrum $S_c(f)$ is obtained by applying Fourier transformation to the compensated signal, as follows:

$$\begin{aligned} S_c(f) &= FT\left\{ rect\left[\frac{t}{T}\right] \cdot s_m(t) \cdot s_h(t) \right\} = FT\left\{ rect\left[\frac{t}{T}\right] \cdot A e^{j2\pi a_0} \exp[j2\pi a_1 t] \right\} \\ &= A e^{j2\pi a_0} T \sin c[T(f - a_1)] \end{aligned} \quad (3)$$

The objective function of the AJTF method is as follows:

$$\begin{cases} \{\hat{a}_n\} = \operatorname{argmax}\left\{ \max\left[FT\left(s_p(t) \cdot s_h(t) \right) \right] \right\} & , \quad n = 2,3,4,\cdots \\ \hat{a}_1 = f_p, \quad S_c(f_p) = \max[S_c(f)] & \end{cases}, \quad (4)$$

where $\{\hat{a}_n\}$ refers to the estimating parameters and f_p is the peak position $S_{c\max} = S_c(f_p)$. The maximum value of the spectrum is at $f = a_1$. The maximum value of the spectrum is at, which is the scattering point image as the $\sin c(\cdot)$ envelope function.

Through optimal estimation, the estimated signal $\widehat{s}_m$ of this component is obtained as:

$$\widehat{s}_m(t) = \widehat{A} \cdot \exp\left\{ j2\pi \sum_{n=1}^{N_p} \widehat{a}_n t^n \right\}, \quad (5)$$

where $\widehat{A}$ is the component intensity of the maximum value of the spectrum $S_c(f)$, and $\{\widehat{a}_n\}$ is the optimal estimated parameter of the signal component.

The time-domain residual signal is updated as follows:

$$y(t) = s(t) - \widehat{s}_m(t) \tag{6}$$

Using the same method, each signal component can be continuously estimated and extracted from the residual signal sequentially. The final estimated signal can be expressed as:

$$\widehat{s}(t) = \sum_{m=1}^{M} \widehat{s}_m(t) + y_M(t), \tag{7}$$

where M is the number of components, $\widehat{s}_m$ is the optimal estimation of the m-th component signal, and $y_M(t)$ is the residual error after extracting M components.

The most critical step in AJTF decomposition processing is the search for the optimum parameters in the multidimensional solution space. Such optimal problems are generally non-convex and require substantial amounts of computation. Therefore, optimization algorithms such as PSO can be applied to AJTF decomposition, thus accelerating the speed of optimal processing and improving the global convergence ability.

3. PSO Algorithm Applied in AJTF Decomposition

3.1. PSO Algorithm

PSO is based on the movement and intelligence of swarms. This approach compares the optimization technique to bird flocking and has become an attractive alternative to other heuristic algorithms [22]. The method is initialized with a population of random solutions. Then, the bird is abstracted as a particle, which is represented by two parameters: position and velocity. The position indicates a feasible solution of the optimization and the velocity means the tendency of particles to move, which is the rate of change of the particle position. The particle tends to move to its historical optimal position and global optimal position during the iteration. The optimal state of each particle in the population during the iteration is evaluated using fitness. The method minimizes the fitness function and obtains the optima by adopting the optimum velocity of each particle toward the local and global particle. From iteration g to $g+1$, the update of the velocity $\vec{V}_i(g+1)$ and position $\vec{P}_i(g+1)$ of the i-th particle is shown as follows:

$$\begin{aligned} &\vec{V}_i(g+1) = \omega\vec{V}_i(g) + r_1c_1\left(\vec{P}_{Lbest} - \vec{P}_i(g)\right) + r_2c_2\left(\vec{P}_{Gbest} - \vec{P}_i(g)\right), \quad (r_1, r_2) \in U(0,1) \\ &\vec{P}_i(g+1) = \vec{P}_i(g) + T\vec{V}_i(g+1) \end{aligned} \tag{8}$$

where ω is the inertial weight; c_1 and c_2 are the local and global attractive coefficients, respectively; and $\vec{P}_{Lbest}$ the $\vec{P}_{Gbest}$ represent the historical best local and global positions, respectively, in the whole swarm. T is the factor of position updating. Therefore, the PSO algorithm may obtain a local optimal solution and retain multiple sub-optimal solutions. The flow chart of PSO is shown in Figure 1.

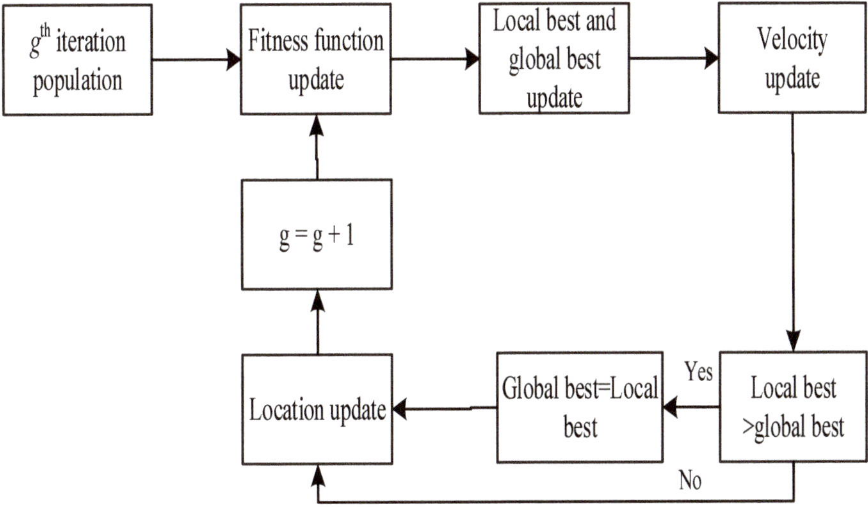

Figure 1. The flow chart of the particle swarm optimization (PSO) algorithm.

In each iteration, the position and velocity of each particle is updated, and each particle has an overall tendency to maintain its local and global best positions. The PSO algorithm retains its individual best value at the end of the algorithm. Compared with other evolutionary algorithms such as GA, PSO avoids complex genetic operations and has a population-based global search strategy. Thus, it is a more efficient search algorithm than other evolutionary algorithms [19]. Unfortunately, PSO compensation processing takes about 20 min, which means it does not meet the requirements of the mission. However, for the mc-PPS in AJTF decomposition, PSO may obtain a locally optimal solution while retaining multiple sub-optimal solutions, which may also be true components. Therefore, PSO easily falls into the local optimal solution, which means the algorithm cannot guarantee convergence to the global optimum during the processing of the mc-PPS signal of the vessel's SAR data.

3.2. Co-Evolutionary PSO Algorithm

To solve the problems of the standard PSO algorithm, Shi Y. devised the co-evolutionary PSO algorithm by reducing the dimension of the fitness function. The solution is divided into multiple sub-groups, each of which is optimized by a separate algorithm, and the fitness value is evaluated and combined into a complete particle. However, the algorithm and subsequent improvements are nonetheless prone to the problem of pseudo-minimums and cannot guarantee global convergence [23]. In theory, the components of the mc-PPS are uncorrelated with each other, which indicates that the extraction of the signal components will not affect other components [18]. Therefore, the improved co-evolutionary PSO algorithm is proposed, based on the division of multiple sub-groups in the original algorithm. This algorithm sets different permissions of the sub-groups and adds a random group. It divides the entire population into several sub-groups, such as an optimal group (Opt-group), a sub-optimal group (Sub-group), and a random group (Ran-group). The term "co-evolutionary" refers to the division of groups in the solution space into several sub-groups; a sub-group represents a sub-goal to be solved, and excellent individuals found in the search migrate between different sub-groups as shared information to guide the progress of evolution. As a result, the global convergence efficiency of this algorithm is significantly improved.

The co-evolutionary PSO algorithm combines the abilities of PSO to explore the search spaces with varied priorities. The Opt-group includes the particle with the global optimal position, and the Sub-group is prohibited from iterating with the neighborhood of the Opt-group. Setting the search forbidden zone to the sub-optimal particle swarm strengthens the global optimal ability of the Opt-group. The Sub-group only limits the search area in the sub-optimal solutions, which means the premature stagnation state problem of the particle is solved. The Sub-group represents a sub-space in the solution space, and also represents one solution to the problem. Based on the azimuth position of the scattering center, the Opt-group and the Sub-group are distinguished and the scattering center corresponds to the maximum value of the spectrum. The random group has no best particle and is updated with the other groups, which improves the randomness of the entire population and the information exchange among the three groups. When the Opt-group falls in the local optimal solution, the Sub-group retains the ability to search for another optimal solution outside of the region. Using this approach, multiple components can be extracted simultaneously, which effectively improves the global convergence speed. The procedure flow of the co-evolutionary PSO algorithm is shown in Figure 2.

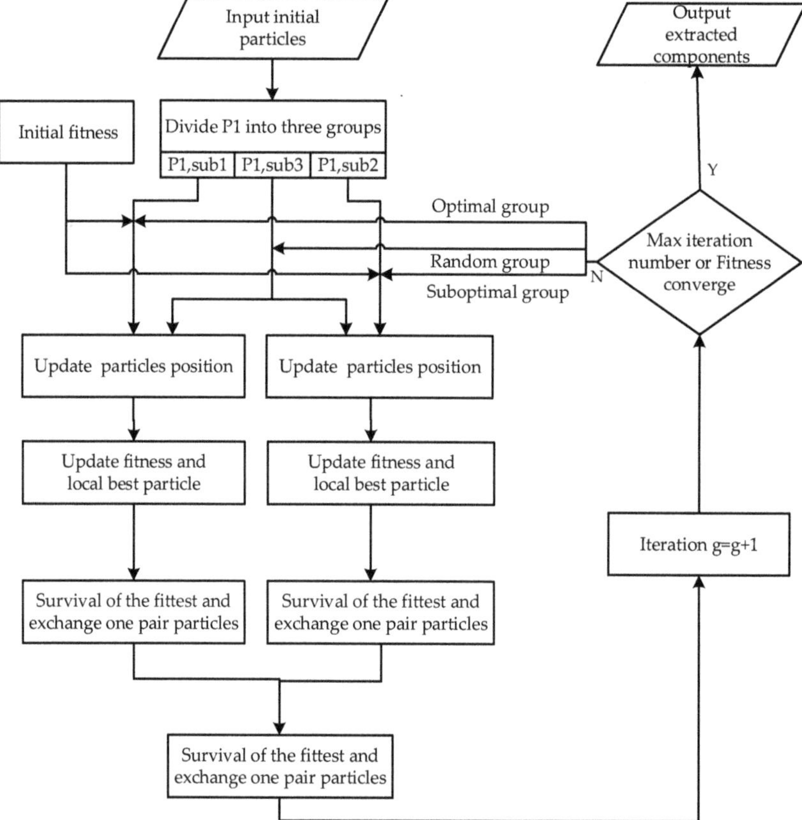

Figure 2. The procedure of the co-evolutionary PSO algorithm.

(a) The input initial particles are the residual signal of one range cell of SAR.

(b) The initial fitness. The fitness of the m-th particle P_m is as follows:

$$F_{fit}(\mathrm{P}_m) = \max\left\{\left|FT\left(s_p(t)\cdot s_h(P_m,t)\right)\right|\right\} \tag{9}$$

where $F_{fit}(\mathrm{P}_m)$ is the fitness function of the m-th component; $FT()$ is the Fourier transform process; $s_p(t)$ is the signal to be processed; and $s_h(t)$ is the phase compensation function. The position of all particles refers to the individual parameters in the multidimensional solution space, including the start and end times of the signal component and the second-order and above polynomial coefficients.

$$\vec{P}_m = \{a_n, \tau_s, \tau_e\}, m = 1, \cdots, N_{pop}, n = 2, 3, \cdots, N_p, \tag{10}$$

where a_n refers to the higher-order parameters and is due to the complex movement; τ_s and τ_e are the start time and end time of the time window, respectively. This indicates that the point scattering characteristics of the complex motion are variable because some signal components will appear and vanish during the dwell time. N_{pop} is the total number of particles, which mainly depends on the vessel size and SAR image resolution, and typically has a value of 50~120. N_p is the polynomial order of the input signal phase, and has a value of 3~4 [18].

(c) Divide P1 into three groups. $P_{1,\,sub1}$ is the Opt-group, $P_{1,\,sub2}$ is the Sub-group, and $P_{1,\,sub3}$ is the Ran-group. Iteration is set to $g = 1$.

(d) Update particle position. By random division, the Ran-group is divided into two sub-groups and combined with the Opt-group and Sub-group in two new mixed groups.

(e) Update fitness and local best particles. According to the fitness function, the global optimal and sub-optimal particle fitness are searched for in the respective ranges, and the optimal particle parameters and scattering center positions are updated. Observing the near zone of the Opt-group scattering center is prohibited to the Sub-group, which has a value of 20~30.

In the g-th iteration, the local and global optimal fitness values of the m-th particle are as follows:

$$\begin{cases} F_{L,g,m} = \max\left\{F_{\text{fit}}\left(\left\{\vec{P}_m(g\prime)\right\}\right)\right\} \\ F_{G,g} = \max\left\{F_{L,g\prime,m}\right\} \end{cases}, g' = 1, 2, \cdots, g, \tag{11}$$

where $\vec{P}_m(g')$ represents all position records from the 1st iteration to the g-th iteration of the m-th particle. Correspondingly, the local and global optimal positions of the m-th particle are as follows:

$$\begin{cases} \vec{P}_{\text{Lbest}}(g) = \text{argmax}\left\{F_{L,g,m}\right\} \\ \vec{P}_{\text{Gbest}}(g) = \text{argmax}\left\{F_{G,g}\right\} \end{cases}, m = 1, 2, \cdots, N_{\text{pop}}, \tag{12}$$

(f) According to local fitness results in the new mixed group of the Opt-group, the survival of the fittest is made by exchanging the two particles inside if the maximum fitness value of the random Sub-group is more than the minimum value of the Opt-group. The same processing is applied to the other mixed group of the Sub-group.

(g) Similarly, according to global fitness results, the survival of the fittest is made by exchanging the roles of the two groups if the maximum fitness value of the Sub-group is more than that of the Opt-group.

(h) The iteration $g = g + 1$. Based on the previous two steps, the new groups become the (g + 1)-th population. The local best position is updated as follows. Furthermore, the global best position can also be obtained, according to Formula (12). When the fitness is stable, the update velocities of the optimal particle become zero, and the position, which is equivalent to the global optimal position, no longer changes.

$$\vec{P}_{\text{Lbest}}(g+1) = \begin{cases} \vec{P}_m(g+1), & F_{\text{fit}}\left(\vec{P}_m(g+1)\right) \geq F_{\text{fit}}\left(\vec{P}_{\text{Lbest}}(g)\right) \\ \vec{P}_{\text{Lbest}}(g) & otherwise \end{cases} \tag{13}$$

(i) Whether the stop conditions are met is determined, including the maximum iteration number and the convergence of the global fitness results. The maximum iteration number is generally 300, and the convergence is expressed by the fitness ratio of the residual signal of the sub-optimal solution before and after the extraction of the global optical component, which is usually set to 0.99. If the conditions are met, step (j) is performed, and if not, then step (e) is performed.

(j) Finally, the global best parameters of Opt-group and Sub-groups are outputted, and the algorithm of one range cell is completed and it continues to process the next range cell. When the traversing of all range cells is completed, the two-dimensional imaging result is obtained with the AJTF decomposition method [1].

4. Experiment Results and Analysis

4.1. Simulation Test

The modeling and simulation system of space-borne SAR was established. The vessel scattering model consists of nine points on the edge of the hull. The sea condition of the complex motion was Level 5 [24]. The main simulation parameters of the moving vessel and imaging condition of the GF-3 satellite are shown in Table 1.

Table 1. Main simulation parameters.

Item	Value
Scattering point model of the vessel with 9 points [X,Y]	[0 0, 100 0, −100 0, 50 50, 50 0, 50 −50, −50 50, −50 0, −50 −50] [m]
Vessel rotation amplitude	[38.4, 3.4, 3.8] [deg]
Vessel rotation period [Roll, Pitch, Yaw]	[12.2, 6.7, 14.2] [s]
Looking Angle	31.2 [deg]
Imaging Mode	Spotlight
Band Width	240 [MHz]
PRF	3725.6 [Hz]
Sample Rate	266.667 [MHz]
Aperture Time	4.0 [s]
Shortest Slant Range	898.388 [Km]

To evaluate the performance, the classical PSO and the proposed algorithm applied to AJTF were compared in the simulation experiment. The imaging results of the scattering point model simulation are shown in Figure 3. Figure 3a,b shows the results of the stationary vessel and the moving vessel, respectively, which were obtained using the Chirp Scaling (CS) algorithm of the conventional SAR ground processing system. Significant differences exist between the results of the model with nine scattering points. Under the stationary condition, all of the points are similar; moreover, the intensity value of the points is also relatively larger, indicating that the focusing effect was better. On the contrary, the imaging result of the vessel in complex motion is defocused and fuzzy, and it is not possible to effectively identify the points of the simulation model. The exception is the central point, which remains focused in the image because it is the center of rotation and is thus not influenced by the rotational motion. The eight surrounding points are distorted and differ from each other. These differences are due to the phase error caused by the rotation movement and the different positions. Figure 3c,d shows the results of the moving vessel using the classical PSO and co-evolutionary PSO, respectively. During the PSO processing, the number of iterations is 300, and the number of particles is 100. The scattering points are focused better than those using the CS algorithm. The nine scattering points of the modeled vessel are obvious and identifiable. The focusing effect in the azimuth (Figure 3d) is more obvious in the detail of its points, which indicates that complex moving compensation is more accurate compared with that using classical PSO.

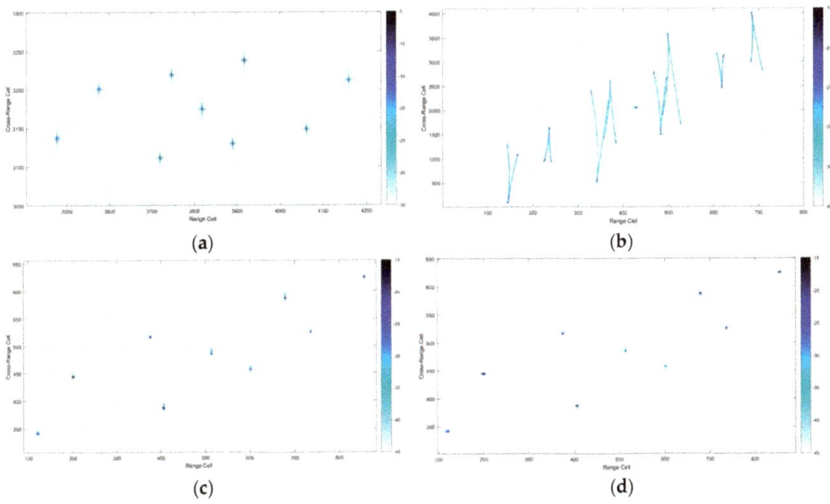

Figure 3. Imaging results of the vessel scattering point model: (**a**) imaging result of a simulated stationary vessel; (**b**) imaging result of a simulated moving vessel; (**c**) imaging result of classical PSO; (**d**) imaging result of co-evolutionary PSO.

To evaluate the effects of point refocus, the peak sidelobe ratio (PSLR) and the integral sidelobe ratio (ISLR) in the azimuth were calculated from the eight surrounding scattering points, and are shown in Table 2. Compared to the imaging results, the PSLR and ISLR of co-evolutionary PSO were superior. The improvement of the average azimuth PSLR was greater than 1.45 dB, and the average azimuth ISLR improvement was greater than 2.1 dB, compared to classical PSO in this simulation test. This indicates the compensation accuracy of the high-order phase error is higher and the algorithm is more robust. Finally, the processing speed of co-evolutionary PSO is about 48.97% faster than that of classical PSO, indicating a significant improvement in the global convergence. Computational complexity relates mainly to iterative processing, including the fitness update and individual parameter update. The fitness update requires the Fourier transform, and the number of calculations in the extracting components is related to the number of population individuals and the total number of iterations. Unfortunately, the PSLR and ISLR in the subsequent experiment were difficult to measure due to the clustered scattering points of the vessel in the real SAR data.

Table 2. Comparison of the algorithms.

	CS	Classical PSO	Co-Evolutionary PSO
PSLR (dB)	−5.02	−10.12	−11.57
ISLR (dB)	−5.41	−13.55	−15.65
Ave computation time (s)	65	1421	725

4.2. Experimental Test

To evaluate the imaging effect and usability of this algorithm, actual GF-3 SAR stationary and moving vessel data were used to verify the experiment, as shown in Figure 4. According to the imaging time and location of the moving vessel in a high sea state taken from the GF-3 satellite, we determined that the wave height was 2.8 m and the ocean velocity was 1.42 m/s at the time the image was captured [25]. In addition, the stationary vessel was selected at other times near the area in a calm sea state to compare the effects of complex motion on SAR imaging. Figure 4a,b show the results of the stationary and moving vessels using the CS algorithm. Figure 4c shows the refocusing result of the moving vessel using ISAR processing algorithms [26]. Figure 4d,e show the imaging results of the moving vessel using the classical PSO and co-evolutionary PSO algorithms, respectively.

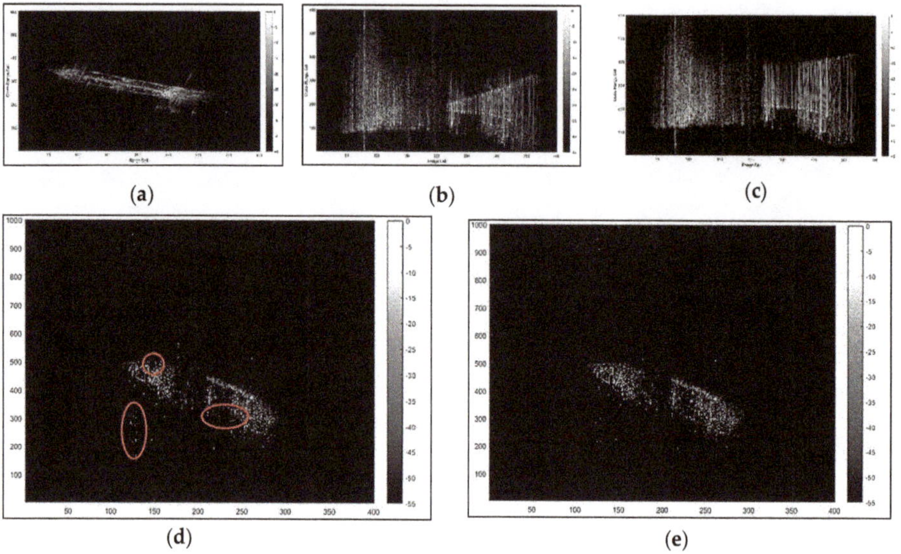

Figure 4. Imaging results of the GF-3 data: (**a**) imaging result of stationary vessel; (**b**) imaging result of moving vessel; (**c**) imaging result of the inverse synthetic aperture radar (ISAR) process; (**d**) imaging result of classical PSO; (**e**) imaging result of co-evolutionary PSO.

In Figure 4a, the edge of the stationary vessel is clear and the inside structure is easy to identify on the calm sea, similar to the case of the SAR imaging result of a stationary ground target. However, as shown in Figure 4b, the vessel image in a high sea state is highly unfocused, and it is impossible to measure the parameters of the vessel and recognize the vessel type. The ISAR principle and the phase gradient autofocus method were used to accomplish the phase compensation of the moving vessel, shown in Figure 4c. Compared with Figure 4b, the focus in the azimuth is improved, indicating that there is a translation component of the vessel. However, previous research has shown that the ISAR method is only applicable to vessels with stable motion. Due to the different movements of the various parts of the vessel, this method is not able to provide complete compensation. Residual uncompensated translation and rotation errors after ISAR processing still have a significant influence on the SAR image. Figure 4d,e shows that the information of the scattering centers of each range unit is effectively extracted. The basic outline of the vessel is maintained, and the length and width of the vessel can be easily measured. In addition, the resolution of each scattering center is improved and the imaging blurring problem caused by phase wrapping is resolved. The edge shape of the vessel in Figure 4e shows more image detail than that of 4d; however, many noise spots are present in the red circles of the classical PSO result image, indicating that it may have fallen into a local optimal solution, and the stability and reliability of the classical PSO algorithm are comparatively worse. In addition, the processing time of co-evolutionary PSO was about 10 min, representing a reduction of about 42.2% compared to classical PSO, which is consistent with the simulation results. Hence, the results proved the efficiency and robustness of this co-evolutionary PSO algorithm by simulation and experimental testing.

5. Conclusions

According to the relevant statistics, the number of defocused vessel images affected by complex motion accounts for 15–20% of GF-3 high-resolution ocean data. The SAR echo of vessels subject to complex motion is a multicomponent polynomial phase signal, which leads to the inability of classical time-frequency analysis methods to process data efficiently. Therefore, a novel co-evolutionary PSO applied to AJTF is proposed that can extract several components in the solution space simultaneously and avoid falling into a local optimal solution. Compared with other algorithms, this method has obvious advantages in image focusing performance, robustness, and efficiency. The results of simulation data and GF-3 satellite SAR data show that the image of a moving vessel in a high sea state was improved and, simultaneously, the processing speed of the algorithm was increased by over 40%. According to our preliminary experiments, this algorithm is also suitable for sub-meter space-borne SAR and airborne SAR data processing, with a processing time of several hours. The existing method performs compensations after traditional SAR imaging and involves redundant calculations. Therefore, we will continue to study methods for simultaneous image processing and phase estimation compensation to improve computing efficiency for processing sub-meter resolution SAR data. To summarize, the use of the proposed and ISAR methods [26] in existing GF-3 SAR ground processing systems could have significant benefits for marine users who need high-precision images of moving marine vessels in applications such as vessel identification and intelligent feature extraction.

Author Contributions: Conceptualization C.L. and J.C.; methodology: L.Y. and P.W.; writing—original draft preparation, L.Y.; writing—review and editing, P.W. and Z.M.; funding acquisition, C.L. All authors have read and agreed to the published version of the manuscript.

Funding: This research received no external funding.

Conflicts of Interest: The authors declare no conflict of interest.

References

1. Lao, G.; Yin, C.; Ye, W.; Sun, Y.; Li, G.; Han, L. An SAR-ISAR Hybrid Imaging Method for Ship Targets Based on FDE-AJTF Decomposition. *Electronics* **2018**, *7*, 46. [CrossRef]
2. Crisp, D.J. *The State-of-the-Art in Ship Detection in Synthetic Aperture Radar Imagery*; Defence Science and Technology Organisation: Edinburgh, Australia, 2004.
3. Gao, G.; Liu, L.; Zhao, L.; Shi, G.; Kuang, G. An adaptive and fast CFAR algorithm based on automatic censoring for target detection in high-resolution SAR images. *IEEE Trans. Geosci. Remote Sens.* **2008**, *47*, 1685–1697. [CrossRef]
4. Lin, M.; He, X.; Jia, Y.; Bai, Y.; Ye, X.; Gong, F. Advances in marine satellite remote sensing technology in China. *Haiyang Xuebao* **2019**, *41*, 99–112.
5. Hwang, S.I.; Ouchi, K. On a novel approach using MLCC and CFAR for the improvement of ship detection by synthetic aperture radar. *IEEE Geosci. Remote Sens. Lett.* **2010**, *7*, 391–395. [CrossRef]
6. Wang, C.; Zhang, H.; Wu, F.; Jiang, S.; Zhang, B.; Tang, Y. A Novel Hierarchical Ship Classifier for COSMO-SkyMed SAR Data. *IEEE Geosci. Remote Sens. Lett.* **2014**, *11*, 484–488. [CrossRef]
7. An, Q.; Pan, Z.; You, H. Ship detection in Gaofen-3 SAR images based on sea clutter distribution analysis and deep convolutional neural network. *Sensors* **2018**, *18*, 334. [CrossRef]
8. Kang, M.; Leng, X.; Lin, Z.; Ji, K. A modified faster R-CNN based on CFAR algorithm for SAR ship detection. In Proceedings of the 2017 International Workshop on Remote Sensing with Intelligent Processing (RSIP), Shanghai, China, 18–21 May 2017; pp. 1–4.
9. Yu, L.; Li, C.; Wang, P.; Fang, Y. Imaging of maritime rotation targets in spaceborne SAR image. In Proceedings of the 2017 Progress in Electromagnetics Research Symposium-Fall (PIERS-FALL), Singapore, 19–22 November 2017; pp. 503–507.
10. Noviello, C.; Fornaro, G.; Martorella, M.; Reale, D. ISAR add-on for focusing moving targets in very high resolution spaceborne SAR data. In Proceedings of the 2014 IEEE Geoscience and Remote Sensing Symposium, Quebec City, QC, Canada, 13–18 July 2014; pp. 926–929.
11. Liu, P.; Jin, Y.-Q. A study of ship rotation effects on SAR image. *IEEE Trans. Geosci. Remote Sens.* **2017**, *55*, 3132–3144. [CrossRef]
12. Martorella, M.; Pastina, D.; Berizzi, F.; Lombardo, P. Spaceborne radar imaging of maritime moving targets with the Cosmo-SkyMed SAR system. *IEEE J. Sel. Top. Appl. Earth Obs. Remote Sens.* **2014**, *7*, 2797–2810. [CrossRef]

13. Lu, G.; Bao, Z. Range instantaneous Doppler algorithm in ISAR based on instant frequency estimation. *J. Xidian Univ.* **1998**, *3545*, 198–201.
14. Hu, C.; Ferro-Famil, L.; Kuang, G. Ship discrimination using polarimetric SAR data and coherent time-frequency analysis. *Remote Sens.* **2013**, *5*, 6899–6920. [CrossRef]
15. Djurović, I.; Simeunović, M. Review of the quasi-maximum likelihood estimator for polynomial phase signals. *Digit. Signal Process.* **2018**, *72*, 59–74.
16. Djurović, I.; Simeunović, M.; Wang, P. Cubic phase function: A simple solution to polynomial phase signal analysis. *Signal Process.* **2017**, *135*, 48–66. [CrossRef]
17. Li, J.; Hao, L.; Chen, V. An Algorithm to Detect the Presence of 3D Target Motion from ISAR Data. *Multidimens. Syst. Signal Process.* **2003**, *14*, 223–240. [CrossRef]
18. Yu, L.; Lao, G.; Li, C.; Sun, Y.; Li, Y. A Novel Multicomponent PSO Algorithm Applied in FDE–AJTF Decomposition. *Electronics* **2019**, *8*, 51. [CrossRef]
19. Brinkman, W.; Thayaparan, T. Focusing ISAR images using the AJTF optimized with the GA and the PSO algorithm-comparison and results. In Proceedings of the 2006 IEEE Conference on Radar, Verona, NY, USA, 24–27 April 2006; p. 8.
20. Li, Y.; Fu, Y.; Li, X.; Le-Wei, L. ISAR imaging of multiple targets using particle swarm optimization–adaptive joint time-frequency approach. *IET Signal Process.* **2010**, *4*, 343–351. [CrossRef]
21. Liu, Q.Y.; Li, Z.S.; Li, H.Y.; Liang, H. Parameter estimation of multicomponent Polynomial Phase Signal. *Acta Electronic. Sinica.* **2004**, *12*, 2031–2034.
22. Shi, Y.; Eberhart, R. A Modified Particle Swarm Optimizer. In Proceedings of the 1998 IEEE International Conference on Evolutionary Computation Proceedings. IEEE World Congress on Computational Intelligence (Cat. No.98TH8360), Anchorage, AK, USA, 4–9 May 1998.
23. Shi, Y.; Krohling, R.A. Co-evolutionary particle swarm optimization to solve min-max problems. In Proceedings of the 2002 Congress on Evolutionary Computation CEC'02 (Cat. No.02TH8600), Honolulu, HI, USA, 12–17 May 2002.
24. Wehner, D.R. *High Resolution Radar*; Artech House, Inc.: Norwood, MA, USA, 1987; p. 484.
25. Global Weather Visual Simulation Network. Available online: https://earth.nullschool.net (accessed on 10 September 2019).
26. Yu, L.; Li, C.; Wang, P.; Fang, Y.; Feng, X. Research into SAR imaging of moving ship target based on ISAR technology. In Proceedings of the International Geoscience and Remote Sensing Symposium (IGARSS 2018), Valencia, Spain, 22–27 July 2018; pp. 2290–2293.

Article

Eddy Detection in HF Radar-Derived Surface Currents in the Gulf of Naples

Leonardo Bagaglini [1,*], Pierpaolo Falco [2] and Enrico Zambianchi [2,3]

1 Institute of Atmospheric Sciences and Climate- National Research Council (ISAC-CNR), 00133 Roma, Italy
2 Università degli Studi di Napoli Parthenope and CoNISMa, 80143 Napoli, Italy; enrico.zambianchi@uniparthenope.it (P.F.); pierpaolo.falco@uniparthenope.it (E.Z.)
3 Institute of Marine Sciences–National Research Council (ISMAR-CNR), 00133 Roma, Italy
* Correspondence: Leonardo.Bagaglini@artov.isac.cnr.it

Received: 4 November 2019; Accepted: 23 December 2019; Published: 27 December 2019

Abstract: Submesoscale eddies play an important role in the energy transfer from the mesoscale down to the dissipative range, as well as in tracer transport. They carry inorganic matter, nutrients and biomass; in addition, they may act as pollutant conveyors. However, synoptic observations of these features need high resolution sampling, in both time and space, making their identification challenging. Therefore, HF coastal radar were and are successfully used to accurately identify, track and describe them. In this paper we tested two already existing algorithms for the automated detection of submesoscale eddies. We applied these algorithms to HF radar velocity fields measured by a network of three radar systems operating in the Gulf of Naples. Both methods showed shortcomings, due to the high non-geostrophy of the observed currents. For this reason we developed a third, novel algorithm that proved to be able to detect highly asymmetrical eddies, often not properly identified by the previous ones. We used the results of the application of this algorithm to estimate the eddy boundary profiles and the eddy spatial distribution.

Keywords: surface currents; HF radar; eddy detection algorithms

1. Introduction

Transport in the ocean develops over an extremely wide range of scales, from the basin to the dissipation scale (e.g., [1]). Our ability to observe and/or to model processes at smaller and smaller scales has greatly increased over the last few decades. Phenomena that in the very recent past could not be detected or described, and thus needed to be parametrized in terms of larger scales (e.g., [2,3]), are now subjects of consolidated research, as is the case of mesoscale features. Now, our focus has shifted to smaller dynamics, such as submesoscale motions. In the wide range of turbulent processes in the ocean, submesoscale eddies are the most volatile ones, due to their short lifetime (few hours) and length scale (below 10 km).

Submesoscale eddies principally act as energy conveyors from the mesoscale to the microscale, and play a crucial ecological role: they may influence the state of health of ocean regions through their ability to carry heat, inorganic matter, nutrients and biomass ([4,5]), ensuring the connectivity between different ecosystems ([6]). They are particularly important for phytoplankton, as they develop over timescales similar to those of phytoplankton growth ([7]), moreover, they may act as carriers of pollutants (see, e.g., [8]). Consequently the detection of eddies, behind its inherent interest, is crucial also for environmental applications.

As more and more synoptic, high resolution data on mesoscale and submesoscale eddies has become available, thanks to remote sensing techniques at different resolutions, automatic eddy detection methods have gained importance and interest. In the recent past, several eddy detection algorithms ([9–13]) have been developed and applied to velocity fields derived from altimeter data, numerical model outputs and HF radar observations. They can be divided into three families: geometrical, dynamical and hybrid ones. The definition depends on the flow characteristics extrapolated by the algorithm itself, that can be geometrical, dynamical or both (e.g., [9–11] and respectively). However, the existing methods have been mainly conceived for meso and larger scale recirculations, which display different kinematic characteristics than submesoscale eddies, in particular in terms of divergence and (a)symmetry of the flow field.

Mesoscale eddies scale with the first internal Rossby radius ([14,15]), which is of the order of 10 km in the Mediterranean (5 to 12 km according to [16]) and four to ten times as large in the north Atlantic ([17]). Differently, submesoscale surface eddies have characteristic lengths starting from 0.1 km up to the mesoscale ([8,18]). They are completely confined in the surface mixed layer, within depths going from tens to hundreds of meters. Therefore, since their relative Rossby and Froude numbers, Ro and Fr, are not small, these structures show highly non-geostrophic behavior, high divergent flow patterns and strong asymmetries. As a consequence, the aforementioned algorithms, typically designed to capture the features of vortices in geostrophic balance, may fail in detecting submesoscale eddies as they are often unable to characterize highly deformed, divergent or convergent motions. For this reason, we have designed a novel algorithm, presented in this paper, that has proved to be able to capture the noncircular symmetry and the divergent character of submesoscale recirculations.

High resolution data is needed in order to identify submesoscale motions. In this framework, HF radars are proving to be an almost irreplaceable tool: They are land-based remote sensing instruments which allow to observe surface currents at very high spatial and temporal resolution, thus suitable to monitor such small scale phenomena ([19,20]). Other remote sensing techniques are available, even with much higher spatial resolution, and are thus able to detect submesoscale flow features ([21–23], but they have very long revisit periods with respect to the hourly sampling provided by coastal radars, thus allowing for detecting but not for tracking such features.

In this study we have utilized HF radar observations of surface currents in the Gulf of Naples (GoN), a semi-enclosed area of the Tyrrhenian, a sub-basin of the western Mediterranean Sea. The GoN is surrounded by a coast characterized by a quite uneven orography, dominated by the presence of the Vesuvius volcano and of Mount Faito in the East, both exceeding 1000 m altitude, and of a number of lower hills very close to the northern coastline. It has a complex bathymetry, with an average depth of 170 m which reaches down to more than 800 m in correspondence of two major canyons, the Magnaghi and the Dohrn, which carve the shelf across the threshold connecting the Gulf with the open Tyrrhenian Sea. Its surface circulation is mainly wind driven, with a strong seasonal regime ([24,25]), even though the offshore circulation of the Southern Tyrrhenian may occasionally affect the current pattern in the interior of the GoN ([25] and references therein). The Gulf represents a very complex system: It hosts a heavily anthropized coastline, with industrial settlements in the immediate vicinity of the coast, side by side with four marine/natural protected areas. Moreover, oligotrophic and eutrophic characteristics coexist in the Gulf. Its outer portion is dominated by Tyrrhenian, oligotrophic waters, while the coastal part is typically eutrophic, as can be expected ([26,27]). Water exchange inside the Gulf is ruled by mechanisms acting at different spatial and temporal scales, triggered by external (local and remote) driving as well as by bottom topography and coastal constraints ([25,28–31]). Fixed-point long term investigations of the local plankton community composition have shown a strong variability of species, alternatively coming from the coast or from offshore ([32]). Recent investigations have pointed out the different roles of physical transport and biological processes, demonstrating in particular the effect of transient current patterns (e.g., [33]). For the above reasons we believe that the Gulf may well represent a universal example of a coastal area facing and intensively interacting with the open sea, but more importantly characterized by the coexistence of different subsystems. In such a framework, submesoscale eddies may act as an extremely powerful exchange mechanism among those subsystems for water and its biogeochemical content, and are therefore worthy of the maximum consideration.

The article is structured as follows. In Section 2 we describe our dataset and the dynamical fields that allow us to identify recirculating structures. Then in Section 3 we accurately describe the chosen detection algorithms, and in Section 4 we describe the algorithm tuning procedure and we provide a method for estimating eddy boundaries and radii. In Section 5 we discuss the results obtained by two algorithms and we analyze the spatial distribution of the detected eddies. Finally, in Section 6, we summarize our results and highlight some possible research directions.

2. Materials

2.1. Dataset

For this study we used the HF radar observations of surface currents in the GoN collected by a CODAR (Coastal Ocean Dynamics Application Radar) SeaSonde system. The product consists of a two-dimensional velocity field with a spatial resolution of 1 km over an area of approximately 20–30 km alongshore by 15–20 km offshore, and with an hourly frequency. The specifics of the radar network operating in the GoN can be found in [30]; see [20] for a review on HF radar theory and its applications to coastal current observations; [34] for a recent utilization of HF radar-detected transport to fisheries. Specific applications to the GoN in terms of description of the dynamics, data validation, as well as their use in conjunction with numerical models, can be found in [24,25,33,35–38]. The data utilized in this study refers to the late fall period 24 November through 8 December 2008. Since the number of eddies was clearly detectable in radar observations, we selected this period among many others, as a sort of training dataset for our algorithm, necessarily limited to a relatively short timespan for validation issues.

Since the observed GoN eddies have radii in a range between 0.5 and 5 km (we found a mean *equivalent radius* of approximately 0.8 km, with extrema reaching 4 km) we decided (following [10]) to refine the grid to approximately 0.5 km, by means of a cubic interpolation, as illustrated in Figure 1.

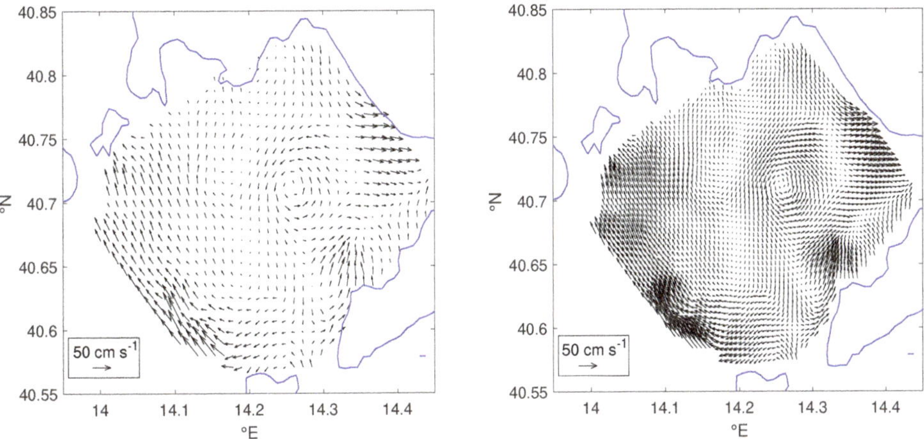

Figure 1. Surface currents data provided by the HF radar system in the Gulf of Naples (**on the left**) and the interpolated data (**on the right**). Black arrows denote the velocity field whereas the blue line represents the coastline.

With a reference velocity scale U of $10\,\text{cm}\,\text{s}^{-1}$, a length scale L of 1 km and a Coriolis parameter $f \sim 9.5 \cdot 10^{-5}\,\text{s}^{-1}$, the Rossby number is Ro ~ 1. It is thus evident that the quasi-geostrophic equations are not accurate for describing the GoN dynamics.

2.2. Dynamical Parameters Characterizing Recirculations

At a first glance, eddies of two-dimensional turbulent flows can be described as flow regions characterized by a rigid-body rotation. In this approximation, many local and semi-local parameters can be adopted to decide whether vortices exist or are likely to develop. As eddies are extensive structures, it is natural to consider integral quantities, rather than pointwise ones, to identify them. Nevertheless, the choice of the appropriate computational regions, specifically their shape and area, is completely arbitrary. For this reason these parameters naturally depend on a scale coefficient.

2.2.1. Okubo–Weiss and Local Okubo–Weiss Parameters

The Okubo–Weiss parameter (OW) is a local dynamical field which, loosely speaking, measures the relative dominance of the rate-of-strain tensor s over the vorticity ω of the velocity field (here $|.|$ denotes the euclidean module)

$$\mathrm{OW} = |s|^2 - |\omega|^2.$$

It was independently introduced by [39,40]. For a two-dimensional flow $\underline{u} = (u, v)$ it turns out that

$$\mathrm{OW} = \left(\frac{\partial}{\partial x}u\right)^2 + \left(\frac{\partial}{\partial y}v\right)^2 + 2\left(\frac{\partial}{\partial y}u\right)\left(\frac{\partial}{\partial x}v\right).$$

By definition $\mathrm{OW} < 0$ whenever the rotation tendency exceeds the strain one.

The local version of the OW parameter, called the *local* Okubo–Weiss parameter (LOW) (see [10]), depends on a positive distance $a > 0$ and is defined as the integral of OW over the disk of radius a:

$$\mathrm{LOW}(\underline{x}) = \int_{B_a(\underline{x})} \mathrm{OW}(\underline{x}')d\underline{x}'.$$

2.2.2. Local Normalized Angular Momentum and Momentum Flux Fields

In the rotating rigid-body analogy the angular momentum of a fluid particle has to be maximized about the eddy center, as pointed out by [41]. This consideration suggested to define the *local normalized angular momentum field* (LNAM):

$$\mathrm{LNAM}(\underline{x}) = \frac{\hat{z} \cdot \int_{B_a(\underline{x})} (\underline{x}' - \underline{x}) \times \underline{u}\, d\underline{x}'}{\int_{B_a(\underline{x})} \left(|\underline{u}||\underline{x}' - \underline{x}| + |\underline{u} \cdot (\underline{x}' - \underline{x})|\right) d\underline{x}'},$$

which assumes extreme values ± 1 at the centers of circular symmetric eddies: $+1$ for cyclonic rotations and -1 for anticyclonic ones (in [41] the term $\underline{u} \cdot (\underline{x}' - \underline{x})$ appears with its sign; we added the modulus to get $|\mathrm{LNAM}| \leq 1$.)

Analogously, the *local normalized momentum flux* field (LNMF) can be defined as follows:

$$\mathrm{LNMF}(\underline{x}) = \frac{\int_{B_a(\underline{x})} \underline{u} \cdot (\underline{x}' - \underline{x}) d\underline{x}'}{\int_{B_a(\underline{x})} \left(|\underline{u}||\underline{x}' - \underline{x}| + |\underline{u} \times (\underline{x}' - \underline{x})|\right) d\underline{x}'}.$$

It is clear that LNMF identically vanishes on centers of rotating eddies, while it assumes extreme values ± 1 at the symmetric sources and sinks; so it can be adopted to distinguish these various types of recirculating structures.

3. Methods

3.1. *Eddy Detection Algorithms*

We implemented two versions of two different existing detection algorithms for our study. The first method, the angular momentum eddy detection and tracking algorithm (AMEDA), developed by [10], was tested on several products such as altimeter data, numerical simulations and laboratory experiment. The second, proposed by [11], the 'Nencioli et al. algorithm' (NEAL), was specifically designed for certain HF radar derived datasets.

It is worth noting that in both cases above the velocity fields utilized for testing and application were geostrophic or quasi-geostrophic. On the other hand the surface flow observed in the GoN is highly non-geostrophic and significant variations of the divergence field frequently occur, often associated to recirculating sources or sinks. So, to distinguish similar structures in our study area, it was necessary to modify those algorithms, and yet, as discussed in the following, our proposed refinements led to just moderate improvements. Therefore, in order to specifically address the aforementioned classification problems, we defined a third method, yet another eddy detection algorithm (YADA), inspired by [10,12].

3.2. Ameda

The AMEDA algorithm ([10]) determines the eddy centers accordingly with the following procedure:

1. Identifies grid points which are local extrema of LNAM satisfying LNAM $>$ K and LOW <0, for a chosen threshold K $\in (0,1)$;
2. Verifies the existence of at least one *closed* streamline around each extremum.

However, as already pointed out, GoN eddies may have hyperbolic orbits, in contrast with the geostrophic flows found in [10,11]. In such cases the second assumption is never verified, so we decided to adopt the following alternative criterion (described in [11]):

2′. Confirms that the velocity field constantly rotates along the perimeter of the square domain of edge $2b$ and centered at the extremum, for a chosen distance b.

The modified version of AMEDA, obtained by substituting 2 with 2′, is here denoted by AMEDAmod.

3.3. Neal

The eddy detection algorithm developed in [11], and here denoted by NEAL, identifies the eddy centers in several steps, namely:

1. Identifies couples of adjacent grid points $(\underline{x}_1, \underline{x}_2)$ such that the meridional component of the velocity field changes sign going westward along the zonal segment of length $2a$, centered at $\underline{x}_i$, and increases its magnitude away from this point. This computation also provides the expected sign of rotation;
2. Verifies that, at any such grid point $\underline{x}_i$, the zonal component of the velocity field changes sign going northward along the meridional segment of length $2a$, centered at $\underline{x}_i$, and increases its magnitude away from this point. This change must be compatible with the expected rotation;
3. Identifies the KE (kinetic energy) local minima inside a square domain of edge $2b$, centered at $\underline{x}_i$, which are global minima in a square neighborhood Q_b of the same size;
4. Confirms that the velocity field constantly rotates along the perimeter ∂Q_b.

3.4. Yada

The YADA algorithm searches for potential eddy centers in two steps:

1. Identifies the local extrema of a dynamical field like LNAM, KE or OW;
2. Analyzes the streamline geometry within some neighborhood Q_b of each extremum, ensuring the existence of either bounded hyperbolic orbits (characterizing eddies with sink-like cores) or elliptic orbits (in presence of eddies having stable orbits).

Note that the second step is precisely designed to distinguish different eddy geometries. During this classification procedure, as we will see in the next section, the YADA algorithm computes quantities that are strongly related to the eddy shape, and therefore provide useful information about its character, which may be either hyperbolic or elliptic, depending on the streamline behavior.

3.5. Tuning Strategy

Each algorithm depends on some parameters, specifically the LNAM threshold K and the neighborhood radii a and b, which have to be tuned in order to maximize the probability of detection. In principle, such a training phase should be carried out with a set of completely characterized observations, for which the real eddy population and its spatio-temporal distribution is perfectly known. This is never the case for eddy detection studies. In [10] the authors, in order to cross-validate their parametric algorithm, considered the number of detected eddies as a score function depending on the algorithm parameters. The best model was then chosen by looking for parameters stabilizing the score function. The reasoning behind this approach can be heuristically described as follows. One starts with an inaccurate model which predicts too few (or too many) eddies. However, by randomly exploring different parameter values, one may observe an increase (decrease) of the score function until reaching a stable region in the parameter space. Then, elements within the stable region can be considered optimal assuming that the observed local fluctuations are caused by the existence of eddies, that randomly fall in (or escape from) the detection range as parameters vary. In our study we chose to adopt the same tuning strategy, better described in the next section.

4. Results

In this section we first describe the tuning procedure for each chosen algorithm and then we discuss the results. Before doing this, some observations about the algorithm definitions are needed. Firstly, for numerical convenience, we substituted the disk $B_a(\underline{x})$ in the definition of LNAM and LOW with the square domain centered at $\underline{x}$ of edge $2a$, which we denote by Q_a. Secondly, we note that in both algorithms AMEDA and NEAL the final step concerns the rotation of the velocity vector along a boundary profile. This was explicitly done by following the path counter-clockwise and verifying that any velocity vector at a given grid point was rotated to the left of the previous by an angle less than $\pi/2$ radians; note that this criterion does not depend on the sense of rotation of the velocity field along the path.

4.1. Ameda Tuning and Results

Three parameters have to be determined to run this algorithm: a, from the definition of LNAM and LOW, K and b. To obtain all dimensionless parameters we divided a and b by the length scale l of one pixel ($l \sim 0.5$ km): $a_0 = a/l$ and $b_0 = b/l$.

The optimal choice of these parameters depends on the scale analysis of the investigated dynamics: if a_0 is too large then LNAM may sum up the contribution of many eddies inside Q_a, leading to a wrong estimate of the angular momentum. Similarly a large b_0 is not recommended, nor is a small one since the velocity vector may abruptly rotate with an angular velocity greater than $\pi/2$ radians per pixel in proximity of the eddy center. The parameter K, in turn, once a_0 is coherently chosen, represents a lower bound for the detected eddy intensity.

We ran the algorithm on the 10-day dataset described in Section 2.1 for different values of the parameters a_0 and b_0, and analyzed the number of detected eddies Ne as a function of K, varying from 0.1 to 1 with step 0.1. The results are shown in Figure 2.

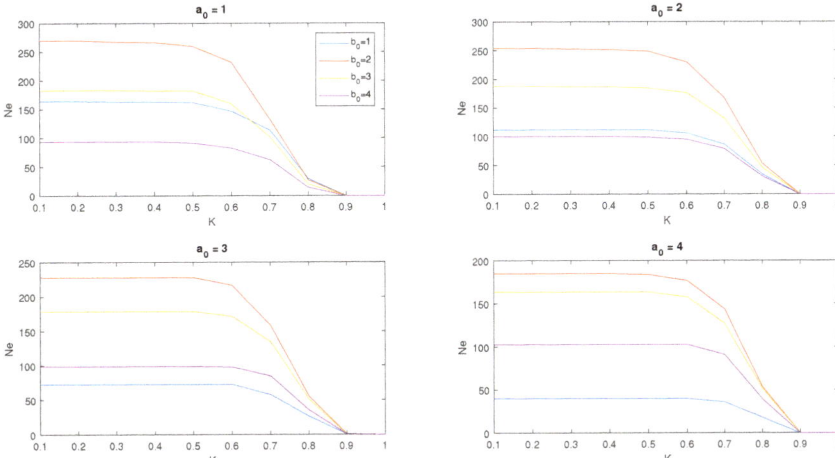

Figure 2. Number of eddies detected in the observation period Ne, obtained with the angular momentum eddy detection and tracking algorithm (AMEDA) for different values of the parameters a_0, b_0 and K. In each figure, corresponding to a value of a_0, the colored curves denote the graphs of Ne as a function of K for different values of b_0 (labeled as in the legend).

For any choice of a_0 and b_0 the values of Ne turned out to be approximately constant for $K < 0.6$, so we set $K = 0.6$. On the other hand for a fixed a_0 the maximum of Ne was achieved at $b_0 = 2$; so we chose this value for b_0. Finally we noted that Ne weakly decreased as a_0 increased, as expected, suggesting to take $a_0 = 1$. In summary, our optimal choice of the parameters turned out be $(a_0, b_0, K) = (1, 2, 0.6)$.

Since we were interested in discriminating diverging structures from converging ones we added a third control to AMEDA (see above, Section 3.4):

3. Discards those extrema satisfying LNMF > 0.2.

In this way we allowed only a little divergence near the eddy core (see the LNMF contour line in Figure 3 for instance). This correction reduced the number of detected eddies Ne by about 16% for K = 0.6, and by 0.4% for K = 0.7; this behavior was expected since strong rotations often imply weak divergences.

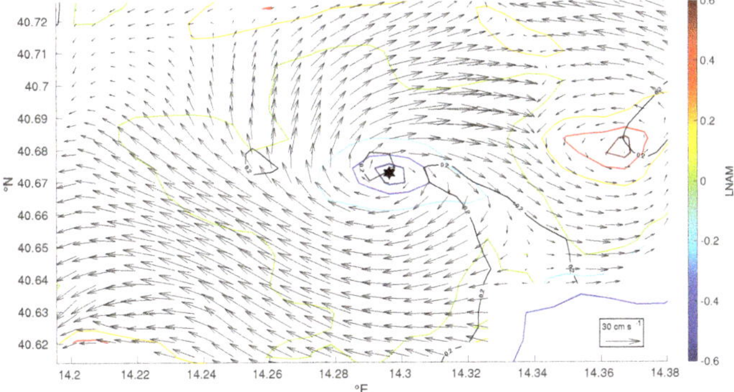

Figure 3. Source-like eddy core detected by the algorithm AMEDA (black star), velocity field (black arrows), local normalized angular momentum field (LNAM) contour lines (colored), local normalized momentum flux (LNMF) = 0.2 contour (black lines) and coastline (blue line)

Unfortunately this criterion is not optimal: It is a pure dynamical control depending on the local behavior of the flow, but eddies are extensive structures which may admit internal divergences. In such a case the eddy center and its real extension is difficult to estimate since it would be necessary to understand the streamline geometry.

4.2. Neal Tuning and Results

By definition NEAL is a purely geometrical method, which is not required to compute any differential quantity: Eddy centers are simply defined as energy minima. Of course this reduces the computation time, making the algorithm fast and efficient. Moreover we note that, as in the previous case, there are two parameters, a and b, to be determined; as before we considered the dimensionless parameters $a_0 = a/l$ and $b_0 = b/l$.

We ran the algorithm on the dataset for $a_0 = 1, \ldots, 8$ and $b_0 = 1, \ldots, 8$. In Figure 4 the number of eddies Ne, discarding the unlikely results obtained for $a_0 = 1$ (Ne > 1000), is shown. We observed that for $a_0 = 2$ there was a weak dependence on b_0, but the values of Ne turned out to be much less than those obtained by AMEDA. For $a_0 = 3, 4$ the number of detected eddies highly depended on b_0, but the results did not converge anywhere; for $a_0 > 4$ we obtained values depending weakly on b_0 but much less than those for $a_0 = 2$. These discrepancies were likely caused by asymmetrical eddies lacking radially increasing velocity components. In conclusion, we were not able to tune NEAL, as no stable regions in the parameter space were identified.

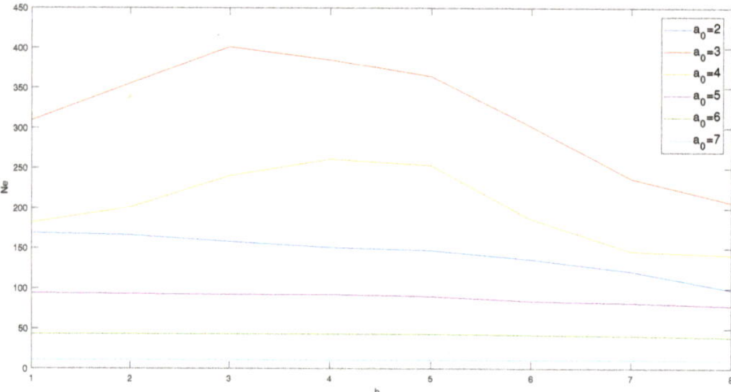

Figure 4. Number of eddies detected in the observation period, Ne, obtained by the 'Nencioli et al. algorithm' (NEAL) for different choices of the parameters a_0 and b_0. Colored lines denote the graphs of Ne as a function of b_0 for different values of a_0 (labeled as in the legend).

4.3. Yada Tuning and Results

The first step of the algorithm coincides with that of AMEDA: it identifies any local extremum $\underline{x}$ of LNAM satisfying LNAM > 0.6 for $a_0 = 1$ (having tested this values in tuning AMEDA).

The second step concerns the study of the streamline geometry in a neighborhood of the extremum. It proceeds as follows: in a square neighborhood Q_b centered at $\underline{x}$ with edge $2b$, where the length b has to be intended as an upper limit for the eddy radius (which, in this study, has been overestimated to be $b = 10l$), it draws a circle C_r of radius $r = l$, centered at $\underline{x}$ and composed by 8 points (as many as the grid points on the tangent square perimeter). It then computes the streamlines originated from these points (each streamline is built by means of a fourth order Runge–Kutta method, with a step of 5 points per pixel. It is composed by up to 1000 points), collecting their mean points (geometric means) and end points.

Then the algorithm performs a selection of all the streamlines such that:

(1) The end points belong to the square domain Q_{b-2l} (that is: they stay away from the boundary of the reference domain);

(2) Each streamline completes at least one revolution.

The second control consists in looking at the *cumulative winding-angle* (given an oriented piece-wise linear curve, its cumulative winding-angle is the sum, over all the angular points, of the angle, with positive sign going counter-clockwise and negative going clockwise, between the two intersecting segments, considered as vectors) of the streamline, as defined in [12]: it has to be, in modulus, equal to or grater than 2π. If no such streamline exists we increase the radius of C_r by l until at least one streamline satisfying (1) and (2) is found; the maximum allowed r will be $b/2$ (at any step we increase the number of points in the circle to match the amount of grid points in the tangent square perimeter).

Note that if one such streamline exists it means that either it converges to some point inside the domain or it definitely stays inside the domain without converging anywhere (at least for the first 1000 points). Of course some diverging streamline, which rotates without reaching the boundary of the domain, could exist and be identified by the algorithm. However, a path starting from $\underline{x}$ and which rotates around it at least three times before reaching the boundary, and having the same step-size of the drawn streamlines, counts approximately 300–400 points. Then, if a spiral-like streamline diverging from the center stays inside the domain without reaching the boundary, it has to complete at least eight revolutions; even in this case we can safely affirm that an eddy exists.

Once the algorithm has selected all the streamlines satisfying (1) and (2) for the first allowable r, it compares the distributions of the mean points with that of the end points. If the eddy core behaves as a sink all the end points will accumulate near it. On the other hand if the orbits around the eddy are elliptic the mean points will be close to the orbits' common center of mass. So the algorithm chooses the distribution with less variance and choose its mean point as eddy center of mass, or eddy symmetry center (ESC); by contrast the extremum will be called the eddy extreme point (EEP). However, to ensure it is not selecting another eddy in the square domain relative to a different extremum, it is required that the expected ESC must belong to the disk bounded by C_r, otherwise the point is discarded and the algorithm moves toward the next extremum; some examples of this procedure can be found in Figures 5 and 6.

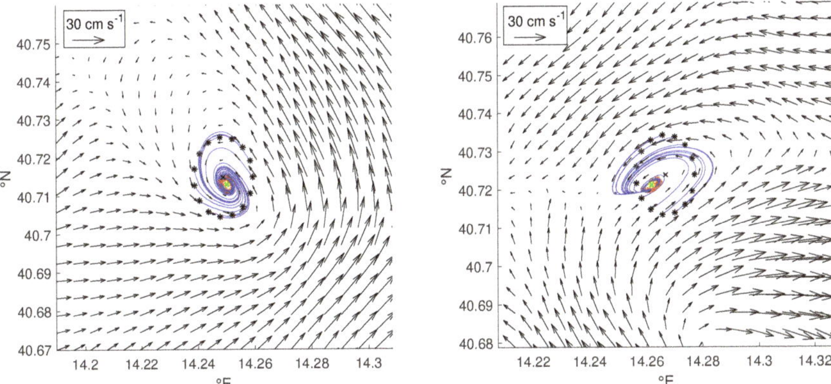

Figure 5. Maps showing the functioning of 'yet another eddy detection algorithm' (YADA) for two eddies with sink-like cores. Once the eddy extreme point (EEP) (black crosses) is detected, YADA identifies a circle (black stars), centered at the extremum, which emanates streamlines (blue lines) with the following property: The streamline has to complete up to a revolution without reaching the domain boundary. Then it evaluates the mean points (yellow stars) and end points (red stars) of such streamlines, choosing the mean point of the second distribution as eddy symmetry center (ESC) (green stars). Black arrows denote the velocity field. In both panels the mean point and the ESC coincide.

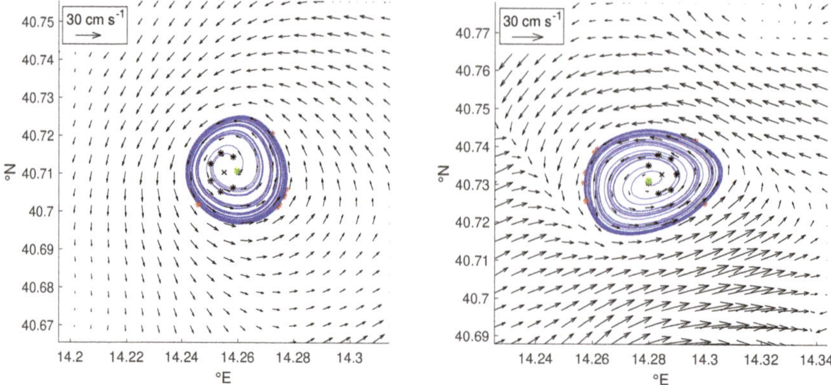

Figure 6. Maps showing the functioning of YADA for two eddies having elliptic orbits. Once the eddy extreme point EEP (black crosses) is detected, YADA identifies a circle (black stars) emanating streamlines (blue lines) with the following property: Each streamline has to complete up to a revolution without reaching the domain boundary. Then it evaluates the mean points (yellow stars) and end points (red stars) of such streamlines, choosing the mean point of the first distribution as eddy symmetry center ESC (green stars). Black arrows denote the velocity field. In both panels the mean point and the ESC coincide.

Generally the ESC does not coincide with the eddy center even though it provides a better approximation of the true eddy core than the EEP; e.g., Figures 5–8. As a consequence the distance between the ESC and the EEP can be considered as a measure of the eddy asymmetry.

Following the procedure just described, the algorithm detected Ne $= 255$ eddies, about 30% more than the value obtained with AMEDA. Eddies such as those in Figure 5, for instance, were missed by AMEDA due to their small extension and asymmetry, whereas they were detected by YADA. However there were still structures detected by AMEDA and missed by YADA, see for instance Figure 9. In some of these cases we noted that the divergence around the LNAM extremum was so weak (LNMF < 0.2) that some orbits complete up to three revolutions before leaving the region.

In conclusion we can affirm that YADA was able to detect and distinguish multiple kinds of eddies and, as it will be shown in the next part, it can be refined to estimate their boundaries.

4.4. Eddy Boundaries

There is no universal definition of eddy boundary: many authors adopted OW or ω contour lines, as well as closed streamlines or closed stream-function contours (not equivalent at all) to locate them.

Based on YADA architecture, we propose a different definition, which aims to distinguish eddies with sink-like cores from those having elliptic orbits. Of course we can not expect to identify the true boundary profile, so we assume it to be in general elliptic (rather than circular).

4.4.1. Sink-Like Cores

We considered the set S_r of all the streamlines originated from C_r and satisfying conditions (1) and (2) as explained in the definition of YADA. We then evaluated the variance ellipse of this distribution of points; let e be its eccentricity. Then we drew the ellipse E_d of eccentricity e, centered at the ESC, with major semi-axis $d = l$. As we did for C_r we consider the streamlines emanated by E_d and if all such streamlines belong to the ellipse interior we increment d by l, repeating the step up to reach $d = b$. Further, in analogy with [10], we also control that the circulation along E_d does not decrease by increasing d. The largest ellipse E_d satisfying this criterion will define the eddy boundary, as shown in Figure 7.

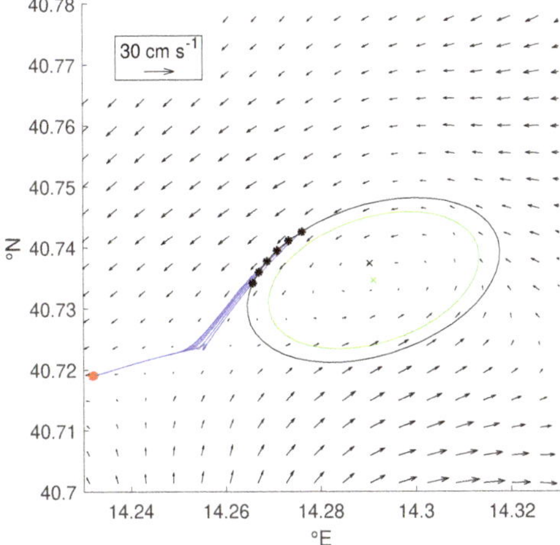

Figure 7. Map showing the YADA boundary computation of an eddy having a sink-like core. Once the eddy extreme point EEP (black cross) and the eddy symmetry center ESC (green cross) are detected the algorithm draws the ellipses centered at the ESC with increasing radii. The cycle breaks when the black ellipse is drawn due to the existence of inadmissible streamlines (blue lines) leaving the domain. The last computed ellipse (green line) will be considered as boundary. Black arrows denote the velocity field.

4.4.2. Eddies Having Elliptic Orbits

For such eddies we also started by building the variance ellipse of the admissible streamlines S_r. Then we drew the ellipses E_d with eccentricity e, centered at the ESC and having semi-major axis $d = l, 2l, \ldots, d'$, where d' was the maximum distance for which the circulation around E_d was a non-decreasing function of d, and we moved each E_d following the flow, thus collecting all the end points of the streamlines emanated by E_d. We denoted this set by $\varepsilon(E_d)$.

We expected that, if E_d approximated the eddy boundary, it had to be close to an elliptic orbit, and therefore $\epsilon(E_d)$ had to be a small deformation of E_d. However, in order to ignore the effects of translating motions, which could occur, we centered the two sets on the same reference point. Then we evaluated the Hausdorff distance $\delta(d)$ between them (see the Appendix A for details).

Finally we took d^* satisfying $\delta(d^*) = \min\{\delta(d)\}$, and E_{d^*} as eddy boundary; we chose E_{d^*} to keep the elliptic symmetric, though $\epsilon(E_{d^*})$ would provide a better approximation. In Figure 8 we plotted the various steps just described; in each panel the ellipse of semi-major axis $d = l, 2l, \ldots, d'$ is drawn.

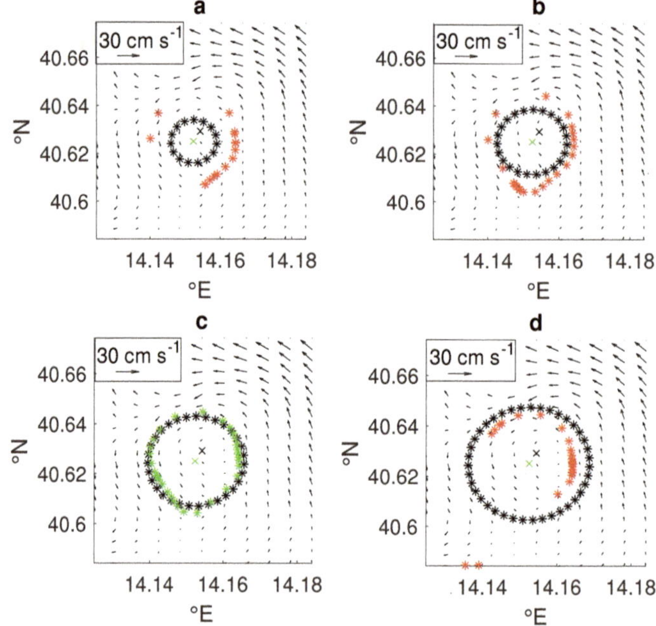

Figure 8. Maps showing the YADA boundary computation of an eddy having stable orbits. From panel (**a–d**) ellipses of increasing semi-major axes are drawn; the temporal frame is unchanged. Once the EEP (black crosses) and the ESC (green crosses) are detected the algorithm draws the ellipses centered at the ESC with increasing semi-major axis d (black stars); from panels (**a–d**) the semi-major axis increases from 2 to 5 pixel lengths. It then evaluates the end points (red stars in panels (**a**,**b**,**d**), and green stars in (**c**)) of the streamlines emanated by these ellipses. The algorithm selects the semi-major axis d^* for which the relative end points (green stars in (**c**)) form the closest deformation of the associated ellipse. Black arrows denote the velocity field.

5. Discussion

5.1. Detected Eddies

In Section 4 we tried to tune the three chosen algorithms by following a stability criterion. We succeeded for AMEDA, but failed for NEAL. Indeed, in the latter case, no stable parameter regions were identified. The algorithm YADA, instead, was indirectly tuned by using the AMEDA common parameters, namely K and a; in fact both these parameters served to set a lower bound to the eddy rotational energy, which was independent on the algorithm itself.

We then compared the tuned version of the two algorithms, AMEDA and YADA. It turns out that AMEDA detected 195 eddies within the time period, whereas YADA detected 255 eddies. However, among these, 157 eddies have been identified by both, leaving 38 eddies detected by AMEDA but missed by YADA and 98 seen by YADA but lost by AMEDA. Mismatches between AMEDA and YADA detections were expected, as already observed: eddies like that in Figure 9 were missed by YADA due to their large extension and weak, but still positive, divergence. On the other hand, deformed recirculations, as in Figure 5, were easily hidden to AMEDA but not to YADA.

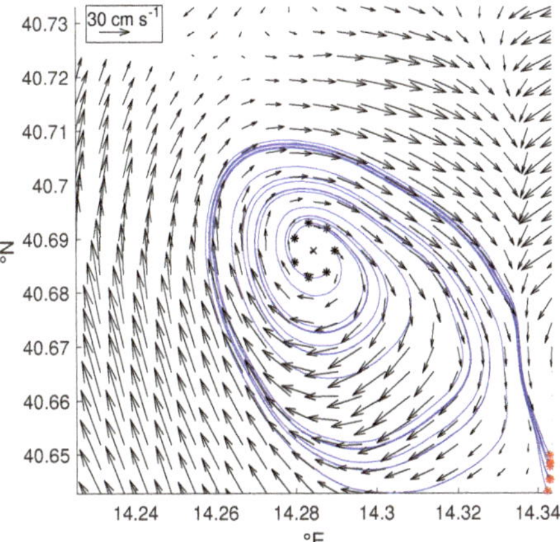

Figure 9. Eddy detected by AMEDA and missed by YADA. The local normalized angular momentum field LNAM extremum x (black cross) corresponds to an eddy core, but any circle centered at x (black stars) emanates streamlines (blue lines) which complete up to 3 revolutions before reaching the domain boundary (contact points in red). Black arrows denote the velocity field.

Finally we checked by visual inspection all the available time frames, in order to determine the existence of false positive detections. Interestingly, no such detections were found, testifying the reliability of both the algorithms, at least in terms of false alarms. The same inspection showed that volatile and higly asymmetrical structures were still missed by both. However, our algorithm was able to detect long-lived eddies for longer time periods, as shown in Figure 10.

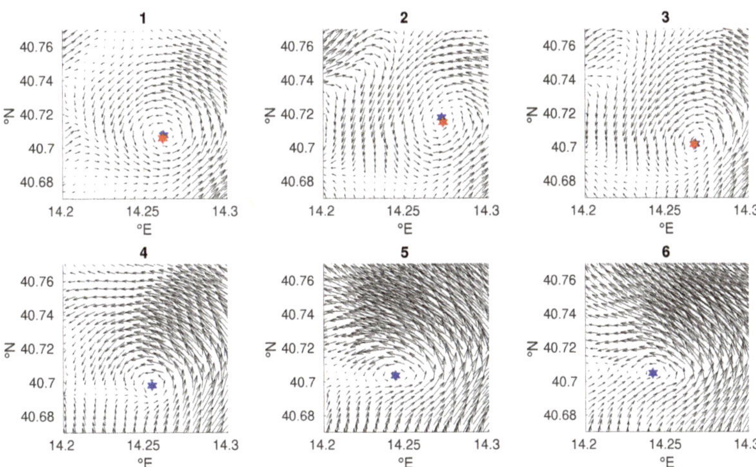

Figure 10. Sequence of time frames (from panel 1 to 6) showing the evolution of a detected eddy. Red stars denote eddy centers identified by AMEDA, whereas blue stars indicate ESCs computed by YADA. As the eddy changes shape and becomes less centrosymmetric AMEDA misses it (panel 3 to 4). Black arrows denote the velocity field (not in scale).

5.2. Equivalent Radii

Following [10] we computed the *equivalent radius* ρ for each detected eddy. It is defined as the radius of the circle bounding an area equivalent to that delimited by the eddy boundary. For elliptic contours it equals $d\sqrt[4]{1-e^2}$. The mean radius $\bar{\rho}$ turned out to be 0.87 km, with a standard deviation of 0.84 km and values of 2.8 and 3.6 km for the 95th and 99th percentile respectively. Similarly we computed a mean eccentricity $\bar{e}$ of 0.71 with standard deviation of 0.02. It turned out that the mean equivalent radius was merely 1–2 times the pixel length scale of the dataset, and therefore we could expect that our method would not accurately describe some kinematic and dynamic features of eddies having ρ close to $\bar{\rho}$. It may have been possible to obtain a more accurate description by increasing the spatial resolution up to reach $l \sim 0.3$ km; however this would have implied performing interpolations at a much higher resolution.

5.3. Spatial Distribution

The hourly sampling frequency of the HF radar allowed to track eddies having longer lifetime. We identified such long-lived structures by looking for eddies encircling an EEP, coming from the previous temporal frame, within their own boundary. The spatial distribution of all the detected long-lived eddies, counted without repetitions, can be found in Figure 11.

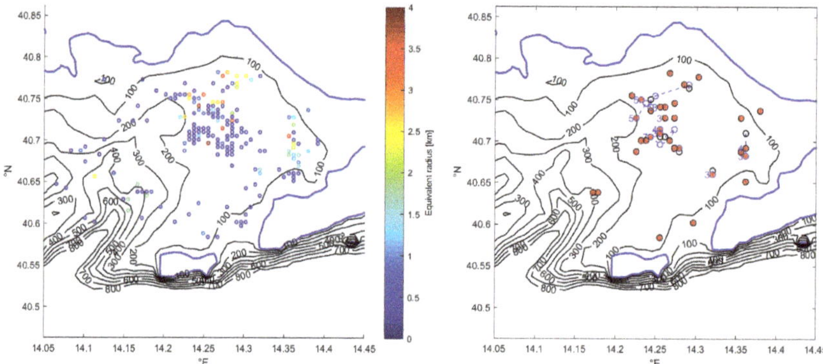

Figure 11. Left panel: Spatial distribution of the detected eddies by means of YADA (colored circles); different colors denote different sizes. Right panel: Detected long-lived eddies (circles) with lifetime $T \geq 2$ h. Initial, mid and final EEPs (black circles, blue circles and red stars respectively) with their relative eddy trajectories (blue dashed lines) and eddy lifetimes $T \geq 3$ h (blue numbers). Shoreline (blue contour) and bathymetric contour lines (black lines) between 100 m and 800 m of depth.

A larger density can be noted in correspondence of a relatively flat plateau located in 40.73° N, 14.27° E, between 120 and 160 m of depth, excluding topographic wakes, that need steep bathymetric slopes, as primary instability causes (for a comprehensive analysis of several submesoscale eddy generation mechanisms see [5]). To understand the instability sources generating the GoN eddies, therefore, it would be necessary to investigate the flow behavior within the SBL (surface boundary layer). Unfortunately there are neither wind observations nor density profiles relative to the GoN SBL. However, there is a work in progress, funded by the Science and Technology department of the Parthenope University of Naples, aiming to investigate the vertical water profile through numerical simulation. Such a study could provide more information to understand the instability sources within the GoN.

Finally we observed that the detected long-lived eddies, namely those with lifetimes greater than 1 h, were 36, distributed as shown in Figure 11. They usually persisted for few hours, 5 or 7 h in some cases, distributed in agreement with the entire density population. We also noted that, except for few examples, they were almost stationary.

6. Conclusions

Submesoscale motions play an important role in the transfer of energy from the mesoscale down to the dissipative range ([5,42]) as well as in the transport of pollutants, of biomass, of organic and inorganic matter ([43,44]). At present, they represent one of the frontiers of the study of transport in the ocean. On the other hand, as recently pointed out by several authors (see, e.g., [5,8,45]), the observation of submesoscale eddies in a synoptic way is challenging, feasible with remote sensing techniques which are typicaly limited by available resolution. HF radars are land-based remote sensing tools that can be very suitable for such investigations, given their temporal and spatial sampling characteristics. In this paper we have tackled the issue of devising an algorithm for submesoscale eddy detection in a high resolution surface velocity field provided by a network of 25 Mhz coastal radar antennas active in the Gulf of Naples. We started by applying two different eddy detection algorithms, here denoted by AMEDA and NEAL, based on the studies of [10,11] respectively, but they both displayed some weaknesses. The application of AMEDA to the selected surface current dataset demonstrated to be unable to distinguish submesoscale eddies entrapping fluid masses from the others. So we refined the algorithm by measuring the divergence occurring in the eddy core. The number of detected eddies then decreased. Differently, we did not succeed to tune the algorithm NEAL.

To obtain a more efficient detection method, able to distinguish asymmetric eddies entrapping fluid masses, we developed a novel, modified algorithm, named YADA, which detected 255 eddies (about 30% more than the refined AMEDA value). Then we used YADA to estimate the eddy boundaries, assuming an elliptical symmetry, and we found a mean equivalent radius of 0.87 km and a mean eccentricity of 0.71.

YADA's results were validated comparison with the results of the algorithm AMEDA, as well as by visual inspection of all time frames. Having developed a more robust algorithm, this also allowed us to look at the spatial distribution of the detected eddies, and to observe a larger density at the plateau located at 160 m of depth, and led us to exclude topographic wakes as main instability sources. Moreover, we obtained estimates for their spatial scales taking into account the noncircular geometry of the vortices.

As mentioned above, submesoscale eddies represent a relevant transport mechanism for waters and their biogeochemical characteristics; their influence is particularly important in those coastal areas, such as the Gulf of Naples, characterized by the coexistence of different subsystems, whose mutual exchanges may strongly affect the whole functioning of the area. For this reason an accurate identification of such structures is a necessary first step for the quantitative assessment of their role, which we plan to further investigate in terms of their specific transport properties both in the horizontal and in the vertical.

Author Contributions: Conceptualization L.B., P.F. and E.Z.; methodology L.B., P.F. and E.Z.; formal analysis L.B., P.F. and E.Z.; writing–review and editing L.B., P.F. and E.Z.; data curation L.B.; funding acquisition E.Z. and P.F. All authors have read and agreed to the published version of the manuscript.

Funding: E.Z. acknowledges support from the Parthenope University of Naples internal research fund; this work was partially supported by "DORA - Deployable Optics for Remote sensing Applications DORA" (ARS01_00653), a project funded by MIUR - PON "Research & Innovation"/PNR 2015-2020.

Acknowledgments: The Department of Sciences and Technologies (formerly the Department of Environmental Sciences) of the Parthenope University of Naples operates the HFR system on behalf of the AMRA consortium (formerly CRdC AMRA), a regional competence center for the analysis and monitoring of environmental risks. Our radar remote sites are hosted by the ENEA Centre of Portici, the 'Villa Angelina Village of High Education and Professional Training', 'La Villanella' resort in Massa Lubrense and the Fincantieri shipyard in Castellammare di Stabia, whose hospitality is gratefully acknowledged. The authors would like to thank Angelo Perilli for useful conversations and Teresa Hann for her help in reviewing the manuscript.

Conflicts of Interest: The authors declare no conflict of interest.

Appendix A. The Hausdorff Distance

The Hausdorff distance $\delta_H(A, B)$ between two compact subsets A and B of the euclidean plane is defined by the formula

$$\delta_H(A, B) = \max \left\{ \sup_{a \in A} d(a, B), \sup_{b \in B} d(b, A) \right\},$$

where $d(a, B)$ and $d(b, A)$ are the usual point-set distances:

$$d(a, B) = \inf_{b \in B} |a - b|, \quad d(b, A) = \inf_{a \in A} |b - a|.$$

The Hausdorff distance δ_H makes the set of all compact subsets a metric space; in particular $\delta_H(A, B) = 0$ if and only if $A = B$.

References

1. Cushman-Roisin, B.; Gualtieri, C.; Mihailovic, D.T. Environmental Fluid Mechanics: Current issues and future outlook. In *Fluid Mechanics of Environmental Interfaces*; Taylor & Francis: Abingdon UK, 2008; pp. 17–30.
2. Zambianchi, E.; Griffa, A. Effects of finite scales of turbulence on dispersion estimates. *J. Mar. Res.* **1994**, *52*, 129–148. [CrossRef]
3. Haza, A.C.; Özgökmen, T.M.; Griffa, A.; Garraffo, Z.D.; Piterbarg, L. Parameterization of particle transport at submesoscales in the Gulf Stream region using Lagrangian subgridscale models. *Ocean. Model.* **2012**, *42*, 31–49. [CrossRef]
4. McWilliams, J.C. Submesoscale, coherent vortices in the ocean. *Rev. Geophys.* **1985**, *23*, 165–182. [CrossRef]
5. McWilliams, J.C. Submesoscale currents in the ocean. *Proc. R. Soc. A* **2016**. [CrossRef] [PubMed]
6. Lévy, M.; Franks, P.J.; Smith, K.S. The role of submesoscale currents in structuring marine ecosystems. *Nat. Commun.* **2018**, *9*, 1–16. [CrossRef] [PubMed]
7. Mahadevan, A. The impact of submesoscale physics on primary productivity of plankton. *Annu. Rev. Mar. Sci.* **2016**, *8*, 161–184. [CrossRef]
8. Poje, A.C.; Özgökmen, T.M.; Lipphardt, B.L.; Haus, B.K.; Ryan, E.H.; Haza, A.C.; Jacobs, G.A.; Reniers, A.; Olascoaga, M.J.; Novelli, G.; et al. Submesoscale dispersion in the vicinity of the Deepwater Horizon spill. *Proc. Natl. Acad. Sci. USA* **2014**, *111*, 12693–12698. [CrossRef]
9. Doglioli, A.M.; Blanke, B.; Speich, S.; Lapeyre, G. Tracking coherent structures in a regional ocean model with wavelet analysis: Application to Cape Basin eddies. *J. Geophys. Res. Ocean.* **2007**, *112*, C5. [CrossRef]
10. Le Vu, B.; Stegner, A.; Arsouzea, T. Angular Momentum Eddy Detection and tracking Algorithm (AMEDA) and its application to coastal eddy formation. *Atmos. Ocean. Technol.* **2017**, *35*, 739–762. [CrossRef]
11. Nencioli, F.; Dong, C.; Duckey, T.; Washburn, L.; McWilliams, J.C. A Vector Geometry Based Eddy Detection Algorithm and Its Application to a High-Resolution Numerical Model Product and High-Frequency Radar Surface Velocities in the Southern California Bight. *Atmos. Ocean. Technol.* **2008**, *27*, 564–579. [CrossRef]
12. Post, F.H.; Sadarjoen, I.A. Detection, quantification, and tracking of vortices using streamline geometry. *Comput. Graph.* **2000**, *24*, 333–341.
13. Pessini, F.; Olita, A.; Cotroneo, Y.; Perilli, A. Mesoscale eddies in the Algerian Basin: do they differ as a function of their formation site? *Ocean. Sci.* **2018**, *14*, 5. [CrossRef]
14. Chelton, D.B.; DeSzoeke, R.A.; Schlax, M.G.; El Naggar, K.; Siwertz, N. Geographical variability of the first baroclinic Rossby radius of deformation. *J. Phys. Oceanogr.* **1998**, *28*, 433–460. [CrossRef]
15. Lévy, M.; Klein, P.; Treguier, A.M. Impact of sub-mesoscale physics on production and subduction of phytoplankton in an oligotrophic regime. *J. Mar. Res.* **2001**, *59*, 535–565. [CrossRef]
16. Grilli, F.; Pinardi, N. The computation of Rossby radii of deformation for the Mediterranean Sea. *MTP News* **1998**, *6*, 4–5.
17. Pinardi, N.; Masetti, E. Variability of the large scale general circulation of the Mediterranean Sea from observations and modelling: a review. *Palaeogeogr. Palaeoclimatol. Palaeoecol.* **2000**, *158*, 153–173. [CrossRef]
18. Corrado, R.; Lacorata, G.; Palatella, L.; Santoleri, R.; Zambianchi, E. General characteristics of relative dispersion in the ocean. *Sci. Rep.* **2017**, *7*, 46291. [CrossRef]

19. Rubio, A.; Mader, J.; Corgnati, L.; Mantovani, C.; Griffa, A.; Novellino, A.; Quentin, C.; Wyatt, L.; Schulz-Stellenfleth, J.; Horstmann, J.; et al. HF radar activity in European coastal seas: next steps toward a pan-European HF radar network. *Front. Mar. Sci.* **2017**, *4*, 8. [CrossRef]
20. Paduan, J.D.; Washburn, L. High-Frequency Radar Observations of Ocean Surface Currents. *Annu. Rev. Mar. Sci.* **2013**, *5*, 115–136. [CrossRef]
21. Karimova, S.; Gade, M. Eddies in the Western Mediterranean seen by spaceborne radar. In Proceedings of the 2016 IEEE International Geoscience and Remote Sensing Symposium (IGARSS), Beijing, China, 10–15 July 2016; pp. 3997–4000.
22. Karimova, S.; Gade, M. Improved statistics of sub-mesoscale eddies in the Baltic Sea retrieved from SAR imagery. *Int. J. Remote. Sens.* **2016**, *37*, 2394–2414. [CrossRef]
23. Karimova, S.S.; Gade, M. Eddies in the Red Sea as seen by satellite SAR imagery. In *Remote Sensing of the African Seas*; Springer: Berlin/Heidelberg, Germany, 2014; pp. 357–378.
24. Menna, M.; Mercatini, A.; Uttieri, M.; Buonocore, B.; Zambianchi, E. Wintertime transport processes in the Gulf of Naples investigated by HF radar measurements of surface currents. *Nuovo Cimento C* **2007**, *30*, 605–622.
25. Cianelli, D.; Falco, P.; Iermano, I.; Mozzillo, P.; Uttieri, M.; Buonocore, B.; Zambardino, G.; Zambianchi, E. Inshore/offshore water exchange in the Gulf of Naples. *J. Mar. Syst.* **2015**, *145*, 37–52. [CrossRef]
26. Carrada, G.C.; Hopkins, T.S.; Bonaduce, G.; Ianora, A.; Marino, D.; Modigh, M.; Ribera D'Alcalà, M.; di Scotto, C.B. Variability in the hydrographic and biological features of the Gulf of Naples. *Mar. Ecol.* **1980**, *1*, 105–120. [CrossRef]
27. Cianelli, D.; Uttieri, M.; Buonocore, B.; Falco, P.; Zambardino, G.; Zambianchi, E. Dynamics of a very special Mediterranean coastal area: the Gulf of Naples. In *Mediterranean Ecosystems: Dynamics, Management and Conservation*; Nova Science: Hauppauge, NY, USA, 2012; pp. 129–150.
28. Moretti, M.; Sansone, E.; Spezie, G.; Vultaggio, M.; De Maio, A. Alcuni aspetti del movimento delle acque del Golfo di Napoli. *Annali Ist Univ Nav XLV-XLVI* **1976**, 207–217.
29. De Maio, A.; Moretti, M.; Sansone, E.; Spezie, G.; Vultaggio, M. Outline of marine currents in the Bay of Naples and some considerations on pollutant transport. *Il Nuovo Cimento C* **1985**, *8*, 955–969. [CrossRef]
30. Uttieri, M.; Cianelli, D.; Nardelli, B.B.; Buonocore, B.; Falco, P.; Colella, S.; Zambianchi, E. Multiplatform observation of the surface circulation in the Gulf of Naples (Southern Tyrrhenian Sea). *Ocean. Dyn.* **2011**, *61*, 779–796. [CrossRef]
31. De Ruggiero, P.; Esposito, G.; Napolitano, E.; Iacono, R.; Pierini, S.; Zambianchi, E. Modelling the marine circulation of the Campania Coastal System (Tyrrhenian Sea) for the year 2016: Analysis of the dynamics. *J. Mar. Syst.* **2019**, Submitted.
32. D'Alelio, D.; Mazzocchi, M.G.; Montresor, M.; Sarno, D.; Zingone, A.; Di Capua, I.; Franzè, G.; Margiotta, F.; Saggiomo, V.; Ribera d'Alcalà, M. The green-blue swing: plasticity of plankton food-webs in response to coastal oceanographic dynamics. *Mar. Ecol.* **2015**, *36*, 1155–1170. [CrossRef]
33. Cianelli, D.; D'Alelio, D.; Uttieri, M.; Sarno, D.; Zingone, A.; Zambianchi, E.; d'Alcalà, M.R. Disentangling physical and biological drivers of phytoplankton dynamics in a coastal system. *Sci. Rep.* **2017**, *7*, 15868. [CrossRef]
34. Sciascia, R.; Berta, M.; Carlson, D.F.; Griffa, A.; Panfili, M.; Mesa, M.L.; Corgnati, L.; Mantovani, C.; Domenella, E.; Fredj, E.; et al. Linking sardine recruitment in coastal areas to ocean currents using surface drifters and HF radar: A case study in the Gulf of Manfredonia, Adriatic Sea. *Ocean. Sci.* **2018**, *14*, 1461–1482. [CrossRef]
35. Iermano, I.; Moore, A.; Zambianchi, E. Impacts of a 4-dimensional variational data assimilation in a coastal ocean model of southern Tyrrhenian Sea. *J. Mar. Syst.* **2016**, *154*, 157–171. [CrossRef]
36. Kalampokis, A.; Uttieri, M.; Poulain, P.M.; Zambianchi, E. Validation of HF radar-derived currents in the Gulf of Naples with Lagrangian data. *IEEE Geosci. Remote Sens. Lett.* **2016**, *13*, 1452–1456. [CrossRef]
37. Ranalli, M.; Lagona, F.; Picone, M.; Zambianchi, E. Segmentation of sea current fields by cylindrical hidden Markov models: a composite likelihood approach. *J. R. Stat. Soc. Ser. Appl. Stat.* **2018**, *67*, 575–598. [CrossRef]
38. Saviano, S.; Kalampokis, A.; Zambianchi, E.; Uttieri, M. A year-long assessment of wave measurements retrieved from an HF radar network in the Gulf of Naples (Tyrrhenian Sea, Western Mediterranean Sea). *J. Oper. Oceanogr.* **2019**, *12*, 1–15. [CrossRef]

39. Okubo, A. Horizontal dispersion of floatable trajectories in the vicinity of velocity singularities such as convergencies. *Deep-Sea Res.* **1970**, *17*, 445–454.
40. Weiss, J. The dynamics of enstrophy transfer in 2-dimensional hydrodynamics. *Phys. D* **1991**, *48*, 273–294. [CrossRef]
41. Mkhinini, N.; Coimbra, A.L.S.; Stegner, A.; Arsouze, T.; Taupier-Letage, I.; Béranger, K. Long-lived mesoscale eddies in the eastern Mediterranean Sea: Analysis of 20 years of AVISO geostrophic velocities. *J. Geophys. Res. Oceans* **2014**, *119*, 8603–8626. [CrossRef]
42. Ferrari, R.; Wunsch, C. Ocean circulation kinetic energy: Reservoirs, sources, and sinks. *Annu. Rev. Fluid Mech.* **2009**, *41*, 253–282. [CrossRef]
43. Lévy, M.; Ferrari, R.; Franks, P.J.; Martin, A.P.; Rivière, P. Bringing physics to life at the submesoscale. *Geophys. Res. Lett.* **2012**, *39*, 14. [CrossRef]
44. Smith, K.M.; Hamlington, P.E.; Fox-Kemper, B. Effects of submesoscale turbulence on ocean tracers. *J. Geophys. Res. Ocean.* **2016**, *121*, 908–933. [CrossRef]
45. Schroeder, K.; Chiggiato, J.; Haza, A.; Griffa, A.; Özgökmen, T.; Zanasca, P.; Molcard, A.; Borghini, M.; Poulain, P.M.; Gerin, R.; et al. Targeted Lagrangian sampling of submesoscale dispersion at a coastal frontal zone. *Geophys. Res. Lett.* **2012**, *39*, 11. [CrossRef]

Article

Influence of Tropical Cyclone Intensity and Size on Storm Surge in the Northern East China Sea

Jian Li [1,2,3,4], Yijun Hou [1,2,3,5,*], Dongxue Mo [1,2,3], Qingrong Liu [4] and Yuanzhi Zhang [6]

1. Institute of Oceanology, Chinese Academy of Sciences, Nanhai Road, 7, Qingdao 266071, China; lijian_bhybzx@ncs.mnr.gov.cn (J.L.); modongxue13@mails.ucas.ac.cn (D.M.)
2. Key Laboratory of Ocean Circulation and Waves, Institute of Oceanology, Chinese Academy of Sciences, Nanhai Road, 7, Qingdao 266071, China
3. University of Chinese Academy of Sciences, Yuquan Road, 19A, Beijing 100049, China
4. North China Sea Marine Forecasting center of State Oceanic Administration, Yunling Road, 27, Qingdao 266061, China; liuqingrong@ncs.mnr.gov.cn
5. Laboratory for Ocean and Climate Dynamics, Qingdao National Laboratory for Marine Science, Qingdao 266061, China
6. Nanjing University of Information Science and Technology, Pukou District, Nanjing 210044, China; yuanzhizhang@cuhk.edu.hk

* Correspondence: yjhou@qdio.ac.cn; Tel.: +86-532-82898516

Received: 5 November 2019; Accepted: 10 December 2019; Published: 16 December 2019

Abstract: Typhoon storm surge research has always been very important and worthy of attention. Less is studied about the impact of tropical cyclone size (TC size) on storm surge, especially in semi-enclosed areas such as the northern East China Sea (NECS). Observational data for Typhoon Winnie (TY9711) and Typhoon Damrey (TY1210) from satellite and tide stations, as well as simulation results from a finite-volume coastal ocean model (FVCOM), were developed to study the effect of TC size on storm surge. Using the maximum wind speed (MXW) to represent the intensity of the tropical cyclone and seven-level wind circle range (R7) to represent the size of the tropical cyclone, an ideal simulation test was conducted. The results indicate that the highest storm surge occurs when the MXW is 40–45 m/s, that storm surge does not undergo significant change with the RWM except for the area near the center of typhoon and that the peak surge values are approximately a linear function of R7. Therefore, the TC size should be considered when estimating storm surge, particularly when predicting marine-economic effects and assessing the risk.

Keywords: storm surge; tropical cyclone size (TC size); ideal test; marine-economic effects; Northern East China Sea (NECS)

1. Introduction

1.1. Background

Storm surge is an abnormal increase in seawater level, which can be induced by typhoons and atmospheric pressure disturbances [1–3], and also may be affected by factors such as interaction with the seiches of (semi) enclosed basins, coastal configuration, coastal bathymetry, and the extend of the continental shelf [4]. Storm surges caused by hurricanes or typhoons (two types of tropical cyclones) are among the most disastrous marine/coastal hazards in the world, resulting in significant property damage and loss of life [5–9]. Weisberg [10], Orton [11], Rey [4], and others have studied the effects of factors such as typhoon landing points, direction, and maximum wind speed (MXW) on storm surges. Condon [12] uses the radius of maximum wind (RMW) to indirectly discuss the impact of TC size on storm surge. However, RMW only represents a single distance parameter and can only reflect the wind speed in the area near the typhoon center, which is not enough to completely describe TC size and consequently, it is necessary to introduce other distance parameters to describe TC size. At present, TC size is often described by the distance from the center of the typhoon to the wind speed of level 7 or 10 in the measurement. However, analysis of the historical hurricane record showed no clear correlation between surge and TC size and consequently, little attention has been given to the role of storm size in surge generation [13]. Researchers have ignored the effect of changes in TC size in the context of global warming [14], which has resulted in a significant underestimation of the TC-driven storm surge destructive potential [13].

Due to the lack of accurate data, there has been relatively less research on the impacts of TC size on storm surge. At present, there are many ways to define typhoon dimensions, due to differences in the means of observation. Brand [15] and Merrill [16], using the mean radius of the outermost closed isobar (ROCI) to measure TC size, and analyzing surface weather charts, reported that there were seasonal and regional variations in TC size, and Northwest Pacific (NWP) TCs had a mean size twice as large as that of the TCs in the Atlantic Ocean. For this reason, more attention should be paid to the impact of different typhoon scales on storm surge in the Pacific. Based on aircraft and other normal surface observations, TC size was investigated using the radii of a specific fixed isobar (1004 hPa) [17]. However, in situ observations are not routinely available and there is still a lack of wind structure observations for TCs in the open ocean [18,19]. The dearth of observations makes TC size estimation heavily dependent upon satellite observations and techniques [19]. Satellite observations include cloud/feature-tracked winds [20,21] and scatterometry [22]. The US Joint Typhoon Warning Center (JTWC) has published TC critical wind radii (mean azimuth radii of 34-, 50-, and 64-kt surface winds—R34, R50, and R64, respectively) in the TC best-track dataset since 2001 [23]. The quick scatterometer (QuikSCAT) surface wind speed data have been used to study the relationship between TC size and weather variables, for which TC size was defined as either R15 [24] or R34 [25]. Knaff [18,19] measured TC size in terms of R5 (the radius of the mean tangential surface wind speed of 5 kt) and presented a relatively simple method for estimating the TC wind radii from two different sources: Infrared satellite imagery and global model analyses. Data from European remote sensing (ERS) satellites [26] and Japan meteorological satellites [23] have also been extensively analyzed in TC size investigations. This study compares the size of tropical cyclones using data from two Japanese satellites, GMS-5 and MTSAT-2. Synthesizing the above, the typhoon size is described in terms of two main aspects: Wind profile and air pressure profile. Different TC sizes exert different effects on the weather and ocean. The timely determination of TC size, intensity, and track is important in weather forecasts, as well as predictions of the potential impacts of TCs [18,19,26–30].

TC size has a significant impact on weather and climate, and the corresponding effects on the ocean can also lead to large storm surges and wave disasters, particularly for very intense storms making landfall in mildly sloping regions [13,31]. For example, Hurricane Katrina's large size contributed to its massive storm surge, enabling it to generate a higher storm surge than hurricane Camille, even though Camille produced stronger winds when it struck the same area in 1969 [13]. Statistical analysis reveals an inverse correlation between storm surge magnitudes and maximum wind speed radius (RMW), while positive correlations exist between storm surge heights and TC sizes [32]. According to Irish [13], storm surge varies by as much as 30% over a reasonable range of TC sizes for a given storm intensity, which means that TC size has a significant impact on storm surge and is worth studying.

1.2. Study Area

The northern East China Sea (NECS), particularly the Bohai Sea and the northern Yellow Sea, is one of the areas in the world most vulnerable to storm surge [33]. The Bohai Sea is a semi-enclosed sea with three bays: Liaodong Bay, Bohai Bay, and Laizhou Bay (Figure 1). The terrain is mostly flat and leaning from bays to the Bohai Strait. The surface area of the Bohai Sea is approximately 77,000 km^2, with a mean depth of 18 m below mean sea level (MSL) [34]. The main industries in the flat (susceptible to inundation) coastal areas include mariculture, salt-drying, oil and gas exploitation, and tourism.

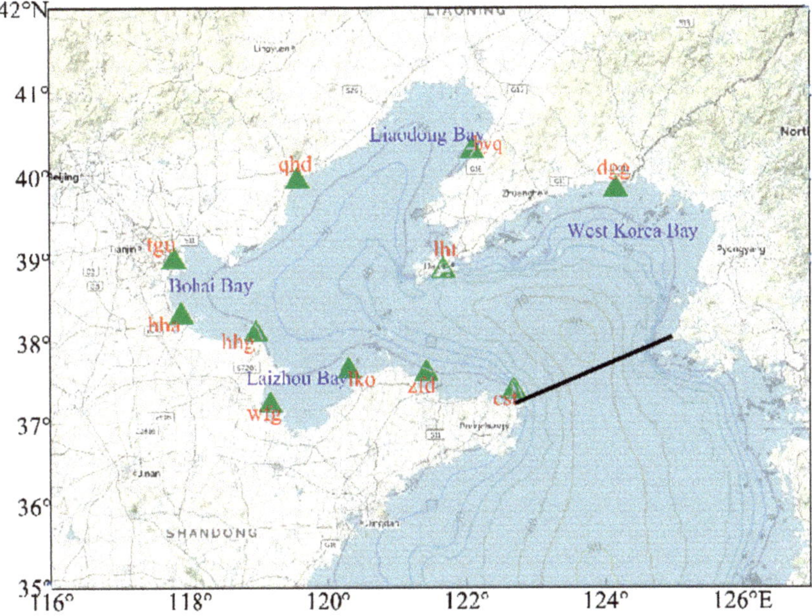

Figure 1. The northern East China Sea, showing names, locations, and bottom topography (in meters with mean sea level (MSL)) as the vertical datum. The black line is the dividing line between northern and southern Yellow Sea.

The Yellow Sea is a marginal sea of the western North Pacific located between the Chinese mainland and the Korean peninsula [35]. The average depth of the northern Yellow Sea is about 38 m below MSL [36]. Many previous studies have divided the Yellow Sea into northern and southern parts, with a dividing line that runs between Chengshan Cape of the Shandong Peninsula and Changyon of the Korean Peninsula [33].

Typhoons in the north Pacific are often serious marine disasters. For example, Typhoon Winnie (TY9711) resulted in 133 people dead or missing and caused more than 28 billion yuan in direct economic losses. Typhoon Winnie (TY9711) and Typhoon Damrey (TY1210) were selected to compare the effects of different TC sizes on storm surge, because the difference in size between typhoons Winnie and Damrey is similar to the difference in size between Hurricane Katrina and Hurricane Camille. As part of this study, ideal tests were conducted by modeling different sizes of typhoons to examine the role of TC size in storm surge. In addition, TC size should be considered when estimating storm surge, particularly when predicting marine-economic effects and assessing the risk.

2. Data and Methods

2.1. Typhoon Information and Data Processing

2.1.1. Typhoon Information

Pacific typhoons Winnie (TY9711) and Damrey (TY1210) made landfall north of NECS. TY9711 and TY1210 were selected because they had similar paths and MXWs but different TC size. The paths of TY9711 and TY1210 are shown in Figure 2. TY9711 made landfall on the coast of Wenling, Zhejiang Province at 20:00 on 18 August 1997. Its MXW was 20 m/s and its seven-level wind circle range (R7) was about 400,000 m in the Shandong area (about 35° north latitude). TY1210 made landfall on the coast of Yancheng City, Jiangsu Province at 21:30 on 2 August 2012. Its MXW was 29 m/s and its R7 was 170,000 m in the Shandong area.

The typhoon data (including TC size) used for comparison and simulation is from the Wenzhou Typhoon Net (http://www.wztf121.com/history.html), where TC size corresponds to the seven-level wind circle range (R7); R7 specifically refers to the radius corresponding to v = 13.9 m/s.

2.1.2. Satellite Data

Since measured data on the typhoon scale is limited, in order to study impacts of typhoon-scale weather variables on storm surge, it is necessary to determine typhoon size based on satellite inversion data. The TY9711 typhoon-scale inversion data were provided by the GMS-5 meteorological satellite, while the TY1210 data were derived from the MTSAT-2 satellite. The TC size inversion method and dataset were provided by Lu [23], using infrared images to determine the typhoon scale based on R34 (defined as the mean azimuth radius of 34-kt surface winds). The TC size data were from the China Meteorological Administration Tropical Cyclone Data Center (http://tcdata.typhoon.org.cn/tcsize.html).

The Japanese geostationary meteorological satellite (GMS) is located at 140° E and has been the most important source of meteorological satellite information for Asian weather forecasting since its launch in July 1977. The GMS-5 satellite was launched in March 1995 and officially positioned at 140° E in June, at which time it commenced operation. The main detection instrument on this satellite is the visible and infrared spin-scanning radiometer (VISSR). The visible light resolution is 1250 m, and the spatial resolution of the infrared and water vapor channels is 5000 m. The GMS has obtained valuable data since its inception, prompting extensive inversion algorithms and application studies. For example, Hyangsun [37] used machine learning to correct GMS data and Broomhall [38] expanded the Australian database with GMS data pairs.

The MTSAT-2 satellite is a geostationary-orbit satellite launched by Japan on 18 February 2006; meteorological observation began in 2010. The imager mounted on the MTSAT-2 satellite has one visible light band and four infrared bands. The imager relies on an internal scanning mirror to capture the image of the Earth's surface. The light collected by the scanning mirror is divided into visible light by a lens and a filter. There is a total of five bands of light. The MTSAT-2 captures images of the Northern Hemisphere every 30 min; this temporal frequency helps to better grasp the movement of typhoons and clouds. At the same time, the horizontal resolution of the satellite reaches 1000 m in the visible light band and 4000 m in the infrared band, and image quality is expected to improve significantly. Since it began accepting meteorological observation missions, the MTSAT-2 has been widely used in meteorological research [23,39].

2.1.3. Tide Station Data Processing

Data from 11 tide stations in NECS were used for model verification and typhoon storm surge comparison studies. The time resolution of the data was 1 h. The station locations are shown in Figure 1. The astronomical tide was calculated and analyzed by harmonic analysis of more than one year of measured data from the tide gauge stations. The remaining tide level was obtained by observing the tide level minus the astronomical tide. The base of the remaining tide level is sea level, which is identical to the base of the numerical model.

2.2. Wind Formula

For the relationship between TC size and storm surge, the radial distribution of wind relative to the MXW is specified following DUAL (dual-exponential) formula [40]. The DUAL formula is the same as the Jelesnianski2 [41] and Holland [42] formulas, but differs in that it is a summary formula based on multiple measurements of the cross-section of a tropical cyclone. The advantage of the DUAL formula is that the parameters can be calculated from measured or predicted wind speed in order to study the effects of different TC sizes on storm surges, which is why this study uses this formula.

The DUAL formula and the numeric coefficients (Formula 1-11) are both derived from the paper of Willoughby [40], as follows:

$$v = V_c = V_s + V_{max}\left(\frac{r}{R_{max}}\right)^n, (0 \le r \le R_1), \tag{1}$$

$$v = V_s + V_c(1-w) + V_o w, (R_1 \le r \le R_2), \tag{2}$$

$$v = V_o = V_s + V_{max}\left[(1-A)exp\left(-\frac{r-R_{max}}{X_1}\right)\right] + Aexp\left(-\frac{r-R_{max}}{X_2}\right),\ (R_2 \le r). \tag{3}$$

Related parameters:

$$X_1 = 317.1 - 2.026V_{max} + 1.915\varphi, \tag{4}$$

$$n = 0.406 + 0.0144V_{max} - 0.0038\varphi, \tag{5}$$

$$w = \frac{nX_1}{nX_1 + R_{max},}, \tag{6}$$

$$A = 0.069 + 0.0049V_{max} - 0.0064\varphi, \tag{7}$$

$$R_{max} = 46.4ex\,p(-0.0155V_{max} + 0.0169\varphi), \tag{8}$$

where r is the radial distance from the typhoon center; v is the wind velocity as functions of r; V_s is the forward speed of the typhoon; V_c and V_o are the tangential wind component in the eye and beyond the transition zone, which lies between $r = R_1$ and $r = R_2$; V_{max} and R_{max} are the MXW and the RMW; X_1 and X_2 are the exponential decay length in the outer vortex; A is the coefficient representing the scale of the E exponential function related to X_1 and X_2 (Formula (3)); φ is the latitude of the typhoon center and n is the exponent for the power law inside the eye.

Note: In order to ensure that both V_c and V_o are equal to V_{max} at $r = R_{max}$, a simple correction to the DUAL formula is as follows.

$$v = V_s + V_c(1-w) + V_{max}w, (R_1 \le r \le R_{max}), \tag{9}$$

$$v = V_s + V_{max}(1-w) + V_o w, (R_{max} < r \le R_2). \tag{10}$$

The weighting function, w, is expressed in terms of a nondimensional argument $\varepsilon = (r - R_1)/(R_2 - R_1)$. When $\varepsilon \le 0, w = 0$; when $\varepsilon \ge 1, w = 1$. In the subdomain $0 < \varepsilon < 1$, the weighting is defined as the polynomial.

$$w(\varepsilon) = 126\varepsilon^5 - 420\varepsilon^6 + 540\varepsilon^7 - 315\varepsilon^8 + 70\varepsilon^9, \tag{11}$$

which ramps up smoothly from zero to one between R_1 and R_2 [40].

TC size is represented by the seven-level wind circle range (R7). The variable v is equal to 13.9 m/s at $r = R_7$, where R_7 is the radial distance between the typhoon center and the seven-level wind circle. The exponential decay length X_2 will be calculated by using R_7 data from the Typhoon Network (http://www.wztf121.com/history.html). Specifically, the measured seven-stage wind speed and R7 are brought into Equation (3) to determine X2, which is the control model wind field cross-section curve.

Wind stress is computed from the following equation:

$$\vec{\tau} = C_d\rho_a\left|\vec{V_W}\right|V_W, \tag{12}$$

where C_d, a drag coefficient dependent on wind velocity, is given by the fitting curve (Figure 2), which is fitted to the mid-air pressure profile of Moon [43] based on previous research [44,45]:

$$C_d \times 10^3 = \begin{cases} 0.040v^4 - 0.5241v^3 + 2.4631v^2 - 5.3025v + 6.1763; \\ v \leq 6 \\ -1.3405 \times 10^{-5}v^4 + 0.0010v^3 - 0.0264v^2 + 0.3428v - 0.0755 \\ 6 < v \leq 31 \\ 1.8701 \times 10^{-7}v^4 - 4.3336v^3 + 0.0043v^2 - 0.2308v + 6.8709 \\ v > 31 \end{cases} . \quad (13)$$

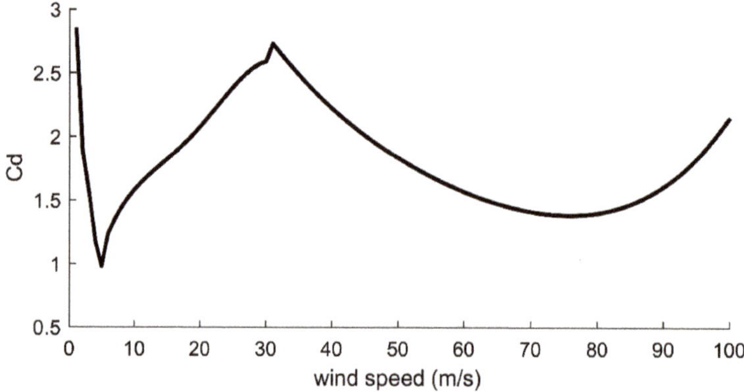

Figure 2. Drag coefficient as a function of maximum wind speed (MXW).

3. Model and Validation

3.1. Model Description and Configuration

The time-dependent, three-dimensional, primitive equation, finite-volume coastal ocean model (FVCOM) [46] is used to model storm surge. A non-overlapping unstructured triangular grid is used at the horizontal to accurately fit complex coastlines.

Storm surge is a cumulative water level effect, so the model must be large enough to contain the spatial extent of the storm and the accumulation of surge response due to nonlocal excitation [10]. The model domain extends from north of Bohai Sea to near Taiwan Strait, with an open boundary arching south. The grid resolution increases from the open boundary toward the NECS, with the highest resolution (about 100 m) on the coast of the Bohai Sea (Figure 3). A total of 168,373 triangular cells with 89,541 nodes comprise the horizontal, and 10 uniformly distributed s-coordinate layers comprised the vertical. The shoreline and bathymetry used in the model were obtained from the ETOPO-1 (http://www.ngdc.noaa.gov/mgg/global/global.html) dataset and were corrected near the NECS using an electronic chart with a scale of 1:50,000 (Figure 4). Based on the Courant—Friedrichs—Levy numerical stability condition, the computational time steps of 1.5 and 15 s are used for the external and internal modes, respectively. Temperature and salinity are estimated and specified to be constant at 18 °C and 35 psu, respectively. Storm surge in this model is only controlled by wind stress. The effects of atmospheric pressure, tides, rivers, and wave run up are not modeled. The conditions applied to the open boundaries were a combination of free surface and close the wet and dry grid.

The wind stress is provided by the above DUAL formula and corresponding typhoon data (Figure 5), but does not consider fusion with the background wind field, given the ideal nature of the design. The wind field established by the DUAL model shows that the typhoon is slightly larger on the right side, an asymmetrical characteristic that is consistent with the actual typhoon.

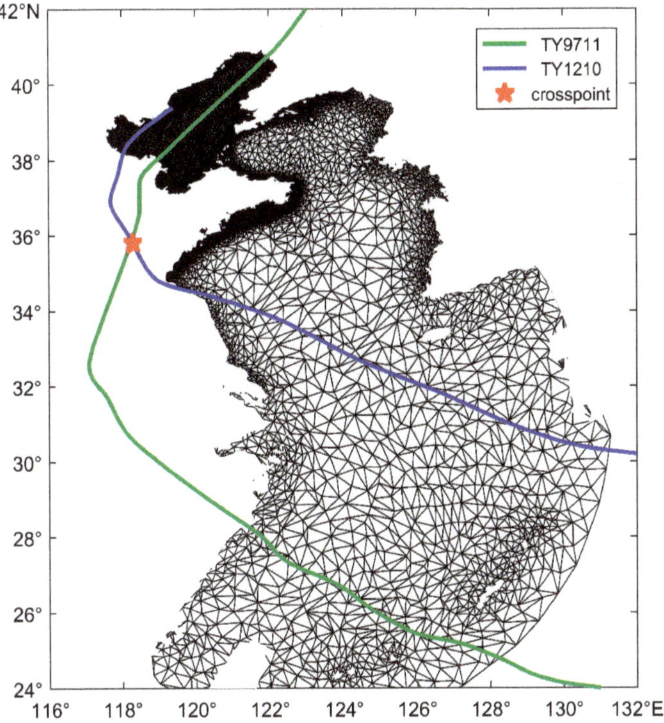

Figure 3. Computed model domains of finite-volume coastal ocean model (FVCOM). The paths of Typhoon Winnie (TY9711) and Typhoon Damrey (TY1210) are indicated by different color lines. The star is the cross point of the two typhoons and is also the time and location of the space map below.

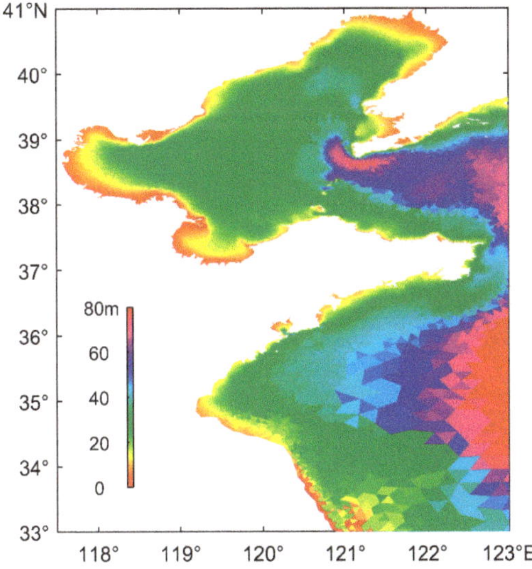

Figure 4. Shoreline and bathymetry of the northern East China Sea (NECS) after correction of the 1:50,000 electronic chart.

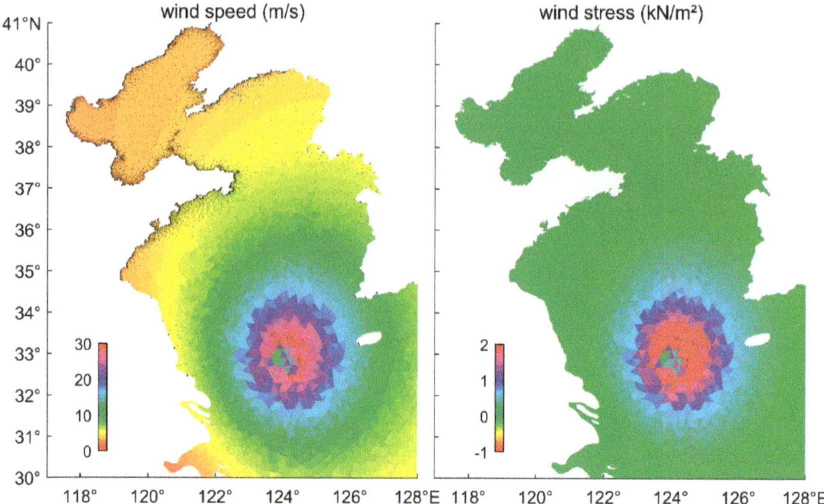

Figure 5. The dual-exponential (DUAL) model wind field and corresponding wind pressure diagram.

3.2. Model Validation

Since Typhoon Winnie (TY9711) and Typhoon Damrey (TY1210) were selected to investigate the relationship between TC size and storm surge, it is necessary to validate the reliability and accuracy of the simulation. The surge observation data were collected from six tide stations: Zfd, lko, wfg, tgu, byq, and lht. The locations are shown in Figure 1. Due to the lack of data, only the TY1210 simulation results are verified here (Figure 6). The timeframe is from 1–3 August 2012 (UTC, the same hereinafter). The results of the model agree well with observations not only in magnitudes but also in phases. However, there are slight differences between the two curves. The probable reasons are as follows: 1) The wind variations calculated by the DUAL formula differs from the actual wind change; 2) there is still some gap between model bathymetry data and actual bathymetry (which brings local effects to storm surges and leads to undulation at the curves); and 3) other factors causing storm surge have not been fully considered in the model, such as the effect of waves and tides. In addition, the FVCOM mesh and model parameters have been validated using additional storms besides TY9711 and TY1210. Therefore, it is given that only two storms are shown in the paper in order to then represent a verified model for a larger experimental design of perturbations to an idealized TY1210. Nevertheless, FVCOM simulates the storm surge elevations well enough to study storm surges induced by typhoons.

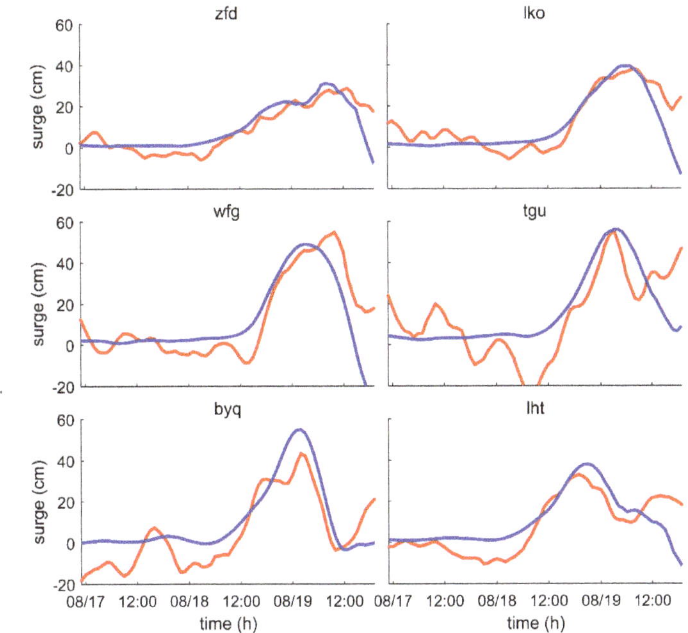

Figure 6. Simulated (blue lines) and observed (red lines) time series of storm surges at tide stations zfd, lko, wfg, tgu, byq, and lht induced by Typhoon Damrey (TY1210) in August 2012.

4. Results

4.1. Comparison of Similar Typhoon Storm Surges

TY9711 and the TY1210 were selected to compare the effects of different TC sizes on storm surge. The maximum envelope of storm surge (MESS) of TY9711 is far greater than MESS of TY1210 (Figure 7). At the zfd, lko, tgu, and lht tide stations, the storm surge curve over time (Figure 8) shows that the storm surge caused by TY9711 is greater than or equal to the storm surge caused by TY1210, but the MXW of TY9711 is less than the MXW of TY1210. One of the possible causes of this phenomenon is TC size, which is illustrated by the size of the cloud circle inversion shown on satellite imagery (see the cloud map at http://www.wztf121.com/history.html).

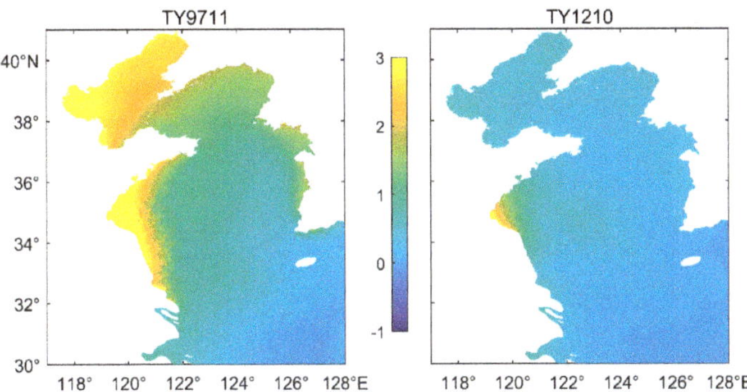

Figure 7. The maximum envelope of storm surge (MESS) of Typhoon Winnie (TY9711) and Typhoon Damrey (TY1210) simulated spatial distribution. The unit of the storm surge is meters (m).

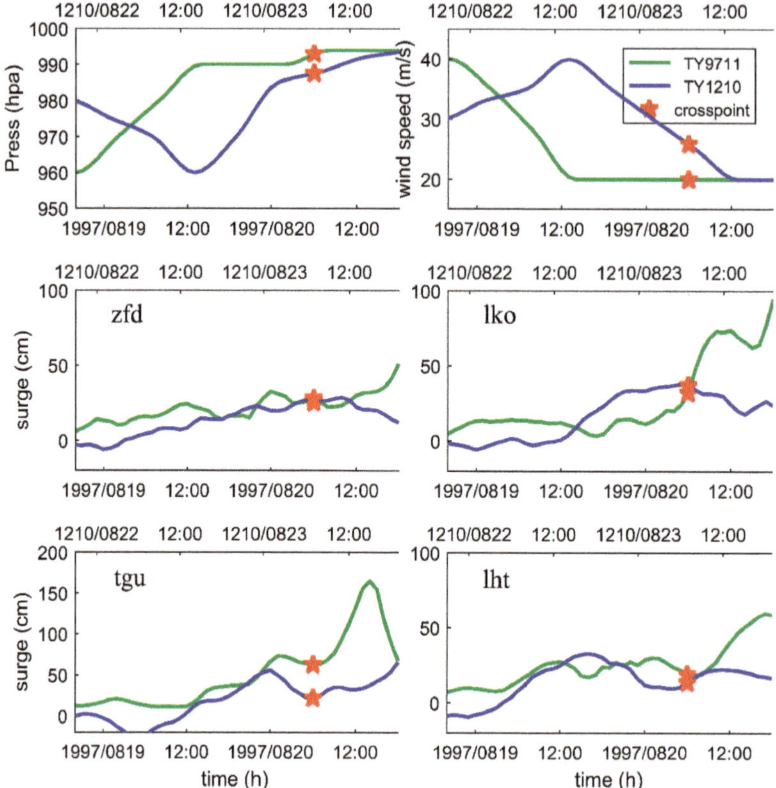

Figure 8. The minimum pressure and MXWs of Typhoon Winnie (TY9711) and Typhoon Damrey (TY1210) curves over time are displayed at the top. Remaining tide levels calculated from observations of storm surges at the tide stations are displayed at the bottom. The star symbol represents the location at which the paths of the two typhoons crossed.

At the same time, the size of the typhoon measured by R34 was inverted by the GTM-5 and MTSAT-2 satellites. The size of TY9711 is obviously larger than that of TY1210 (Figure 9). However, the amount of cloud map and inversion contour data is very small, making it difficult to accurately describe the chronological change of typhoon size and the relationship with the storm surge. Since further research is needed, we designed ideal tests to study the effects of different TC sizes on storm surge (see Section 3.2).

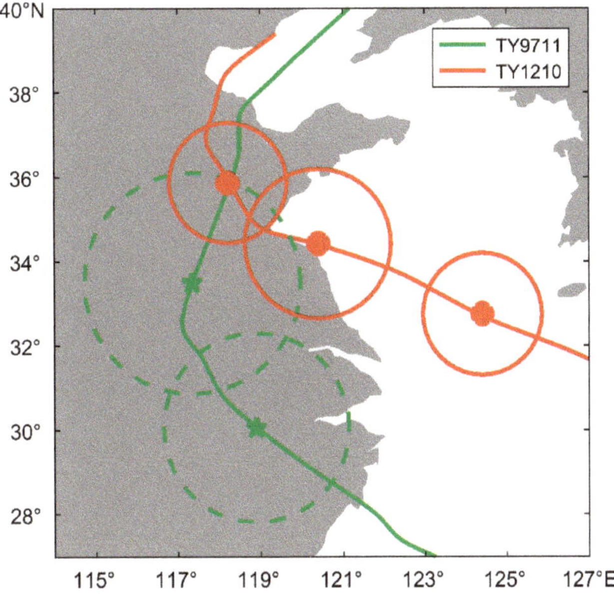

Figure 9. Map of the satellite inversion of TY9711 and TY1210 size.

4.2. Ideal Tests

There are many factors affecting storm surges, such as topography and external environmental impacts, but wind causes the largest proportion of storm surges [11]. In order to further explore the influence of tropical cyclone intensity and size on storm surge, we designed an ideal test based on the path of TY1210.

4.2.1. Maximum Wind Speed

Based on the path of TY1210, six contrast tests (ideal tests) were designed by varying the MXW and RMW, but fixing R7. The constant R7 is equal to 200,000 m. The MXW is constant for each ideal test, at values of 20, 30, 35, 40, 45, and 50 m/s (Table 1). The RMW is calculated from the MXW and the latitude of the center of TY1210 using Equation (8).

Table 1. Maximum wind speed ideal test parameters.

	MXW (m/s)	RMW (m, Latitude = 36°N)	R7 (m)
test 1	20	62,533	200,000
test 2	30	53,555	200,000
test 3	35	49,561	200,000
test 4	40	45,865	200,000
test 5	45	42,445	200,000
test 6	50	39,279	200,000

Using the DUAL formula wind fields, the ideal tests are computed using the FVCOM model. Figure 10 shows time series of simulated surges induced by different MXWs. The time span is one day before and after the time of maximum surge. The wind cross-section (Figure 11) is the curve of wind as a function of the distance from the center of the typhoon. It is intercepted from the center of the typhoon to periphery along longitude. The maximum envelope of storm surge (MESS) is given by calculating the extreme value of storm surge in the tests (Figure 12).

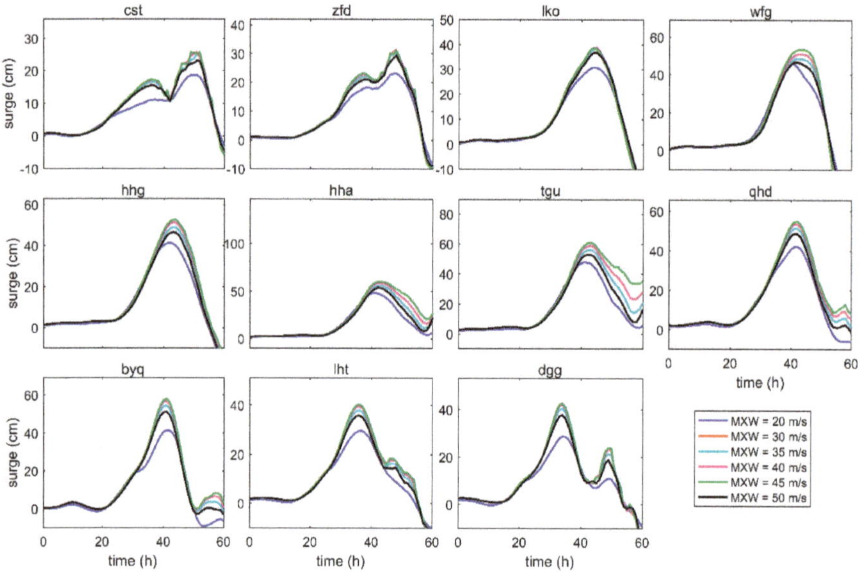

Figure 10. Time series (at representative spots for the tide stations) of simulated surges (cm) induced by different MXWs.

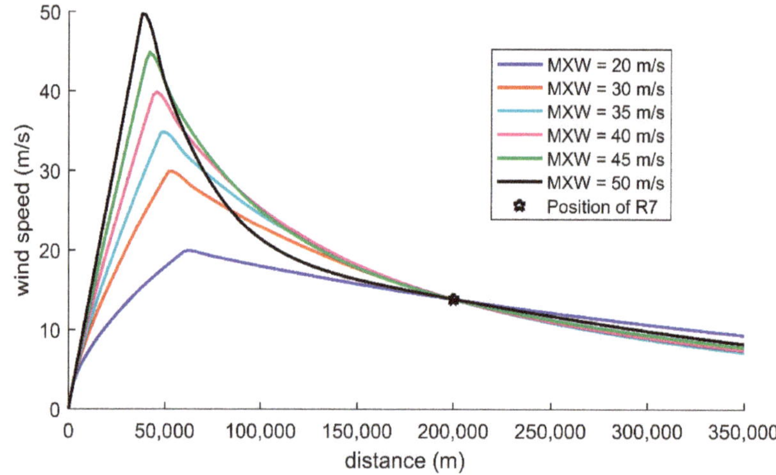

Figure 11. Wind cross-sections from the center of the typhoon at different MXWs. The black star represents the position of the seven-level wind circle range (R7).

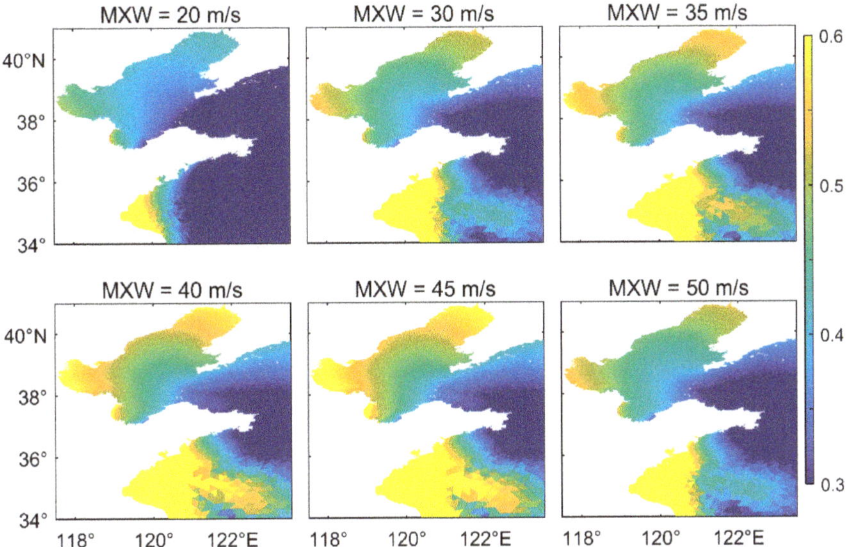

Figure 12. The maximum envelope of storm surge (MESS) of the typhoon at di_erent MXWs. The unit of the storm surge is meter (m).

The maximum value of storm surge occurs at an MXW of 40–45 m/s rather than at an MXW of 50 m/s (Figure 10). The reason for the phenomenon is related to the fact that the model only changes MXW and RMW, which can be seen from the wind cross-section (Figure 9). In addition, the storm surge does not undergo significant change with the MXW, probably due to the small TC size (Figure 11).

The MESS spatial distribution map of the ideal MXW test (Figure 12) reveals that the spatial distributions of storm surge caused by different wind intensities are slightly different, and the water increases in each bay mouth area are larger. For example, the storm surge in Bohai Bay is larger than the storm surge in other areas. Most of the storm surge extremes occur in the MXW = 45 m/s test.

4.2.2. TC Size and RMW

Based on the path of TY1210, a series of idealized storm surge tests with varying TC sizes and RMW were designed. The TC size is represented by the seven-level wind circle range (R7). There are three ideal tests (test 1, test 3, test 6) for which the MXW is constant and RMW varies with the latitude of the center of TY1210. The constant MXW is equal to 35 m/s. R7 in each test is a constant value of 200,000, 400,000, and 600,000 m (Table 2).

Table 2. Tropical cyclone (TC) size ideal test parameters.

	MXW (m/s)	RMW (m, Latitude = 36°N)	R7 (m)
test 1	35 m/s	49,561	200,000
test 2	35 m/s	25,000	400,000
test 3	35 m/s	49,561	400,000
test 4	35 m/s	75,000	400,000
test 5	35 m/s	100,000	400,000
test 6	35 m/s	49,561	600,000

In the other three tests (test 2, test 4, test 5), MXW and R7 values are fixed, but RMW fixed values are different in different tests (Table 2). The constant MXW is equal to 35 m/s and the constant R7 is equal to 400,000 m. The RMW is constant for each ideal test, at value of 25,000, 75,000, and 100,000 m (Table 2). The cross-section of all the tests wind is shown in Figure 13.

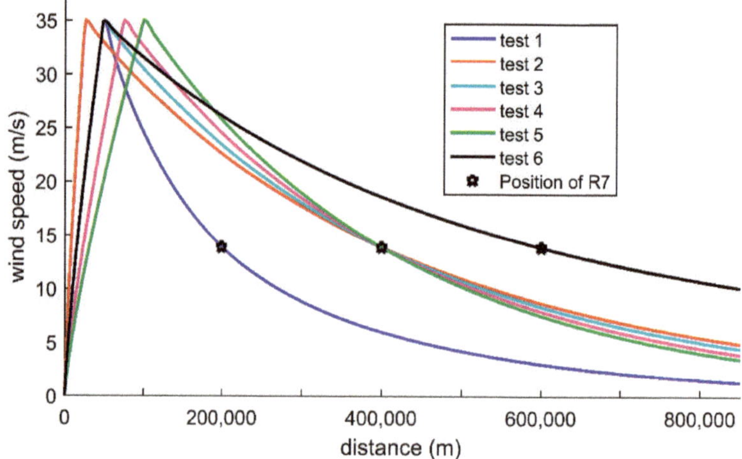

Figure 13. Wind cross-sections from the center of the typhoon at different TC sizes and radius of maximum wind (RMW). The black star represents the position of the seven-level wind circle range (R7).

Using the DUAL formula wind fields, the ideal tests are computed by the FVCOM model. The wind cross-section of the different tests (Figure 13) are represented as function curves of distance from the center of a typhoon. The maximum envelope of storm surge (MESS) is given by calculating the extreme value of storm surge in the tests (Figure 14). Time series of surges induced by different tests are shown Figure 15.

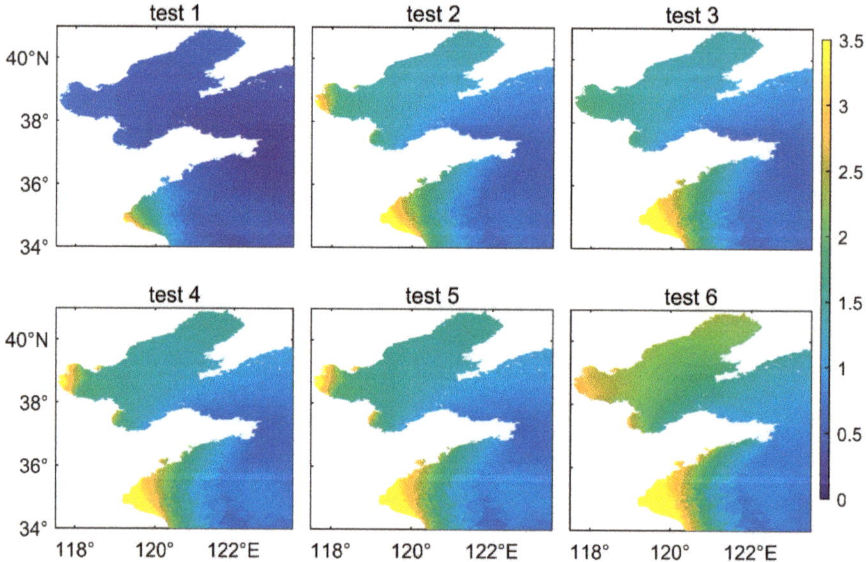

Figure 14. Maximum envelope of storm surge (MESS) induced by the ideal test at different TC sizes and RMWs. The unit of the storm surge is meter (m).

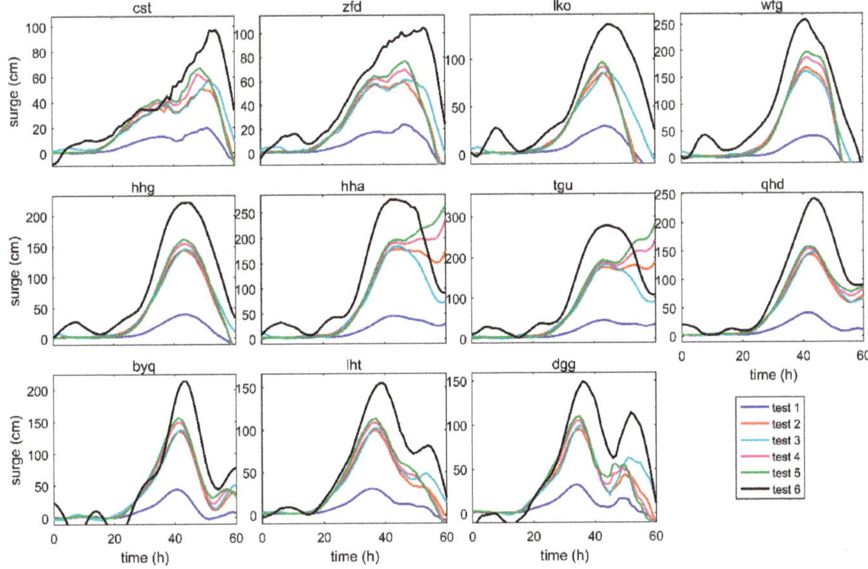

Figure 15. Time series (at representative spots from the tide stations) of simulated surges (cm) induced by different TC sizes and RMWs.

The maximum value of the storm surge increases as R7 increases, and the maximum value is greatest when R7 is equal to 600,000 m (Figures 14 and 15). These results are the same as reported by Irish [13]. In addition, we found that the peak surge values are approximately a linear function of R7. This is consistent with conclusions from Condon [12].

Tests 2–5 show that the storm surge does not undergo significant change with the RWM except for the area near the center of typhoon (Figures 14 and 15, location of hha and tgu), which is related to the change of wind field profile caused by the fixing of R7 (Figure 13).

5. Discussion

As shown above, the numerical ideal tests with different TC sizes indicate that in addition to storm intensity, TC size can dramatically change maximum storm surge. TC size data are lacking, however. Field observations, such as ground reports and buoy observations, can provide high-quality surface conditions, but these observations are not routinely available. Aircraft reconnaissance can also provide detailed spatial distributions of low-level or surface winds, but these missions rarely occur in the Northwest Pacific Ocean [18,19]. A weather center's estimate of the wind radius is based on subjective analysis of existing information, such as the 2007 publication of the global Hurricane Satellite (HURSAT) dataset by the National Oceanic and Atmospheric Administration (NOAA) (available online at: www.ncdc.noaa.Gov/hursat/index.php) [23].

Due to the lack of in situ observations, conventional TC size estimates are heavily dependent upon satellite observations and techniques. Satellites observations typically employ infrared light, cloud visible light, etc. [20,21]. In addition, satellite inversion relies on good inversion methods. Predecessors used a variety of techniques to transform measured data, such as data fusion [19] and machine learning [37,47]. The size and cloud top brightness temperature profiles of TCs are used for parameter inversion [23]. In the future, we should continue to study satellite inversion methods, and develop new observation methods to ensure timely access to typhoon-scale intensity and movement trajectory data to support weather system assessment and storm surge forecasting.

In the study of the impact of TC size on storm surge, Weisberg [10] investigated the storm surge in the Tampa Bay area of Florida, in the United States, and found that as the intensity of the wind increases, the maximum value of the storm surge increases, and it is suggested that as the RMW increases, the maximum value of the storm surge will increase. Irish [13] used a coupled hurricane vortex–planetary boundary layer (PBL) model to establish a wind field to simulate the entire Gulf of Mexico and found that storm surge increases with storm size. However, these studies have certain regional characteristics and rely on the establishment of wind fields.

Rey [4] and the NHC [48] built a database from the perspective of hypothetical hurricanes, using similarly anticipated typhoon storm surges that could not fully respond to the upcoming typhoon storm surge due to the complexity and variability of the typhoon storm surge. In this paper, using the predicted typhoon path, MXW and TC size, the entire wind field is inverted by the DUAL formula to simulate the storm surge. The approach can supplement the historical database and increase the accuracy of forecasting storm surges in order to reduce disaster losses. Therefore, it is necessary to establish the wind field by TC size and study the tropical cyclone intensity and size in the NECS.

In order to more clearly study the impact of TC size on storm surges, we have parameterized RMW according to the Weisberg [10] research. It is well known that typhoon wind field can be is described by at least two curves, that is, the wind speed rising curve with distance from the typhoon center to RWM and the wind speed decreasing curve with distance from RWM to the periphery. R7 or the peripheral location of fixed wind speed can be used to compensate for the inaccuracy of the wind field profile by RMW and typhoon center location alone. Our research found the small TC size limits the impact of wind field intensity changes on storm surges to some extent. This phenomenon occurs because with a fixed R7, the cross-section of the typhoon changes as the storm's strength varies, with the maximum value of storm surge occurring at an MXW of 40–45 m/s rather than 50 m/s. It is usually assumed that TC size increases as RWM increases, while other TC size changes are ignored, in many studies of storm surge changes through different RWM [4,12]. Our research found the storm surge does not undergo significant change with the RWM except for the area near the center of typhoon, probably due to the fixing of R7 (TC size). The maximum value of the storm surge increases as TC size increases. Therefore, we suggest that at least three positional parameters (typhoon center, RMW, TC size) are needed to accurately describe the wind field profile of a typhoon, especially TC size, which is more important in storm surge simulation. However, an ideal test with varying MXWs, RMWs, and TC sizes demonstrated that the MXW, RMW, and pressure of a typhoon will change with different TC sizes. There is no clear relationship between MXW, RMW, and typhoon pressure with TC size [49]. What is certain, however, is that the relationship is not simply linear. In the future, we will continue to investigate these relationships and their impacts on storm surge.

Typhoon scale varies, and the disasters caused by typhoon-scale weather factors vary significantly. According to the Saffir–Simpson hurricane scale, Hurricane Camille was a Category 5 storm, while Katrina was only a Category 3. Hurricane Katrina, however, was a greater natural disaster and resulted in a higher death toll due to its larger size [13].

The same conclusion was gleaned from the disaster losses caused by TY9711 and TY1210 (Table 3). TY9711 affected eight provinces, while TY1210 affected only four. In the same provinces (Jiangsu, Shandong, Hebei, and Tianjin) in which losses occurred, TY9711 destroyed 557 dykes, damaged 1000 ships, and damaged a large amount of farming equipment and homes, causing direct economic losses of 8300 million yuan, while TY1210 destroyed 40 dykes, damaged 690 ships, and destroyed a number of pieces of farming equipment and homes, resulting in a direct economic loss of 4175 million yuan. With the exception of Hebei Province, the disaster losses caused by TY9711 far exceeded those from TY1210, which was also due to the larger scale of TY9711. It can be seen that TC size variation has a significant impact on storm surge disasters. Moreover, the analysis of TC size variation is of great importance in disaster prevention and reduction of marine-economic effects.

Table 3. Direct economic losses.

	Jiangsu	Shandong	Hebei	Tianjin	Total
TY9711 (million yuan)	3000	4500	200	600	8300
TY1210 (million yuan)	537	1599	2044	4	4184

6. Conclusions

Three-dimensional FVCOM is used to simulate storm surge in the NECS region. First, we used observational data from tide stations to test the reliability and accuracy of the model. Through a hindcasting test of TY1210 storm surge, we found that the overall trend of the storm surge is consistent with actual measurements, and the maximum value of the storm surge was also similar to the measured value.

Analysis of the satellite cloud images and inverse typhoon profiles of TY9711 and TY1210 revealed that TY9711 had a larger TC size than TY1210. Based on observations and model simulations, the storm surge generated by TY9711 was far greater than the surge produced by TY1210, even though the paths and MXWs of the typhoons were similar.

Based on the path of TY1210, six contrast tests (ideal tests) were designed by varying the MXW and RMW, but fixing R7. The result of the ideal tests showed that the maximum value of storm surge occurs at a velocity of 40–45 m/s, rather than 50 m/s. Differences in the wind cross-section and the storm surge do not induce significant changes in the MXW, probably due to the small TC size.

In order to study the effect of TC size on storm surge, a series of ideal tests were designed by fixing the MXW, RMW, and R7. The maximum value of the storm surge becomes larger as R7 becomes larger, and the peak surge values are approximately a linear function of R7. The storm surge does not undergo significant change with the RWM except for the area near the center of typhoon.

Using the predicted typhoon path, MXW, RMW, and TC size, the entire wind field was inverted by the DUAL formula to simulate the storm surge. This approach can supplement the historical database and increase the accuracy of forecasting storm surges in order to reduce disaster losses of marine-economic effects. Therefore, it is necessary to establish the wind field by TC size and further study tropical cyclone intensity and size in the NECS.

Author Contributions: Conceptualization, J.L. and Y.H.; methodology, D.M.; software, Q.L.; validation, J.L.; formal analysis, Q.L.; investigation, J.L.; resources, Y.H.; data curation, D.M.; writing—original draft preparation, J.L.; writing—review and editing, Y.Z.; visualization, Q.L.; supervision, Y.Z.; project administration, Q.L.; funding acquisition, Y.H.

Funding: This research was jointly funded by the Natural Science Foundation of China, grant numbers U1706216 and U1901215.

Acknowledgments: The Japanese geostationary meteorological satellite (GMS) data, MTSAT-2 satellite data, and typhoon data from the Wenzhou Typhoon Net are highly appreciated.

Conflicts of Interest: The authors declare no conflict of interest.

References

1. You, S.H.; Seo, J.W. Storm surge prediction using an artificial neural network model and cluster analysis. *Nat. Hazards* **2009**, *51*, 97–114. [CrossRef]
2. Yang, Z.Q.; Wang, T.P.; Leung, R.; Hibbard, K.; Janetos, T.; Kraucunas, I.; Rice, J.; Preston, B.; Wilbanks, T. A modeling study of coastal inundation induced by storm surge, sea-level rise, and subsidence in the Gulf of Mexico. *Nat. Hazards* **2014**, *71*, 1771–1794. [CrossRef]
3. Xu, J.; Zhang, Y.; Cao, A.; Liu, Q.; Lv, X. Effects of tide-surge interactions on storm surges along the coast of the Bohai Sea, Yellow Sea, and East China Sea. *Sci. China Earth Sci.* **2016**, *59*, 1308–1316. [CrossRef]
4. Rey, W.; Mendoza, E.T.; Salles, P.; Zhang, K.; Teng, Y.C.; Trejo-Rangel, M.A.; Franklin, G.L. Hurricane flood risk assessment for the Yucatan and Campeche State Coastal Area. *Nat. Hazards* **2019**, *96*, 1041–1065. [CrossRef]

5. Lin, N.; Emanuel, K.; Oppenheimer, M.; Vanmarcke, E. Physically based assessment of hurricane surge threat under climate change. *Nat. Clim. Chang.* **2012**, *2*, 462–467. [CrossRef]
6. Andrade, C.A.; Thomas, Y.F.; Lerma, A.N.; Durand, P.; Anselme, B. Coastal flooding hazard related to swell events in Cartagena de Indias, Colombia. *J. Coast. Res.* **2013**, *29*, 1126–1136. [CrossRef]
7. Olbert, A.; Nash, S.; Cunnane, C.; Hartnett, M. Tide-surge interactions and their effects on total sea levels in Irish coastal waters. *Ocean Dyn.* **2013**, *63*, 599–614. [CrossRef]
8. Haigh, I.D.; Macpherson, L.R.; Mason, M.S.; Wijeratne, E.M.; Pattiaratchi, C.B.; Crompton, R.P. Estimating present day extreme water level exceedance probabilities around the coastline of Australia: Tropical cyclone-induced storm surges. *Clim. Dyn.* **2014**, *42*, 139–157. [CrossRef]
9. Bonaldo, D.; Antonioli, F.; Archetti, R.; Bezzi, A.; Correggiari, A.; Davolio, S.; Falco, D.G.; Fantini, M.; Fontolan, G.; Furlani, S.; et al. Integrating multidisciplinary instruments for assessing coastal vulnerability to erosion and sea level rise: Lessons and challenges from the Adriatic Sea, Italy. *J. Coast. Conserv.* **2019**, *23*, 19–37. [CrossRef]
10. Weisberg, R.H.; Zheng, L. Hurricane storm surge simulation for Tampa Bay. *Estuar. Coasts* **2006**, *29*, 899–913. [CrossRef]
11. Orton, P.; Georgas, N.; Blumberg, A.; Pullen, J. Detailed modeling of recent severe storm tides in estuaries of the New York City region. *J. Geophys. Res. Oceans* **2012**, *117*. [CrossRef]
12. Condon, A.J.; Sheng, Y.P. Optimal storm generation for evaluation of the storm surge inundation threat. *Ocean Eng.* **2012**, *43*, 13–22. [CrossRef]
13. Iish, J.L.; Resio, D.T.; Ratcliff, J.J. The influence of storm size on hurricane surge. *J. Phys. Oceanogr.* **2008**, *38*, 2003–2013. [CrossRef]
14. Sun, Y.; Zhong, Z.; Li, T.; Yi, L.; Hu, Y.; Wan, H.; Chen, H.; Liao, Q.; Ma, C.; Li, Q. Impact of ocean warming on tropical cyclone size and its destructiveness. *Sci. Rep.* **2017**, *7*, 8154. [CrossRef] [PubMed]
15. Brand, S. Very large and very small typhoon of the Western North Pacific Ocean. *J. Meteorol. Soc. Jpn. Ser. II* **1972**, *50*, 332–341. [CrossRef]
16. Merrill, R.T. A comparison of large and small tropical cyclones. *Mon. Weather Rev.* **1984**, *112*, 1408–1418. [CrossRef]
17. Cocks, S.B.; Gray, W.M. Variability of the outer wind profiles of western North Pacific typhoons: Classifications and techniques for analysis and forecasting. *Mon. Weather Rev.* **2002**, *130*, 1989–2005. [CrossRef]
18. Knaff, J.A.; Longmore, S.P.; Molenar, D.A. An objective satellite-based tropical cyclone size climatology. *J. Clim.* **2014**, *27*, 455–476. [CrossRef]
19. Knaff, J.A.; Slocum, C.J.; Musgrave, K.D.; Sampson, C.R.; Strahl, B.R. Using Routinely Available Information to Estimate Tropical Cyclone Wind Structure. *Mon. Weather Rev.* **2016**, *144*, 1233–1247. [CrossRef]
20. Holmlund, K.; Velden, C.S.; Rohn, M. Enhanced automated quality control applied to high-density satellite-derived winds. *Mon. Weather Rev.* **2016**, *129*, 517–529. [CrossRef]
21. Velden, C.; Daniels, J.; Stettner, D.; Santek, D.; Key, J.; Dunion, J.; Holmlund, K.; Dengel, G.; Bresky, W.; Menzel, P. Recent innovations in deriving tropospheric winds from meteorological satellites. *Bull. Am. Meteorol. Soc.* **2005**, *86*, 205–223. [CrossRef]
22. Jones, W.L.; Grantham, W.L.; Schroeder, L.C.; Johnson, J.W. Microwave scattering from the ocean surface (short papers). *Microw. Theory Tech. IEEE Trans.* **1975**, *23*, 1053–1058. [CrossRef]
23. Lu, X.; Yu, H.; Yang, X.; Li, X. Estimating tropical cyclone size in the northwestern pacific from geostationary satellite infrared images. *Remote Sens.* **2017**, *9*, 728. [CrossRef]
24. Lee, C.S.; Cheung, K.K.W.; Fang, W.T.; Elsberry, R.L. Initial maintenance of tropical cyclone size in the western North Pacific. *Mon. Weather Rev.* **2010**, *138*, 3207–3223. [CrossRef]
25. Chan, K.T.F.; Chan, J.C.L. Size and strength of tropical cyclones as inferred from QuikSCAT data. *Mon. Weather Rev.* **2012**, *140*, 811–824. [CrossRef]
26. Liu, K.S.; Chan, J.C.L. Size of Tropical cyclone as inferred from ERS-1 and ERS-2 Data. *Mon. Weather Rev.* **1999**, *127*, 2992–3001. [CrossRef]
27. Holland, G. An analytic model of the wind and pressure profiles in Hurricanes. *Mon. Weather Rev.* **1980**, *108*, 1212–1218. [CrossRef]
28. Zou, X.L.; Xiao, Q.N. Studies on the initialization and simulation of a mature hurricane using a variational bogus data assimilation scheme. *J. Atmos. Sci.* **1999**, *57*, 836–860. [CrossRef]

29. Knaff, J.A.; DeMaria, M.; Molenar, D.A.; Sampson, C.R.; Seybold, M.G. An automated, objective, multi-satellite platform tropical cyclone surface wind analysis. *J. Appl. Meteorol. Climatol.* **2011**, *50*, 2149–2166. [CrossRef]
30. Ricchi, A.; Miglietta, M.M.; Bonaldo, D.; Cioni, G.; Rizza, U.; Carniel, S. Multi-physics ensemble versus Atmosphere–Ocean coupled model simulations for a tropical-like cyclone in the Mediterranean Sea. *Atmosphere* **2019**, *10*, 202. [CrossRef]
31. Benetazzo, A.; Barbariol, F.; Bergamasco, F.; Torsello, A.; Carniel, S.; Sclavo, M. Observation of extreme sea waves in a space-time ensemble. *J. Phys. Oceanogr.* **2015**, *45*, 2261–2275. [CrossRef]
32. Needham, H.F.; Keim, B.D. An empirical analysis on the relationship between tropical cyclone size and storm surge heights along the US. Gulf Coast. *Earth Interact.* **2014**, *18*, 1–15. [CrossRef]
33. Mo, D.; Hou, Y.; Li, J.; Liu, Y. Study on the storm surges induced by cold waves in the northern east China Sea. *J. Mar. Syst.* **2016**, *160*, 26–39. [CrossRef]
34. Liu, F.; Su, J.; Moll, A.; Krasemann, H.; Chen, X.; Pohlmann, T.; Wirtz, K. Assessment of the summer–autumn bloom in the bohai sea using satellite images to identify the roles of wind mixing and light conditions. *J. Mar. Syst.* **2014**, *129*, 303–317. [CrossRef]
35. Hu, L.; Shi, X.; Bai, Y.; Qiao, S.; Li, L.; Yu, Y.; Yang, G.; Ma, D.; Guo, Z. Recent organic carbon sequestration in the shelf sediments of the Bohai Sea and Yellow Sea, China. *J. Mar. Syst.* **2015**, *155*, 50–58. [CrossRef]
36. Bao, X.; Na, L.; Yao, Z.; Wu, D. Seasonal variation characteristics of temperature and salinity of the North Yellow Sea. *Period. Ocean. Univ. China* **2009**, *39*, 553–562.
37. Hyangsun, H.; Sanggyun, L.; Jungho, I.; Miae, K.; Myong-In, L.; Myoung, A.; Sung, C. Detection of convective initiation using meteorological imager onboard communication, ocean, and meteorological satellite based on machine learning approaches. *Remote Sens.* **2015**, *7*, 9184–9204. [CrossRef]
38. Broomhall, M.; Grant, I.; Majewski, L.; Willmott, M.; Jones, D.; Kuleshov, Y. Improving the Australian tropical cyclone database: Extension of the GMS satellite digital image archive. In *Indian Ocean Tropical Cyclones and Climate Change*; Springer: Haarlem, The Netherlands, 2010; pp. 199–206. [CrossRef]
39. Yi, L.; Zhang, S.P.; Thies, B.; Shi, X.M.; Trachte, K.; Bendix, J. Spatio-temporal detection of fog and low stratus top heights over the yellow sea with geostationary satellite data as a precondition for ground fog detection—A feasibility study. *Atmos. Res.* **2015**, *151*, 212–223. [CrossRef]
40. Willoughby, H.E.; Darling, R.W.; Rahn, M.E. Parametric representation of the primary hurricane vortex. Part II: A new family of sectionally continuous profiles. *Mon. Weather Rev.* **2006**, *134*, 1102–1120. [CrossRef]
41. Jelesnianski, C.; Chen, J.; Shaffer, W. *SLOSH: Sea, Lake, and Overland Surges from Hurricanes*; NOAA Tech. Rep. NWS 48; United States Department of Commerce, NOAA/AOML Library: Miami, FL, USA, 1992; p. 71.
42. Holland, G.J.; Belanger, J.I.; Fritz, A. A revised model for radial profiles of hurricane winds. *Mon. Weather Rev.* **2010**, *138*, 4393–4401. [CrossRef]
43. Moon, I.J.; Ginis, I.; Hara, T. Impact of the reduced drag coefficient on ocean wave modeling under hurricane conditions. *Mon. Weather Rev.* **2018**, *136*, 1217–1223. [CrossRef]
44. Powell, M.D.; Vickery, P.J.; Reinhold, T.A. Reduced drag coefficient for high wind speeds in tropical cyclones. *Nature* **2003**, *422*, 279–283. [CrossRef] [PubMed]
45. Oey, L.Y.; Ezer, T.; Wang, D.P.; Fan, S.J.; Yin, X.Q. Loop current warming by Hurricane Wilma. *Geophys. Res. Lett.* **2006**, *33*, 153–172. [CrossRef]
46. Chen, C.; Liu, H.; Beardsley, R.C. An unstructured grid, finite-volume, three-dimensional, primitive equations ocean model: Application to coastal ocean and estuaries. *J. Atmos. Ocean. Technol.* **2003**, *20*, 159–186. [CrossRef]
47. Benassai, G.; Migliaccio, M.; Montuori, A. Sea wave numerical simulations with COSMO-SkyMed? SAR data. *J. Coast. Res.* **2013**, *65*, 660–665. [CrossRef]
48. Zachry, B.C.; Booth, W.J.; Rhome, J.R.; Sharon, T.M. A national view of storm surge risk and inundation. *Weather Clim. Soc.* **2015**, *7*, 109–117. [CrossRef]
49. Lin, Y.; Zhao, M.; Zhang, M. Tropical cyclone rainfall area controlled by relative sea surface temperature. *Nat. Commun.* **2015**, *6*, 6591. [CrossRef]

Article

Shoreline Extraction in SAR Image Based on Advanced Geometric Active Contour Model

Xueyun Wei [1,2,*], Wei Zheng [1,2], Caiping Xi [1,2] and Shang Shang [1,2]

1 School of Electronics and Information, Jiangsu University of Science and Technology, Zhenjiang 212100, China; zhengwei@just.edu.cn (W.Z.); xicaiping@just.edu.cn (C.X.); shangshang@just.edu.cn (S.S.)
2 Zhenjiang Smart Ocean Information Perception and Transmission Laboratory, Zhenjiang 212003, China
* Correspondence: xywei@just.edu.cn

Abstract: Rapid and accurate extraction of shoreline is of great significance for the use and management of sea area. Remote sensing has a strong ability to obtain data and has obvious advantages in shoreline survey. Compared with visible-light remote sensing, synthetic aperture radar (SAR) has the characteristics of all-weather and all-day working. It has been well-applied in shoreline extraction. However, due to the influence of natural conditions there is a problem of weak boundary in extracting shoreline from SAR images. In addition, the complex micro topography near the shoreline makes it difficult for traditional visual interpretation and image edge detection methods based on edge information to obtain a continuous and complete shoreline in SAR images. In order to solve these problems, this paper proposes a method to detect the land–sea boundary based on a geometric active contour model. In this method, a new symbolic pressure function is used to improve the geometric active-contour model, and the global regional smooth information is used as the convergence condition of curve evolution. Then, the influence of different initial contours on the number and time of iterations is studied. The experimental results show that this method has the advantages of fewer iteration times, good stability and high accuracy.

Keywords: SAR images; shoreline extraction; geometric active contour model

Citation: Wei, X.; Zheng, W.; Xi, C.; Shang, S. Shoreline Extraction in SAR Image Based on Advanced Geometric Active Contour Model. *Remote Sens.* **2021**, *13*, 642. https://doi.org/10.3390/rs13040642

Academic Editors: Martin Gade and Andrea Buono

Received: 24 December 2020
Accepted: 4 February 2021
Published: 10 February 2021

1. Introduction

The coastal zone is rich in biological, mineral, energy, land and other natural resources. Shoreline is the boundary between sea and land, and is also the outpost of national defense. Accurate and rapid determination of the location, direction and outline of shoreline plays an important role in coastal environmental protection and marine resource management. The traditional field survey methods are labor-intensive, inefficient, long-term, and the data obtained are not easy to be counted. Remote sensing has a strong ability to obtain data and has obvious advantages in shoreline survey [1]. Visible remote sensing is often used in shoreline extraction due to its imaging characteristics in line with human visual characteristics. However, it is limited by illumination and climate conditions. Compared with visible-light remote sensing, synthetic aperture radar (SAR) has the characteristics of all-weather and all-day operation [2]. It can image a large area and record the information of shoreline changes under bad weather conditions. SAR has been well-applied in shoreline extraction [3–6]. However, sometimes the contrast between ocean and land is not strong and the boundary is not always clear, which makes the shoreline extraction of SAR images a challenging problem.

So far, there are two main methods to detect shoreline from SAR images: visual interpretation and automatic interpretation. Usually, the digital manual tracking method is adopted to mark the boundary between ocean and land according to the trend, texture, shape and other interpretation marks of shoreline characteristics in SAR images, as well as the differences between tidal flats and water bodies near the coastal zone. This process is visual interpretation. However, because visual interpretation needs image interpretation, the reader must have rich visual interpretation experience and master all kinds of geoscience knowledge, which requires greater labor. Moreover, this interpretation method is time-consuming in human and material resources, and to a certain extent, it is difficult to ensure the quality of the interpretation image. Based on this, the current research on SAR-image shoreline extraction mainly focuses on the automatic interpretation.

Based on the characteristics of land and water boundary, extracting shoreline from SAR images is actually an image-segmentation process. The most classical image-segmentation methods are edge-differential operators (such as Sobel operator, Canny operator and Roberts operator). Although these edge-differential operators are simple and fast they are sensitive to noise, and the edge location is not accurate enough, so they lack universal applicability. In order to detect the complete and continuous shoreline from SAR images, many researchers have done a lot of active exploration and also put forward some effective extraction methods.

The geometric active contour (GAC) model [7] was developed on the basis of the active-contour model (ACM). ACM was a major breakthrough in the field of extracting image boundary and has very practical research value [8]. In recent years, with the extensive and in-depth study of ACM, the GAC model has wide use in the world, involving more and more fields. The GAC model also shows strong practicability in the field of SAR-image boundary extraction. However, due to problems of SAR images such as fuzzy boundary, low contrast, high gray level and easy interference by noise, the method of the GAC model can still encounter some problems such as weak boundary, the number of iterations and iteration times being easily affected by the initial contour of the image, and influence of image preprocessing on the shoreline extraction from SAR images.

In order to solve the problem of weak boundary in SAR-image shoreline extraction, this paper proposes a method of sea–land boundary detection based on a geometric active-contour model. This method improves the GAC model through combining the global regional smooth information as the convergence condition of curve evolution, which is helpful to solve the problem of weak shoreline boundary. New symbolic pressure function combined with regional information is proposed as the boundary stop condition of the GAC model, and the shoreline is accurately extracted. In this paper, the influence of different initial contours of a SAR image on the iterations of shoreline detection is studied. It shows that the larger the initial contour selection of the image, the fewer number of iterations and the shorter the iteration times. Experiments show that the proposed method cannot only effectively detect the shoreline in SAR images, but also reduce the number of iterations and shorten the iteration times compared with other related shoreline extraction methods, and the detection accuracy is further improved.

The paper is structured as follows. The background and related work, which includes the main methods for shoreline extraction of SAR images, are presented in Section 2. The materials and methods, which include the study area, traditional geometric active contour model, the improvement of geometric active contour model and the method of shoreline extraction in this paper, are described in Section 3. In Section 4, the results are presented. The influence of the selection of a SAR-image initial contour is verified in Section 5. The results and the future research directions are discussed in Section 6. Finally, the conclusions are summarized in Section 7.

2. Background and Related Work

At present, many scholars have carried out research on SAR-image shoreline-extraction technology and achieved many meaningful research results. There are two main methods for shoreline detection of remote sensing image, which are based on edge detection and region segmentation. The main methods include boundary tracking algorithm, Markovian segmentation method, active contour model method, level set algorithm and so on.

The boundary tracking algorithm [9,10] first analyzes the normal distribution of ocean and land pixels in the image, and then sets a threshold value according to the mean value and standard deviation to distinguish the ocean and land in the image to obtain the binary image. Then, the boundary-tracking algorithm is set to send out from a certain shoreline point to plot the boundary contour of ocean and land. The algorithm is intuitive, simple and easy to operate, and can get continuous shoreline. However, the shoreline obtained by this algorithm depends on the separation of land and ocean in the image, that is, smoothing, filtering and threshold selection, so it has great limitations and is generally applied in the case of low-accuracy requirements. In order to solve the problem of edge discontinuity and false edge, a ridge-tracking technique for edge extraction from noisy data was proposed in [11,12].

Markovian segmentation method uses the concept of the Markovian random field and simulated annealing method to extract shoreline [13]. First, the resolution of the image is reduced and the influence of speckle noise is reduced. The minimum value of energy function is solved by a simulated annealing method. The pixels in the image are classified (sea, land, low wave zone, beach). The right angle gradient operator is defined to obtain an approximate rough boundary. Then, the image resolution is restored and the above steps are applied to the high-resolution image. Finally, the shoreline is obtained. However, the method of Markovian random field and simulated annealing still has errors in the classification of pixels in the image, and the amount of calculation is relatively large [14].

ACM is also called the snakes algorithm, which is a kind of algorithm based on human visual characteristics [8]. The algorithm first gives an initial contour in the region of interest in the image and then minimizes an energy function to drive the contour line to move in the image. After several iterations, the contour line is constantly changed, and finally the boundaries of the objects in the image are approached. The active contour method can get the outline of each object in the image. However, because of the poor stability of the active contour method and high requirement for the position of initial contour, it can only be applied to the detection of simple images.

The level-set algorithm [15,16] follows the characteristics of the active-contour method. In this kind of algorithm, it is also necessary to give the initial contour line, and the requirement of the initial contour position is lower than that of the active-contour method. The level-set algorithm has strong topology adaptability, and contour curves can be merged or separated automatically without additional treatment. Given a simple initial contour, the boundary of the object in the image can be obtained. Moreover, the two-dimensional curves are embedded into the three-dimensional surfaces, so that the numerical solution in this method is stable and there is a unique solution. However, due to the iterative algorithm of 3D surface, it leads to a large number of calculations and high complexity.

The GAC model is based on the level-set method and curve-evolution theory [17]. The basic idea of the GAC model for extracting shoreline from SAR images is: (1) Using continuous curve to describe image edge and combining with image information to define energy functional. (2) Then using the Euler Lagrange method to get the curve-evolution equation corresponding to the energy functional. (3) Finally using level set to simulate the evolution process of initial curve along the direction of the fastest energy decline to obtain the optimal boundary-contour curve. The GAC model can be classified as edge-based, region-based and hybrid models.

Above all, the shoreline obtained by the boundary-tracking algorithm depends on the separation of land and ocean in the image, which has great limitations. ACM can obtain continuous shoreline, but it is sensitive to the initial contour and cannot handle the boundary topology adaptively; it is usually used in combination with other methods, such as clustering algorithms [18] and wavelet-edge detection [19]. According to different energy function, ACM can be divided into region-based ACMs and edge-based ACMs (EL-ACMs). Region-based ACMs cannot simulate the heterogeneity of coastal zones with a single probability distribution, especially in high-resolution images [20]. EL-ACMs construct edge indicators based on edge information or gradient, which allows contours to evolve rapidly in homogeneous regions and stop at real boundaries [21]. They have been also used for shoreline extraction from SAR images [18]. The level-set algorithm follows the characteristics of the active-contour method and also needs to give the initial contour line, but the requirement for the position of the initial contour is lower than that of the active-contour method. The existing SAR-image shoreline-detection technology mostly uses the method based on the active-contour model or level set for iterative calculation, which has high computational complexity, and detection accuracy is greatly affected by the initial contour, window size and other factors [8,16,22]. The method of the GAC model to extract the shoreline of a SAR image will still encounter some problems, such as weak boundary, and number of iterations and iteration times, which are easily affected by the initial contour of the SAR image.

Shoreline extraction methods based on multipolarization SAR have also been proposed [14,23]. In [23], radar frequency was shown to have great influence on the method of SAR-based shoreline extraction.

In addition, the classical fuzzy C-means (FCM) method was also applied to shoreline detection, and the Wavelet decomposition algorithm was combined to better suppress the inherent speckle noise of SAR images [24]. In [25], a nonparametric fuzzy approach was proposed for shoreline extraction from Sentinel-1A. In [26], a shoreline extraction method based on spatial pattern analysis was proposed, which includes image decomposition, smoothing, segmentation and shoreline compensation. A learning process that involves spatial patterns was presented in the image-decomposition step. A nonlocal means filter was used to smooth the outline images, and then the graphic cutting technology was applied to segment the images into sea and land areas. The positioning accuracy was determined using the snakes algorithm. In [27], J-Net Dynamic which is an experimental algorithm was applied on a high-resolution Sentinel-1 SAR image for the first time.

Relevanting works, a new diffusion-based method for the delineation of shorelines from space-borne polarimetric SAR imagery, was presented in [28]. The over-segmentation problem is solved by combining neighboring segments with similar radar brightness. In [29], shoreline rotation has been analyzed to provide a better understanding of the morphodynamic processes of natural embayed beaches. In [30], a shoreline monitoring system based on satellite SAR imagery was studied. In this system, a shoreline-extraction technique was developed based on the edge-detection technique, and a simple polynomial function was introduced to represent the shoreline location at arbitrary water level. In [31], a semiautomatic coastline-extracting approach was proposed based on fuzzy connectivity concepts. And an automatic procedure was proposed for the evaluation of results.

Although the above results have improved the performance of shoreline extraction to a certain extent, there are still many problems to be further studied. Generally speaking, the boundary tracking algorithm, Markovian segmentation method and active-contour method are seldom used independently due to their detection effect. For SAR-image shoreline extraction, it is necessary to analyze the extraction effect, antinoise ability and complexity.

3. Materials and Methods

3.1. Study Area

We used the SAR image observed by RADARSAT-2, a Canadian radar satellite series. Compared with the RADARSAT-1 satellite, the RADARSAT-2 satellite has a more powerful imaging function and has become one of the most advanced commercial SAR satellites in the world. First, the RADARSAT-2 satellite can switch between left view and right view according to the command, and all wave velocities can be viewed left or right, which shortens the revisit times and increases the ability to obtain stereo images. Second, RADARSAT-2 retains all imaging modes of RADARSAT-1 and adds spot light mode, hyperfine mode, four polarization (fine, standard) mode and multiview fine mode, giving users more flexibility in imaging-mode selection. Third, the RADARSAT-1 satellite only provides HH polarization mode, while the RADARSAT-2 satellite can provide VV, HH, HV, VH and other polarization modes. The coasts of interest are shown in Figure 1. The study area is near the South China Sea.

Figure 1. Administrative division map of the study area.

3.2. *Theoretical Background*

3.2.1. Traditional Geometric Active Contour Model

The geometric active contour model was developed on the basis of the active-contour model. Compared with the active-contour model, the geometric active contour model has the advantages of natural handling of topological structure changes, insensitivity to initial conditions and simple numerical implementation. These characteristics have attracted more and more attention, and this model is widely used in computer vision and image processing.

Based on the definition of energy functional, the GAC model can fall into boundary model and region model categories. The geodesic active-contour model was the most typical boundary model proposed in 1997 [32]. This model can solve the problems of the sensitivity of the snakes model to initial conditions and the inability to deal with topological changes automatically. The geodesic active-contour model is a special case of the snakes model, and its energy functional E is:

$$E(C(q)) = \int_0^1 g(|\nabla I(C(q))|)|C'(q)|dq \tag{1}$$

where C is the parametric plane curve, I is the known image, and g is the edge stopping function (ESF):

$$g = \frac{1}{1 + |\nabla G_\sigma * I|^2} \tag{2}$$

where G is a Gaussian function with variance σ.

The value of g tends to 0 where the image gradient is large, and tends to 1 where the image gradient is small. The curve evolves to the position where g tends to 0, which can effectively extract the target boundary. The geometric active-contour model based on the level-set method can automatically deal with the topological changes of curves in the evolution process, and a similar boundary model has been proposed in [33]. The curve evolution termination conditions of the above boundary models all depend on the edge-detection operator based on image gradient. In fact, for low-contrast targets, the edge detection operator does not converge to 0, and then the evolution curve can cross the boundary. Moreover, the edge-detection operator is sensitive to noise, which makes the evolution curve of the boundary model easy to fall into local extremum, resulting in redundant contour.

Compared with the boundary model, the region model defines the energy functional by using the global region information inside and outside the active contour, and does not use the edge-detection operator based on image gradient, which is more conducive to SAR-image shoreline detection. Based on this, this paper uses the region-based geometric active-contour model to detect the shoreline of SAR images. Assuming that the image is composed of two homogeneous regions, I is the original image to be segmented, C is the closed contour, and the energy functional is defined as follows [34]:

$$E(C, c_1, c_2) = t_1 \int_{in(C)} |I(x) - c_1|^2 dx + t_2 \int_{out(C)} |I(x) - c_2|^2 dx, x \in \Omega \quad (3)$$

where t_1 and t_2 are constants greater than zero and are used to control the weight of the internal and external energy of the curve, and c_1 and c_2 are the average gray values of the image inside and outside the contour-division area, respectively.

It can be seen that the model combines the global information of the image, and its energy function is independent of the gradient of the image, so it is suitable for the edge extraction of the image with smooth boundary and discontinuous boundary. But it is not suitable for the image whose gray level of target and background is not obvious. In addition, although the initial position of the evolution curve has little effect on the result of edge detection, the evolution speed still depends on the initial position of the evolution curve, and the level-set function must be periodically reinitialized, which increases the time and computational complexity of edge detection to a certain extent.

3.2.2. Improvement of Geometric Active Contour model

The signed pressure function (SPF) is often used as the edge-stopping function in the region-based geometric active-contour model [35]:

$$SPF(I(x)) = \frac{I(x) - \frac{c_1+c_2}{2}}{max\left(\left|I(x) - \frac{c_1+c_2}{2}\right|\right)}, x \in \Omega \quad (4)$$

where I is the original image to be segmented, and c_1 and c_2 are the average gray values of the image inside and outside the contour division area, respectively.

Because c_1 and c_2 are the average gray values of the image inside and outside the contour division area, the SPF function will not be able to segment the weak boundary when the contrast of the image is not high. For solving this problem, we replace $(c_1 + c_2)/2$ in Equation (4) with a weighted function f^{LBF} in the local binary fitting (LBF) model [36], and a new SPF function for the image area Ω can be given by:

$$SPF^{LBF}(I(x)) = \frac{I(x) - f^{LBF}(x)}{max(|I(x) - f^{LBF}(x)|)}, x \in \Omega \quad (5)$$

The weighted function is $f^{LBF}(x) = \frac{1}{(2\pi)^{n/2}\sigma^n} e^{-|x|^2/2\sigma^2}$ with parameter $\sigma > 0$.

Then the corresponding evolution equation of the level-set function region-based can be written as:

$$\frac{\partial \varphi}{\partial t} = SPF^{LBF}(I(x)) \cdot \left(div\left(\frac{\nabla \varphi}{|\nabla \varphi|}\right) + \alpha\right)|\nabla \varphi| + \nabla SPF^{LBF}(I(x)) \cdot \nabla \varphi, x \in \Omega \quad (6)$$

where α is the spherical force controlling the contraction and expansion of the curve, and φ is the level set function.

In this paper, the evolution process of the geometric active-contour model shown in Figure 2 includes the following steps:

Step 1: Initialize the level-set function φ as a binary function;

$$\varphi(x,t=0)=\begin{cases}-k & x\in\Omega_0-\alpha\Omega_0\\ 0 & x\in\alpha\Omega_0\\ k & x\in\Omega-\Omega_0\end{cases} \tag{7}$$

where k is a constant greater than zero, Ω_0 is a subset of the image domain Ω, and $\alpha\Omega_0$ is the boundary of the region Ω_0.

Step 2: The simplest level-set evolution equation is calculated by combining f^{LBF} and SPF^{LBF};

$$\frac{\partial\varphi}{\partial t}=SPF^{LBF}(I(x))\alpha|\nabla\varphi|,x\in\Omega \tag{8}$$

Step 3: If $\varphi>0$, set as $\varphi=1$; otherwise, set as $\varphi=-1$;

Step 4: Selective binary and Gaussian filtering regularized level set (SBGFRLS) method [35,36] is used;

$$\varphi^{n+1}=G_{\sqrt{\nabla t}}*\varphi^{n} \tag{9}$$

where φ^n and φ^{n+1} are the values of φ obtained by the nth and (n + 1)th iterations respectively, and $G_{\sqrt{\nabla t}}$ is the Gaussian kernel function with variance ∇t.

Step 5: Check whether φ converges. If not, return to step 2.

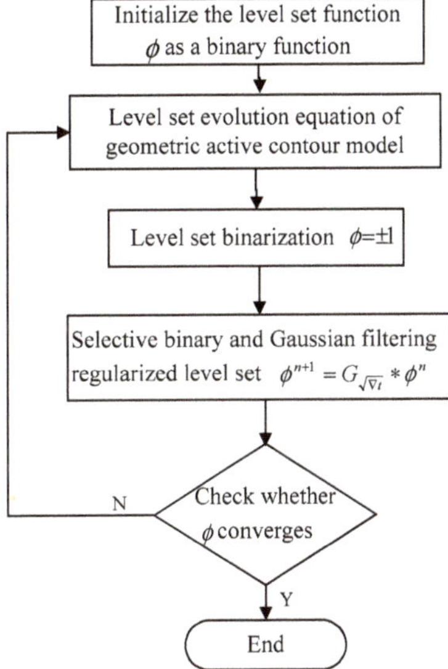

Figure 2. Evolution process of geometric active-contour model.

The geometric active-contour model proved in this paper combines the global regional smooth information as the convergence condition of curve evolution, which can effectively solve the influence of speckle noise on the segmentation of the land–sea boundary line in SAR images. The improvement of the symbolic pressure function can solve the problem of weak shoreline boundary. In addition, the SBGFRLS method can obtain faster convergence speed in the process of level-set evolution. In this paper, we use simple grid sampling points to obtain the initial positioning of the shoreline boundary as the initial contour of curve evolution, which can not only reduce the iterative time of the algorithm, but also reduce the possibility of boundary leakage caused by fuzzy boundary to a certain extent, so as to obtain more accurate detection results.

3.3. Method

3.3.1. Acquisition of Initial Contour

In order to reduce the evolution time of the geometric active-contour model, several small disks are used as the initial contour of shoreline, as shown in Figure 3. First, a numerical matrix is used to convolute the preprocessed target image. In the convolution processed SAR image, the grid sampling points are generated by the grid-sampling point function [37]. A disk with a radius of 9 pixels is created in the grid, and the image is inflated by the method of image expansion to achieve the effect of strengthening the shoreline edge.

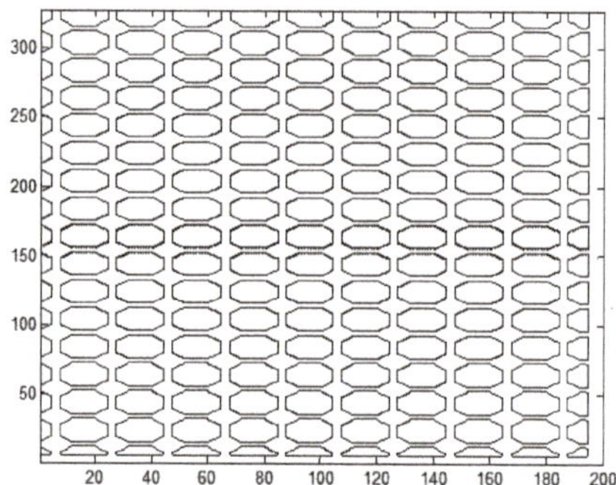

Figure 3. Initial contour.

3.3.2. Accurate Extraction of Shoreline

The accurate extraction process of shoreline is as follows:

Step 1: The SAR image is read and preprocessed;

Step 2: The SAR image is convoluted to generate grid sampling points and then several small disks are drawn as the initial contour of the shoreline;

Step 3: The initial contour of shoreline obtained in step 2 is used as the input of the geometric active-contour model. The improved symbolic pressure function is used as the boundary stop condition of the geometric active-contour model. The binary level-set function is quickly initialized by a Gaussian filter, and the shoreline is vectorized. Finally, a continuous shoreline is obtained.

Figure 4 is the flow chart of this method.

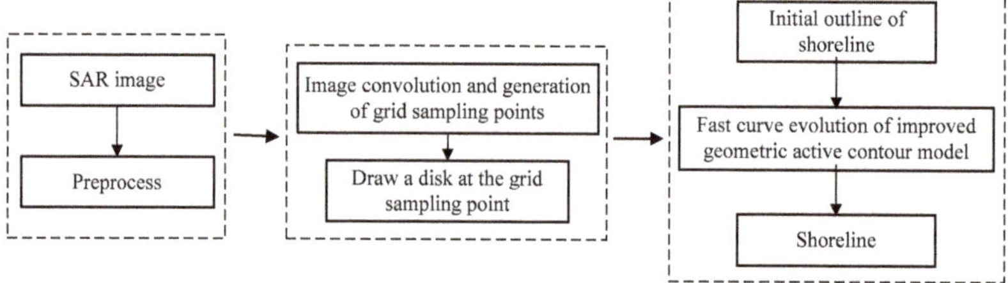

Figure 4. Extraction process of shoreline.

4. Results

In this section, the method is compared with the improved Canny operator method [38], boundary-tracking algorithm [9] and a traditional geometric active-contour model method [7]. At the same time, we label the shoreline manually and define it as follows: Error pixel is the sum of missed pixel and false detection pixel; correct pixel is the difference between detection result and false detection pixel; error rate is the ratio of wrong pixel number to manually labeled pixel number; accuracy is the ratio of correct pixel number to manually labeled pixel number. An ideal detection method must have a high accuracy and a low error rate. Figure 5a shows the original stripmap, select single-polarimetric, HH collected SAR image around the South China Sea area. The spatial resolution is 6.8 × 7.0 m; the number of looks is 1. The acquisition mode is fine.

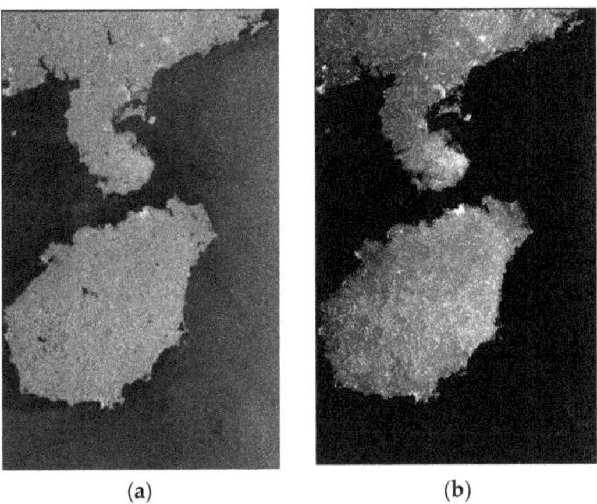

Figure 5. Synthetic aperture radar (SAR) image: (**a**) Original SAR image; (**b**) SAR image after gray transformation.

The gray contrast of the original SAR image used in this paper is low. In order to achieve better shoreline extraction, the gray-level transformation method of piecewise linear transformation is used to adjust the image contrast, and the result is shown in Figure 5b.

In Figure 6, the shoreline is less affected by the external natural conditions, and the clarity is relatively high. The improved Canny operator method can detect continuous edge points, but there are many false detection pixels, resulting in a high error rate. The boundary-tracking algorithm and the traditional geometric-active contour model method have many false detection pixels and missing pixels, which lead to a high error rate. The method in this paper has higher accuracy and a lower error rate, and the extraction results are better. See Table 1 for details.

Table 1. Algorithm extraction performance results.

	Improved Canny Operator Method	Boundary Tracking Algorithm	Traditional Geometric Active Contour Modeling Method	Proposed Method	Shoreline Pixels
Extraction result	707	789	906	681	
False extraction pixel	158	184	311	53	
Missing pixels	114	58	68	35	663
Error rate	0.410	0.365	0.572	0.133	
Accuracy	0.828	0.913	0.897	0.947	

It can be seen from Table 2 that the improved Canny method has 1693 iterations, and the operation time is 363.15 s. The boundary-tracking method has 1389 iterations, with the operation time of 289.63 s, and the traditional GAC model method has 1212 iterations, and the operation time is 267.98 s. However, due to the use of small disks as the initial contour of shoreline, the proposed method has 164 iterations and the operation time is 25.92 s.

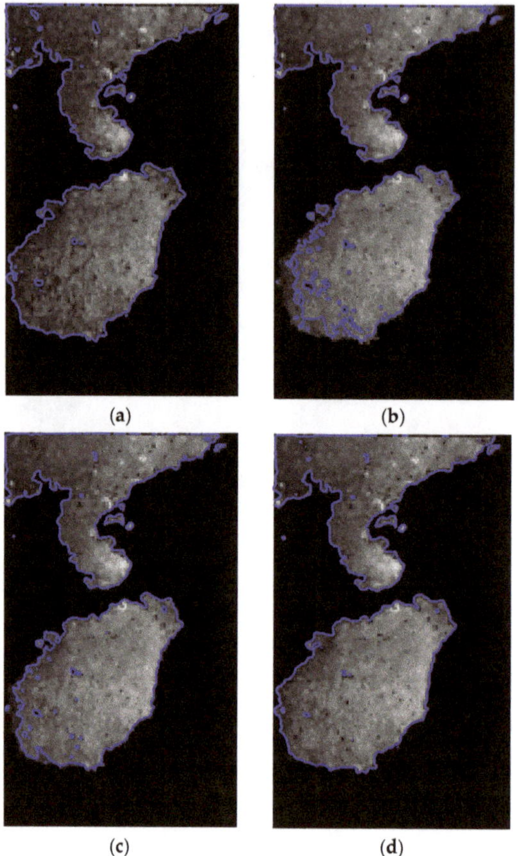

Figure 6. Shoreline extraction results: (a) Improved Canny operator method; (b) Boundary-tracking algorithm; (c) Traditional geometric active-contour modeling method; (d) Proposed method.

Table 2. Comparison of extraction efficiency of algorithm.

	Improved Canny Operator Method	Boundary Tracking Algorithm	Traditional Geometric Active Contour Modeling Method	Proposed Method
Number of iterations/times	1693	1389	1212	164
Operation time/second	363.15	289.63	267.98	25.92

In order to further intuitively observe the detection effect, Figure 7 shows the local enlarged results of shoreline extraction in Figure 6. From the visual point of view, it can be clearly seen that the shoreline extracted by the improved Canny operator method, boundary-tracking algorithm and traditional GAC model method has a large extraction error, while the shoreline extracted from the SAR image by the proposed method is more ideal.

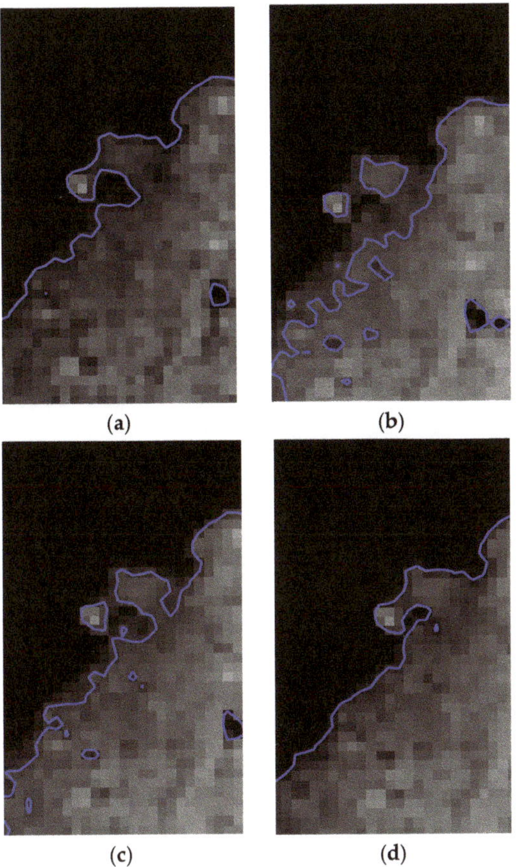

Figure 7. Local shoreline extraction results: (**a**) Improved Canny operator method; (**b**) Boundary-tracking algorithm; (**c**) Traditional geometric active-contour modeling method; (**d**) Proposed method.

5. Validation

In order to further verify the influence of the selection of the SAR image initial contour on the iteration number and operation times, this section selects small initial contour, large initial contour and global image as the initial contour of the image, and sets the maximum iteration number to 6000 times. The specific extraction results are shown in Figures 8–10, in which Figure 8 is the extraction-effect diagram of small initial contour, Figure 9 is the extraction-effect diagram of large initial contour, and Figure 10 is the detection result of global image as initial contour. Among the three extraction result graphs, (a) is the initial contour of the target image, (b) is the position of the initial contour in the target image, (c) is the effect map of extracting the shoreline, and (d) is the binary map of the extraction results of shoreline.

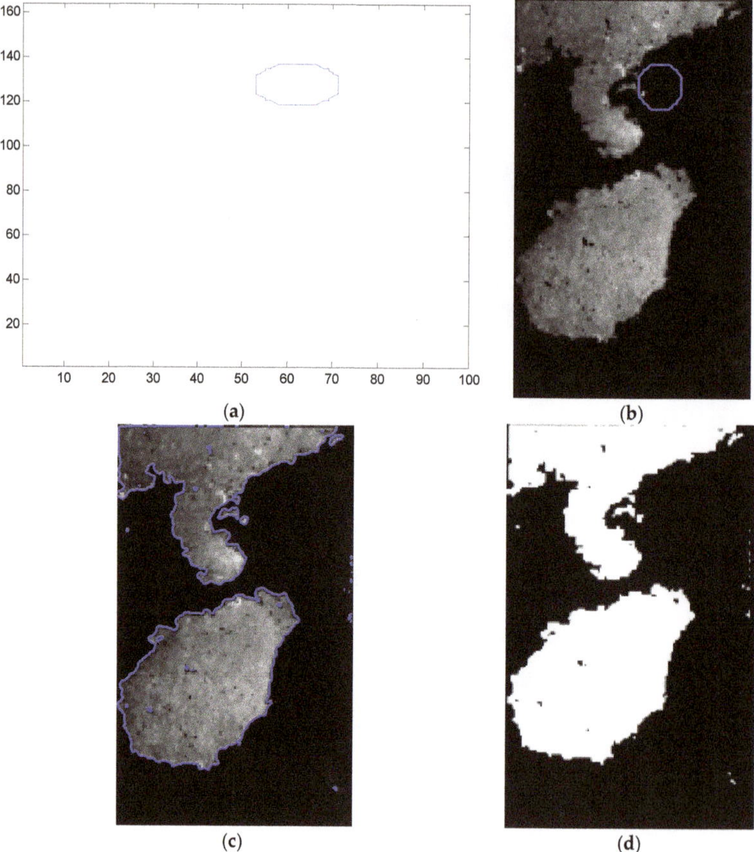

Figure 8. Small initial contour extraction results: (**a**) Small initial contour of target image; (**b**) The position of initial contour in target image; (**c**) Shoreline extraction result; (**d**) Binary map of the extraction results of shoreline.

The details are shown in Table 3. When a small initial contour is selected, the iteration has not been completed when the iteration reaches 6000 times, but the set maximum number of iterations has been reached, so the extraction automatically stops. It is further verified that the size of the initial contour of the SAR image affects the iteration number and operation time of the model: the larger the initial contour, the fewer the iteration number and the shorter the operation time. Moreover, the initial contour combined with the global information of the image has the fewest number of iterations and the shortest operation time.

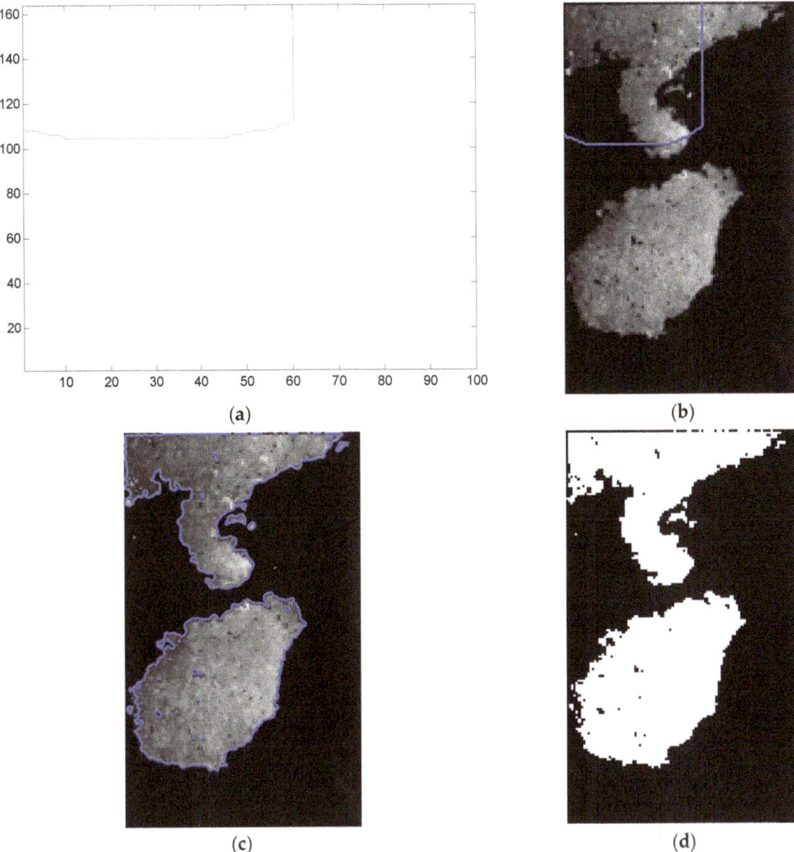

Figure 9. Large initial contour extraction results: (**a**) Large initial contour of target image; (**b**) The position of large initial contour in target image; (**c**) Shoreline extraction result; (**d**) Binary map of the extraction results of shoreline.

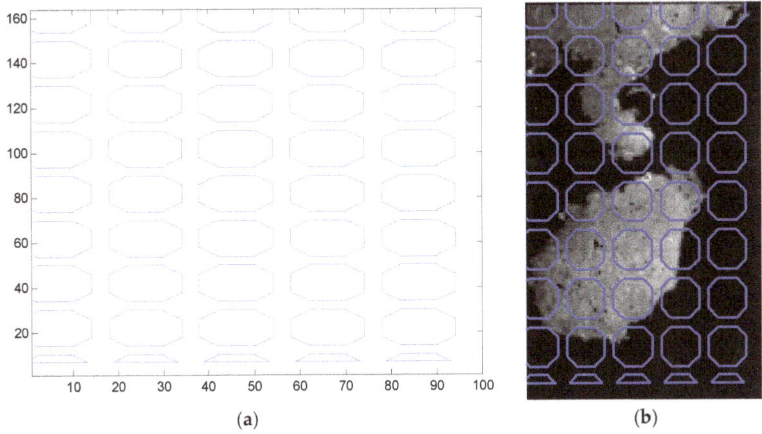

Figure 10. *Cont.*

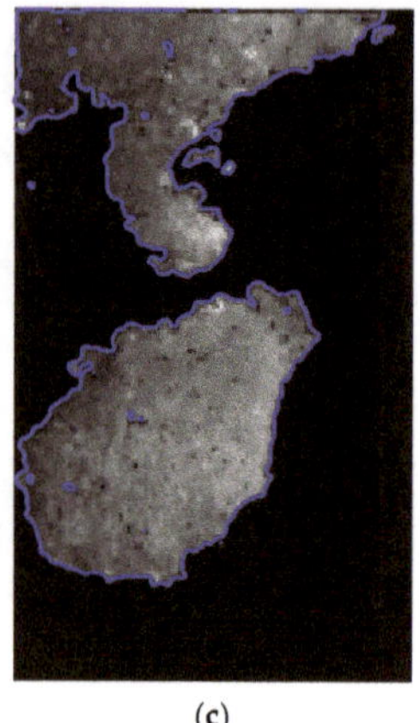

(c)

(d)

Figure 10. Extraction results of global image as initial contour: (**a**) Global initial contour of target image; (**b**) The position of global initial contour in target image; (c) Shoreline extraction result; (**d**) Binary map of the extraction results of shoreline.

Table 3. Comparison of different initial contour extraction results.

	Iteration Number/Times	Operation Time/Second
Small initial contour	6000	161.23
Large initial contour	1593	58.48
Global initial contour	164	25.92

In order to further remove the redundant blocks from the shoreline, this paper uses the method of block tracking. The results are shown in Figures 11 and 12, in which Figure 11 is the comparison chart before and after the block-tracking processing, and Figure 12 is the partially enlarged comparison map before and after the block-tracking processing.

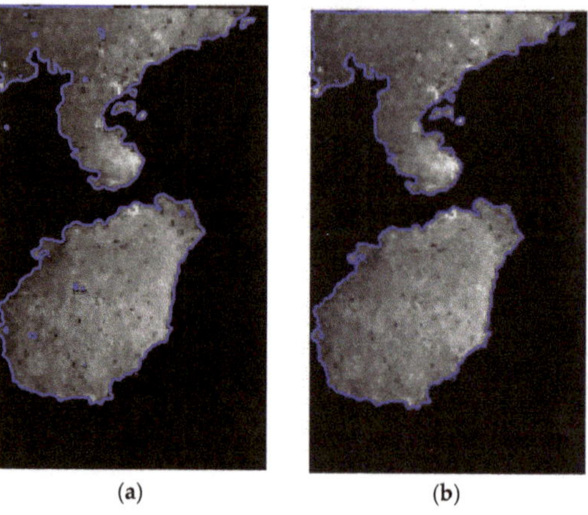

Figure 11. Comparison of whole block tracking: (**a**) Before block tracking (**b**) After block tracking.

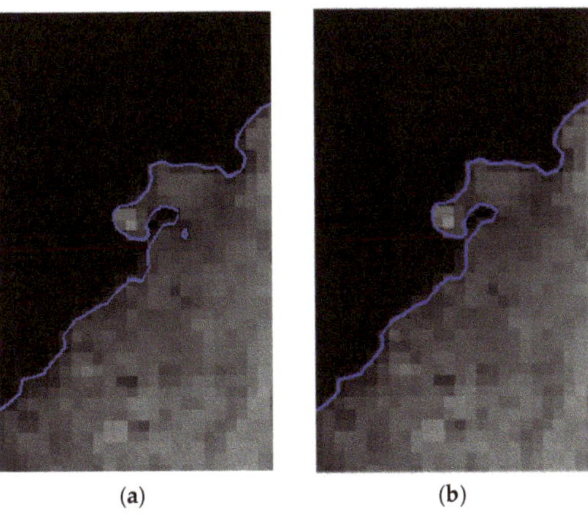

(a) (b)

Figure 12. Comparison of partial block tracking: (**a**) Before block tracking (**b**) After block tracking.

In order to further verify the proposed method, one more stripmap, select single-polarimetric, HH collected SAR image is used in the experiment. The administrative division map and the original SAR image are shown in Figure 13. The spatial resolution is 25.3 × 24.7 m; the number of looks is 1. The acquisition mode is standard. The extraction result is shown in Figure 14. The traditional geometric active-contour modeling method has 532 iterations, and the operation time is 49.06 s. The proposed method has 219 iterations, with the operation time of 38.81 s. It shows that the proposed method can reduce the number of iterations and shorten the iteration time.

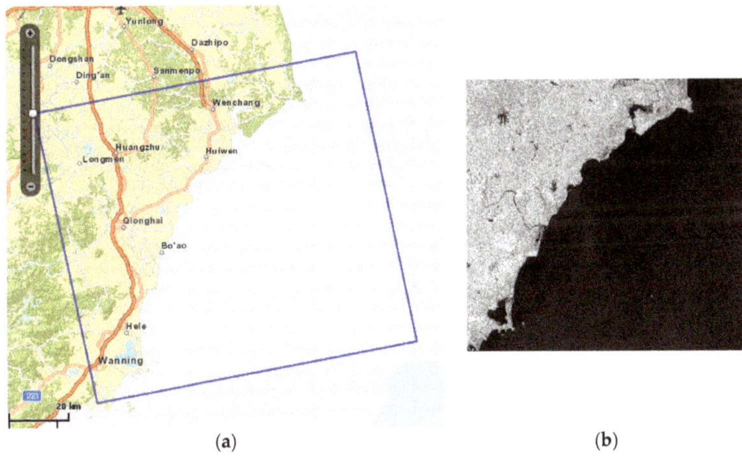

(a) (b)

Figure 13. SAR image: (**a**) The administrative division map (**b**) The original SAR image.

Figure 15 shows the local enlarged results of shoreline extraction. It can be seen that the shoreline extracted by the proposed method is more continuous and accurate, which proves the proposed method can effectively solve the weak boundary problem.

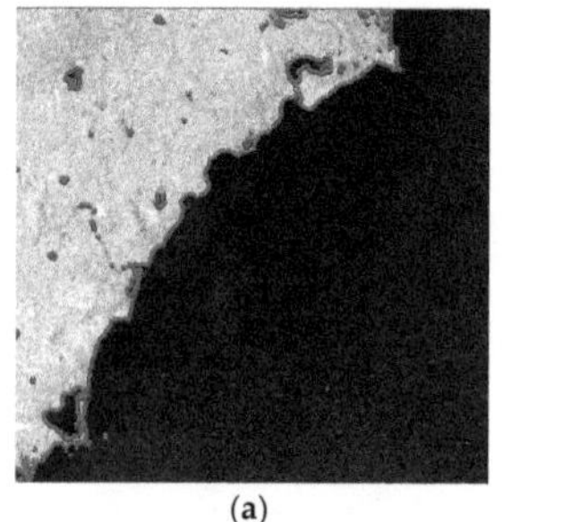
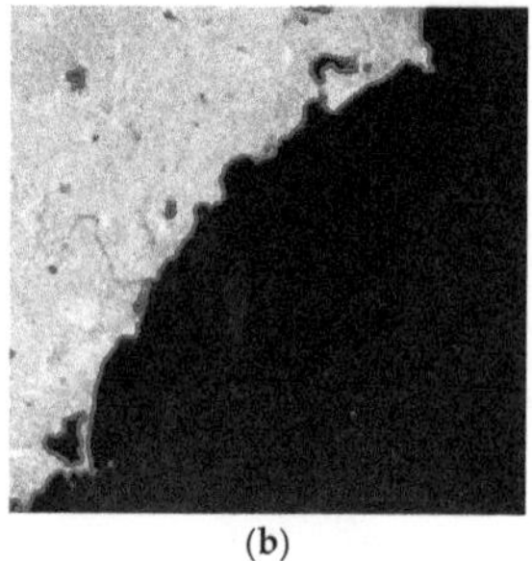

(a) (b)

Figure 14. Shoreline extraction results: (**a**) Traditional geometric active-contour modeling method; (**b**) Proposed method.

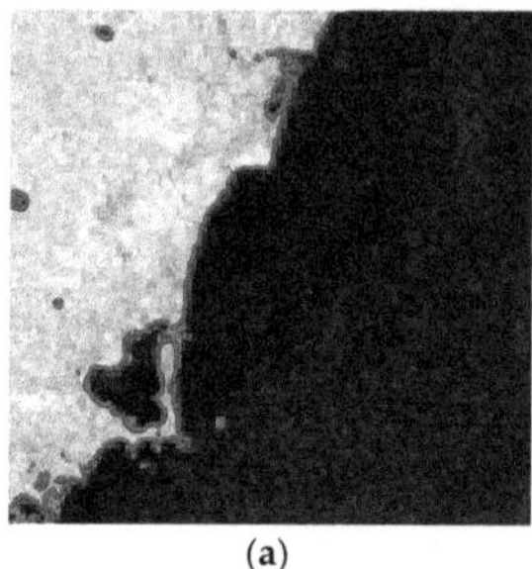
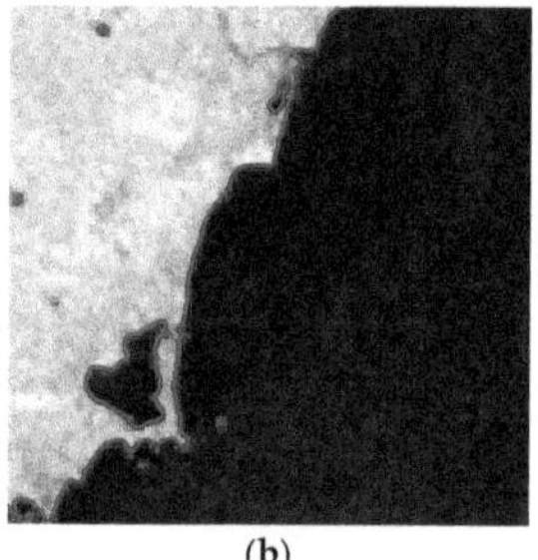

(a) (b)

Figure 15. Local shoreline extraction results: (**a**) Traditional geometric active contour modeling method; (**b**) Proposed method.

6. Discussion

In order to extract shoreline from a SAR image effectively, an improved GAC model was proposed. Although the application of the GAC model in SAR-image shoreline extraction was studied in this paper, much content and many technologies related to this subject can be further studied and explored. The later research can be carried out from the following aspects:

1. Although this paper solves the problem of shoreline extraction from SAR images, there is no theoretical basis for the setting of model parameters. As future studies, further research could be conducted to automatically set the parameters.
2. In this paper, the GAC model was improved by combining the global region smooth information as the convergence condition of curve evolution. In order to further reduce the iteration times and time, it is necessary to determine the initial contour of shoreline accurately and find a way to eliminate the redundant points in the detected shoreline.
3. The energy functional design of the model also needs a set of theories as a guide.

7. Conclusions

According to the characteristics of the automatic navigation process of satellite remote-sensing images, an image edge-extraction algorithm based on the geometric active-contour model was proposed to detect the land—sea boundary in a SAR image. First, the SAR image was convoluted and the grid sampling points were generated. Then, several small disks were drawn in the grid sampling points as the initial contour of the shoreline, which realized the coarse positioning of the shoreline-, and provided conditions for the reduction of the iteration times of the subsequent level-set evolution. Then, the improved symbolic pressure function combined with regional information was used as the boundary stop condition of the geometric active-contour model, and the shoreline was extracted accurately. The experimental results showed that this method reduces the number of iterations and the execution time, and increases the accuracy.

Author Contributions: Conceptualization, W.Z.; methodology, S.S.; validation, X.W., writing—original draft preparation, X.W.; writing—review and editing, X.W.; funding acquisition, C.X. All authors have read and agreed to the published version of the manuscript.

Funding: This research was funded by the National Nature Science Foundation of China (NSFC) under Grant 61901195 and supported by Zhenjiang smart ocean information perception and transmission laboratory project GX2017004.

Institutional Review Board Statement: Not applicable.

Informed Consent Statement: Not applicable.

Data Availability Statement: At https://www.mdpi.com/ethics (accessed on 30 January 2021).

Acknowledgments: The authors declare that there is no conflict of interest regarding the publication of this paper. This work was supported by the project of National Natural Science Foundation of China and the Zhenjiang smart ocean information perception and transmission laboratory project. The above funding did not lead to any conflict of interest regarding the publication of this manuscript.

Conflicts of Interest: The authors declare no conflict of interest. The funders had no role in the design of the study; in the collection, analyses, or interpretation of data; in the writing of the manuscript, or in the decision to publish the results.

Abbreviations

The following abbreviations are used in this manuscript:

SAR	synthetic aperture radar
ACM	active contour model
GAC	geometric active contour
ESF	edge stopping function
SPF	signed pressure function
LBF	local binary fitting
SBGFRLS	selective binary and Gaussian filtering regularized level set

References

1. Buono, A.; Nunziata, F.; Mascolo, L.; Migliaccio, M. A multipolarization analysis of coastline extraction using X-Band COSMO-SkyMed SAR data. *IEEE J. Sel. Top. Appl. Earth Obs. Remote Sens.* **2014**, *7*, 2811–2820. [CrossRef]
2. Al Fugura, A.; Billa, L.; Pradhan, B. Semi-automated procedures for shoreline extraction using single RADARSAT-1 SAR image. *Estuar. Coast. Shelf Sci.* **2011**, *95*, 395–400. [CrossRef]
3. Nunziata, F.; Buono, A.; Migliaccio, M.; Benassai, G. Dual-Polarimetric C- and X-Band SAR Data for Coastline Extraction. *IEEE J. Sel. Top. Appl. Earth Obs. Remote Sens.* **2016**, *9*, 4921–4928. [CrossRef]
4. Ferrentino, E.; Nunziata, F.; Migliaccio, M. Full-polarimetric SAR measurements for coastline extraction and coastal area classification. *Int. J. Remote Sens.* **2017**, *38*, 7405–7421. [CrossRef]
5. Nunziata, F.; Migliaccio, M.; Li, X.; Ding, X. Coastline extraction using dual-polarimetric COSMO-SkyMed PingPong mode SAR data. *IEEE Geosci. Remote Sens. Lett.* **2014**, *11*, 104–108. [CrossRef]
6. Schmitt, M.; Baier, G.; Zhu, X.X. Potential of nonlocally filtered pursuit monostatic TanDEM-X data for coastline detection. *ISPRS J. Photogramm. Remote Sens.* **2019**, *148*, 130–141. [CrossRef]
7. Shen, Q.; Wang, C.Y.; Zhao, B. Automatic waterline extraction in VHR imagery using geometric active contour model. *J. Fudan Univ. (Nat. Sci.)* **2012**, *51*, 77–82.
8. Michael, K.; Andrew, W.; Demetri, T. Snake: Active Contour Models. In Proceedings of the First International Conference on Computer Vision, London, UK, 8–11 June 1987; pp. 259–269.
9. Lee, J.-S.; Jurkevich, I. Coastline detection and Tracing in SAr images. *IEEE Trans. Geosci. Remote. Sens.* **1990**, *28*, 662–668. [CrossRef]
10. Baselice, F.; Ferraioli, G. Unsupervised coastal line extraction from SAR images. *IEEE Geosci. Remote. Sens. Lett.* **2013**, *10*, 1350–1354. [CrossRef]
11. Chu, C.; Siao, J.S.; Wang, L.C.; Deng, W.S. Estimation of 2D jump location curve and 3D jump location surface in nonpar-ametric regression. *Stat. Comput.* **2012**, *22*, 17–31. [CrossRef]
12. Wang, D.; Liu, X. Coastline Extraction from SAR Images Using Robust Ridge Tracing. *Mar. Geod.* **2019**, *42*, 286–315. [CrossRef]

13. Descombes, X.; Moctezuma, M.; Maître, H.; Rudant, J.-P. Coastline detection by a Markovian segmentation on SAR images. *Signal Process.* **1996**, *55*, 123–132. [CrossRef]
14. She, X.; Qiu, X.; Lei, B. Accurate sea–land segmentation using ratio of average constrained graph cut for polarimetric synthetic aperture radar data. *J. Appl. Remote Sens.* **2017**, *11*, 26023. [CrossRef]
15. Margarida, S.; Sandra, H. Separation between water and land in SAR Images using Region-Based level sets. *IEEE Geosci. Remote Sens. Lett.* **2009**, *6*, 471–475.
16. Liu, C.; Yang, J.; Yin, J.J.; An, W. Coastline detection in SAR images using a hierarchical level set segmentation. *IEEE J. Sel. Top. Appl. Earth Obs. Remote Sens.* **2016**, *9*, 4908–4920. [CrossRef]
17. Maciel, L.D.S.; Gomide, F.; Ballini, R. A differential evolution algorithm for yield curve estimation. *Math. Comput. Simul.* **2016**, *129*, 10–30. [CrossRef]
18. Modava, M.; Akbarizadeh, G. Coastline extraction from SAR images using spatial fuzzy clustering and the active contour method. *Int. J. Remote Sens.* **2017**, *38*, 355–370. [CrossRef]
19. Niedermeier, A.; Romaneessen, E.; Lehner, S. Detection of coastlines in SAR images using wavelet methods. *IEEE Trans. Geosci. Remote. Sens.* **2000**, *38*, 2270–2281. [CrossRef]
20. Liu, C.; Xiao, Y.; Yang, J. A Coastline Detection Method in Polarimetric SAR Images Mixing the Region-Based and Edge-Based Active Contour Models. *IEEE Trans. Geosci. Remote Sens.* **2017**, *55*, 3735–3747. [CrossRef]
21. Tirandaz, Z.; Akbarizadeh, G. A two-phase algorithm based on kurtosis Curvelet energy and unsupervised spectral regres-sion for segmentation of SAR images. *IEEE J. Sel. Top. Appl. Earth Obs. Remote Sens.* **2016**, *9*, 1244–1264. [CrossRef]
22. Modava, M.; Akbarizadeh, G.; Soroosh, M. Integration of spectral histogram and level set for coastline detection in SAR im-ages. *IEEE Trans. Aerosp. Electron. Syst.* **2018**, *55*, 810–819. [CrossRef]
23. Kim, D.-J.; Moon, W.M.; Park, S.-E.; Kim, J.-E.; Lee, H.-S. Dependence of Waterline Mapping on Radar Frequency Used for SAR Images in Intertidal Areas. *IEEE Geosci. Remote. Sens. Lett.* **2007**, *4*, 269–273. [CrossRef]
24. An, M.; Sun, Q.; Hu, J.; Tang, Y.; Zhu, Z. Coastline Detection with Gaofen-3 SAR Images Using an Improved FCM Method. *Sensors* **2018**, *18*, 1898. [CrossRef]
25. Demir, N.; Bayram, B.; Şeker, D.Z.; Oy, S.; Erdem, F. A nonparametric fuzzy shoreline extraction approach from Sentinel-1A by integration of RASAT pan-sharpened imagery. *Geo-Mar. Lett.* **2019**, *39*, 401–415. [CrossRef]
26. Fuse, T.; Ohkura, T. Development of Shoreline Extraction Method Based on Spatial Pattern Analysis of Satellite SAR Images. *Remote Sens.* **2018**, *10*, 1361. [CrossRef]
27. Zollini, S.; Alicandro, M.; Cuevas-Gonzalez, M.; Baiocchi, V.; Dominici, D.; Buscema, M. Shoreline Extraction Based on an Active Connection Matrix (ACM) Image Enhancement Strategy. *J. Mar. Sci. Eng.* **2019**, *8*, 9. [CrossRef]
28. Yu, Y.; Acton, S.T. Automated delineation of coastline from polarimetric SAR imagery. *Int. J. Remote Sens.* **2004**, *25*, 3423–3438. [CrossRef]
29. Di Luccio, D.; Benassai, G.; Di Paola, G.; Mucerino, L.; Buono, A.; Rosskopf, C.M.; Nunziata, F.; Migliaccio, M.; Urciuoli, A.; Montella, R. Shoreline Rotation Analysis of Embayed Beaches by Means of In Situ and Remote Surveys. *Sustainability* **2019**, *11*, 725. [CrossRef]
30. Tajima, Y.; Wu, L.; Fuse, T.; Shimozono, T.; Sato, S. Study on shoreline monitoring system based on satellite SAR imagery. *Coast. Eng. J.* **2019**, *61*, 401–421. [CrossRef]
31. Dellepiane, S.; De Laurentiis, R.; Giordano, F. Coastline extraction from SAR images and a method for the evaluation of the coastline precision. *Pattern Recognit. Lett.* **2004**, *25*, 1461–1470. [CrossRef]
32. Caselles, V.; Kimmel, R.; Sapiro, G. Geodesic active contours. *Proc. IEEE Int. Conf. Comput. Vis.* **2002**, *22*, 61–79. [CrossRef]
33. Malladi, R.; Sethian, J.; Vemuri, B. Shape modeling with front propagation: A level set approach. *IEEE Trans. Pattern Anal. Mach. Intell.* **1995**, *17*, 158–175. [CrossRef]
34. Chan, T.F.; Vese, L.A. Active contours without edges. *IEEE Trans. Image Process.* **2001**, *10*, 266–277. [CrossRef]
35. Zhang, K.; Zhang, L.; Song, H.; Zhou, W. Active contours with selective local or global segmentation: A new formulation and level set method. *Image Vis. Comput.* **2010**, *28*, 668–676. [CrossRef]
36. Yang, Y.; Li, C.; Kao, C.-Y.; Osher, S. Split Bregman Method for Minimization of Region-Scalable Fitting Energy for Image Segmentation. *Min. Data Financ. Appl.* **2010**, *2008*, 117–128. [CrossRef]
37. Chan, T.F.; Sandergby, V. Active contours without edges for vector-valued images. *J. Vis. Commun. Image Represent.* **2000**, *11*, 130–141. [CrossRef]
38. Huo, Y.-K.; Wei, G.; Zhang, Y.-D.; Wu, L.-N. An Adaptive Threshold for the Canny Operator of Edge Detection. In Proceedings of the 2010 International Conference on Image Analysis and Signal Processing, Zhejiang, China, 9–11 April 2010; pp. 371–374.

 remote sensing 

Article

A Sensitivity Analysis on the Spectral Signatures of Low-Backscattering Sea Areas in Sentinel-1 SAR Images

Valeria Corcione [1,‡], Andrea Buono [1,*,†,‡], Ferdinando Nunziata [1,†,‡] and Maurizio Migliaccio [1,2,†,‡]

1 Dipartimento di Ingegneria, Università di Napoli Parthenope, Centro Direzionale isola C4, 80143 Naples, Italy; valeria.corcione@uniparthenope.it (V.C.); ferdinando.nunziata@uniparthenope.it (F.N.); maurizio.migliaccio@uniparthenope.it (M.M.)

2 Istituto Nazionale di Geofisica e Vulcanologia, Via di Vigna Murata 605, 00143 Rome, Italy

* Correspondence: andrea.buono@uniparthenope.it

† Current address: Centro Direzionale isola C4, 80143 Napoli, Italy.

‡ These authors contributed equally to this work.

Abstract: Satellite synthetic aperture radar (SAR) is a unique tool to collect measurements over sea surface but the physical interpretation of such data is not always straightforward. Among the different sea targets of interest, low-backscattering areas are often associated to marine oil pollution even if several physical phenomena may also result in low-backscattering patches at sea. In this study, the effects of low-backscattering areas of anthropogenic and natural origin on the azimuth autocorrelation function (AACF) are analyzed using VV-polarized SAR measurements. Two objective metrics are introduced to quantify the deviation of the AACF evaluated over low-backscattering areas with reference to slick-free sea surface. Experiments, undertaken on six Sentinel-1 SAR scenes, collected in Interferometric Wide Swath VV+VH imaging mode over large low-backscattering areas of different origin under low-to-moderate wind conditions (speed ≤ 7 m/s), spanning a wide range of incidence angles (from about 30° up to 46°), demonstrated that the AACF evaluated within low-backscattering sea areas remarkably deviates from the slick-free sea surface one and the largest deviation is observed over oil slicks.

Keywords: SAR; sentinel-1; low-backscattering areas; azimuth autocorrelation function

Citation: Corcione, V.; Buono, A.; Nunziata, F.; Migliaccio, M. A Sensitivity Analysis on the Spectral Signatures of Low-Backscattering Sea Areas in Sentinel-1 SAR Images. *Remote Sens.* **2021**, *13*, 1183. https://doi.org/10.3390/rs13061183

Academic Editor: Martin Gade

Received: 29 December 2020
Accepted: 16 March 2021
Published: 19 March 2021

1. Introduction

Continuous and effective monitoring of the oceans is of paramount importance to improving global marine awareness and the understanding of ocean dynamics, including man-made target surveillance, pollution monitoring, and the impact on climate change [1]. Satellite Earth observation represents a valuable tool that provides extensive data collection over the oceans. An important sensor for ocean observation is the synthetic aperture radar (SAR), an active, band-limited, and coherent microwave imaging sensor that provides day and night imagery in almost all-weather conditions [2]. The exploitation of SAR imagery for marine and maritime applications is now well-established [3–6] and has been further boosted when the European Space Agency (ESA) started providing Sentinel-1 (S1) SAR satellite measurements free of charge in 2014 [7–10].

The general physical modeling that rules sea surface scattering in SAR imagery accounts for both the sea surface roughness, i.e., the sea surface spectrum, and the sea permittivity. In the context of marine pollution, low-backscattering areas, whose normalized radar cross section (NRCS) is lower than the surrounding sea one, represent a broad class of targets of interest since they can be often associated to natural or anthropogenic oil slicks. Even though low-backscattering sea areas may have different origins, e.g., organic films, low-wind areas, etc., most analyses on low-backscattering areas have been focused on oil spill detection and classification [11–17], with important advancements that have been recently accomplished by means of proper physical processing of polarimetric SAR data [18–21].

Low-backscattering areas are due to the damping of the capillary and small gravity sea waves generated by the local wind and responsible for the measured NRCS, i.e., the centimetric Bragg resonant waves, which result in a reduced sea surface roughness. Accordingly, most of the incident energy is scattered in the specular direction rather than toward the SAR antenna. This is why low-backscattering sea areas appear as dark patches in graytones intensity SAR images [22,23]. However, note that the NRCS values depend on incident wavelength, polarization, and incidence angle [22]. In this paper, a new approach is proposed to exploit VV-polarized, i.e., vertical transmit/vertical receive, S1 SAR data, collected in interferometric wide (IW) swath mode, to characterize different kinds of large low-backscattering sea areas of a known origin. The approach, based on the autocorrelation function evaluated along the azimuth (AACF), relies on the spectral information inherently carried on by SAR measurements and on the SAR peculiar imaging mechanism along the azimuth [24]. The experimental analysis, performed under low-to-moderate wind regime, i.e., wind speed ≤ 7 m/s, and in a broad range of incidence angles (approximately 30°–46°), considers slick-free sea surface as the reference scenario. The deviation of the AACF evaluated over low-backscattering areas of both an anthropogenic and natural origin is quantified using two objective metrics, namely the Euclidean distance and the percentage relative difference. Furthermore, the AACF analysis is compared to the conventional contrast, i.e., the slick-free sea surface to low-backscattering area NRCS ratio, and the effect of incidence angle is also discussed. The experimental results show a pronounced sensitivity of the AACF to low-backscattering sea areas. According to the metrics, the oil AACF differs from the slick-free sea surface AACF and from other natural low-backscattering area AACFs. In addition, the Euclidean distance, if compared to the NRCS contrast, is less dependent on the incidence angle.

The remainder of this paper is organized as follows: The theoretical background is presented in Section 2, while the dataset is described in Section 3. The experiments are presented in Section 4, where the results are discussed, while conclusions are drawn in Section 5.

2. Theoretical Background

2.1. Physical Rationale

The SAR estimates the sea reflectivity by means of two different scanning mechanisms. The range or across-track direction imaging is done at the speed of light, therefore insensitive to temporal changes of the sea, while the azimuth or along-track direction imaging is done at satellite velocity, therefore sensitive to temporal changes of the sea [24–26]. The microwave signal scattered by the sea, under low-to-moderate sea state conditions and in a broad range of incidence angles, can be well described by two-scale scattering models [27]. The class of two-scale scattering models assumes that the full-range sea surface roughness spectrum is artificially split into two parts: The larger-scale roughness, mainly associated to longer surface waves, and the smaller-scale roughness, mainly associated to shorter surface waves, which are responsible for the Kirchhoff and Bragg scattering, respectively [28]. The choice of the K_{lim}, i.e., the wavenumber that splits the sea surface spectrum, has been addressed in several papers and is summarized in [28,29] and references therein.

These two contributions are not independent since, for instance, the non-linear interaction between the capillary (shorter) and gravity (longer) waves is the physical mechanism that underpins the energy transfer from the wind to the waves [30]. However, to achieve a good compromise between accuracy, practical implementation, and interpretation, these two contributions are added incoherently [27]:

$$\sigma^0_{pq} = \sigma^0_{pq,0} + \sigma^0_{pq,1}, \tag{1}$$

where σ^0_{pq}, $\sigma^0_{pq,0}$, and $\sigma^0_{pq,1}$ are the total NRCS, the zeroth-order Kirchhoff scattering contribution, and the first-order tilted-Bragg scattering contribution, respectively [28]. In (1), the subscripts "q" and "p" stand for transmitted and received polarization, respectively. In the S1 IW SAR imaging mode, the incidence angle is such that the main term contributing to σ^0_{pq} is the first-order Bragg scattering [28]. The latter, which can be considered as an average of the untilted Bragg scattering over the larger-scale ripple by the long-waves structure, broadens the spectrum of the Bragg resonant waves [28]. Hence, according to the tilted-Bragg model, the microwave signal backscattered to the radar antenna depends directly on the small-scale ripple (through the small-scale sea surface roughness spectrum) and, indirectly, on the longer-wave part of the spectrum through the probability density function of the slopes [28,31,32].

The presence of a surfactant over the sea surface affects both the NRCS and SAR image spectrum [33]. In fact, it modifies the full-range sea surface spectrum [28], i.e., both the short- and the longer-wave part of the sea surface spectrum are affected by the surfactant. The visco-elastic properties of the surfactant have a direct impact on the small-scale part of the spectrum through a damping coefficient that, in the case of monomolecular surface films, can be modeled by the Marangoni damping coefficient [34]. In this case, as expected, the Marangoni damping mainly affects the small-scale part of the sea surface spectrum. The surfactant, reducing the sea surface roughness, also affects the energy transfer from the wind to the sea waves. The latter is typically modeled by the friction velocity, with this phenomenon that is well described by a reduced friction velocity over the surfactant. A reduced friction velocity has a direct implication on the long-wave part of the sea surface spectrum by modifying the peak wavenumber and the significant slope [28].

The above-described physical rationale does not explicitly take into account sea dynamic processes that are of paramount importance when dealing with SAR imaging of sea surface. The effect of the sea dynamics on SAR images is known as the velocity bunching phenomenon [35,36] that is related to the azimuth channel. This imaging is affected by the scene coherence time which, being shorter than the SAR integration time, makes the SAR imaging act as a non ideal filter along the azimuth direction [25,37].

These dynamics processes, which make SAR images appear blurred in the azimuth direction [36], are such that the "optimal" focusing depends on the SAR image patch [36] and it can narrow/broaden according to the sea state [24]. Although all this matter was first seen as a limit in SAR imaging of the oceans [24], it was later considered as a geophysical information to be potentially exploited [36]. This phenomenon is known as azimuth cut-off [38]. In filter theory words, one can say that the azimuth cut-off is a measure of the actual SAR azimuth spatial resolution [24]. In [35] the azimuth cut-off is explained as being due to two main contributions. The radial velocity to the radar of the single moving water particle generates a Doppler shift with respect to a stationary scene and the Doppler shifts of the elementary scatterers in the SAR resolution cell are not all identical, i.e., the single elementary Doppler shift is different from the mean Doppler shift of the SAR resolution cell, producing a "velocity spread" that smears the image of the resolution cell [24,35].

The physical model detailed in [35] shows that the quasi-linear approximation of the SAR image power spectrum $P_{ql}(\boldsymbol{K})$ is modeled as a perturbation of the linear SAR image power spectrum $P_l(\boldsymbol{K})$ modeled in accordance to the linear imaging theory. Hence, the quasi-linear SAR image power spectrum $P_{ql}(\boldsymbol{K})$ can be expressed in function of the linear SAR image power spectrum $P_l(\boldsymbol{K})$ as follows [33,35]:

$$P_{ql}(\boldsymbol{K}) = \exp(-{K_x}^2\xi^2)P_l(\boldsymbol{K}), \tag{2}$$

where the subscript x refers to the azimuth direction and ξ^2 is the total variance of the azimuthal displacements within the SAR integration time.

Equation (2) means that the SAR image spectrum is not able to represent the waves whose wavelength is less than the so-called azimuth cut-off wavelength λ_c, which can be interpreted as a measure of the low-pass filtering, in the wavenumber domain, witnessed by the exponential term. In (2), ξ and consequently λ_c are related to the sea surface spectrum as follows [38]:

$$\lambda_c(\theta,\phi) = \pi\frac{R(\theta)}{V}\sqrt{\int_0^\infty \Omega^2 S(K)F(\boldsymbol{K},\theta,\phi)\mathrm{d}\boldsymbol{K}}, \tag{3}$$

where θ is the incidence angle, ϕ is the wind mean direction relative to the range axis [38], R/V is the ratio between the slant-range distance and the velocity of SAR platform, $S(\cdot)$ is the omni-directional sea spectrum and $F(\cdot)$ is the directional sea spectrum. This reasoning can be exploited over slick-free sea surface, by means of a tailored semi-empirical model and proper estimation procedure, to determine the wind speed [33,39].

2.2. Methodology

In this study, the large degree of heterogeneity resulting from the low-backscattering sea class prevents the use of a compact parameter for the spectral analysis as in the case of wind speed estimation, that relies on λ_c. Hence, here the analysis is focused on the AACF that is obtained as the inverse Fourier transform of the power spectral density (PSD) following a guideline similar to the one proposed in [39] to estimate λ_c. We expect a sensitivity of the AACF to the low-backscattering areas. In fact, the theoretical modeling here outlined shows that the surfactant affects the NRCS and the SAR image spectrum. The relationship between the slick-covered sea surface spectrum and the SAR image spectrum is generally complex and to some extent poorly known. Since the friction velocity reduction affects the dynamic process, it is expected to have an additional impact on the azimuthal SAR spectrum. Hence, we expect that the AACF evaluated over low-backscattering areas exhibits deviation from the reference slick-free sea surface one.

In detail, the methodology proposed to process S1 SAR imagery using the AACF can be summarized as follows, see also the flowchart of Figure 1:

- The VV-polarized uncalibrated intensity image of the SAR scene is divided into non-overlapped square boxes whose size is n = 128 pixels. This value is set according to the S1 pixel spacing, i.e., 10 m, in order to ensure both a satisfactory degree of homogeneity and a consistent number of samples for a reliable AACF estimation;
- For each box, the 2-D PSD is evaluated as the square modulus of the Fourier transform of the VV-polarized uncalibrated intensity image;
- The 1-D azimuth PSD, $\mathrm{PSD_x}$, is evaluated by averaging the PSD along the range direction;
- The AACF is obtained by applying the inverse Fourier transform:

$$\mathrm{AACF} = \frac{1}{n^2}\mathrm{IFFT}(\mathrm{PSD_x}), \tag{4}$$

  where PSD = FFT(X), with X is the VV-polarized uncalibrated intensity SAR image;
- A smoothing 7×1 median filter is applied to the modulus of the azimuth autocorrelation function (AACF) in order to remove the 0-lag contribution.

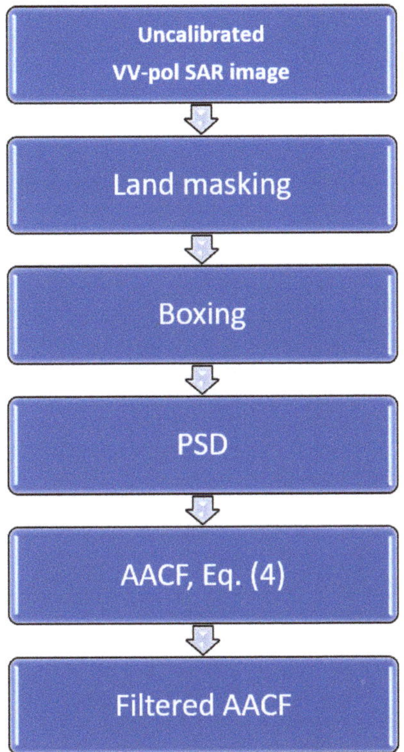

Figure 1. Flowchart of the AACF estimation procedure.

3. Dataset

The Copernicus Earth observation S1 mission consists of two polar-orbiting satellites equipped with a C-band (5.405 GHz) SAR. The latter supports dual-polarimetric imaging modes through a single switchable linear polarization transmission (horizontal—H, or vertical—V) while receiving coherently in a linear orthogonal polarization basis H-V. S1 operates in four different acquisition modes: Stripmap, extra-wide swath, wave mode, and IW mode. The IW mode is composed by three sub-swaths obtaing using the terrain observation with progressive scans SAR (TOPSAR) imaging technique. The TOPSAR technique, together with the electronic steering of the beam, result in a high-quality image characterized by no scalloping and homogeneity throughout the swath.

In this study, a SAR dataset consisting of 6 C-band S1 VV+VH ground range detected images collected in IW mode is considered that allows combining a large swath width (250 km) with a moderate geometric resolution (5 m by 20 m). The swath coverage is around 250 km, the pixel spacing is 10 m, and the incidence angle θ spans from about 30° up to 46°. Although both co- and cross-polarized channels are available, in this study only co-polarized S1 imagery is used since the cross-polarized channel is significantly affected by noise due to the smaller signal-to-noise ratio (SNR) and, in addition, it exhibits a low sensitivity to low-backscattering areas. All the images are characterized by low-to-moderate wind conditions, i.e., wind speed between 2 m/s and 7 m/s, therefore representing an optimal range for studying biogenic surfactants and oil slicks as suggested in [40]. The wind speeds are provided by space/time co-located ancillary European Centre for Medium-range Weather Forecasting (ECMWF) information.

The S1 SAR images include low-backscattering areas of a known origin [41–47] related to both verified oil slicks and other natural phenomena, see Table 1. The corresponding VV-polarized NRCS images are shown in Figure 2, where a graytones dB scale is used.

The regions of interest (ROIs) considered for the experimental analysis are highlighted with dashed boxes that refer to slick-free sea surface (blue and red), oil slick (green), and look-alike (orange).

The image shown in Figure 2a is acquired on 10 August 2017, off the southern coast of Kuwait in the Persian Gulf, where a certified oil spill is present [41]. The oil spill is likely due to an accidental collision between the pipeline laying vessel "DLB 1600" and an old pipeline on the seafloor. A conservative estimation of the oil-covered area is 131 km^2. Another low-backscattering area, due to very low wind conditions (<3 m/s), is also present. The incidence angle relevant to both low-backscattering areas is equal to 32°.

The image shown in Figure 2b was acquired on 8 October 2018, over the northern part of the Tyrrhenian Sea between Corsica and Tuscany coasts. A 20-km-long oil spill due to an accidental collision between two cargo ships the day before is present [42,43]. A large low-backscattering area due to very calm sea state is also present. The incidence angles over the two ROIs are 33° and 41°, respectively.

The image shown in Figure 2c is acquired on 8 March 2017, off the coast of Fujairah in the western coast of the United Arab Emirates (Persian Gulf). Multiple oil slicks due to a seafloor leakage from the jack-up drill rig "Pasargad 100" are present [44]. The polluted sea area is conservatively estimated to be about 334 km^2. The incidence angle over the ROIs is 35°.

The image shown in Figure 2d was acquired on 11 March 2017, over the same area of Figure 2c, during the same oil spill event. Another low-backscattering area due to very low wind conditions (<3 m/s) is also visible [44]. The incidence angle over the ROIs are 35° and 40°, respectively.

The image shown in Figure 2e was acquired on 20 July 2019, around the Gotland island in the Baltic Sea between Sweden and Latvia. A diffuse low-backscattering area due to swirling green algae blooms covers most of the observed sea surface [45]. The incidence angle measured at mid-range is equal to 38°.

The image shown in Figure 2f was acquired on 1 April 2018, over the Balikpapan Bay on the eastern coast of Indonesia. An oil spill due to a 25-m-underwater oil pipeline damaged the day before is present. The oil-affected area was estimated to be about 130 km^2 [46,47]. The incidence angle over the oil slick is 43°.

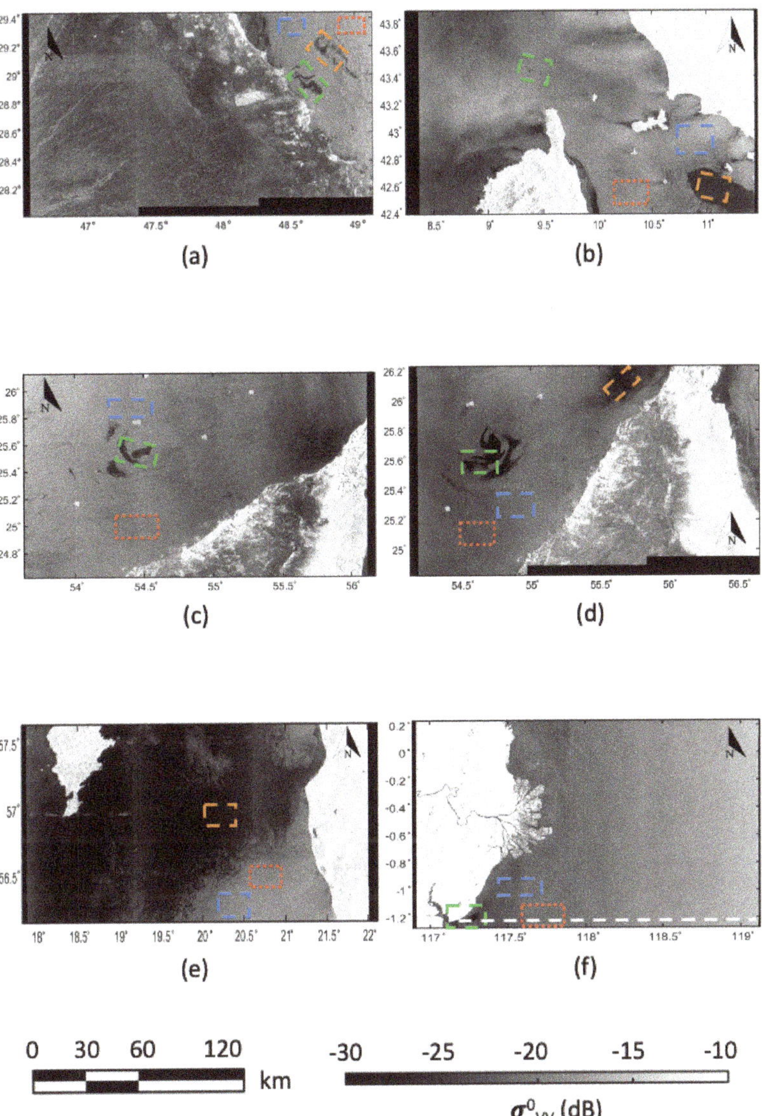

Figure 2. S1 synthetic aperture radar (SAR) dataset: VV-polarized normalized radar cross section (NRCS) graytones images (in dB scale) that include low-backscattering areas. (**a**–**f**) refer to the SAR scenes labeled as data ID 1–6 in Table 1. The regions of interest (ROIs) that refer to slick-free sea surface, oil slick, and look-alike are highlighted with blue-, green-, and orange-dashed boxes, respectively. An extra slick-free sea surface ROI, highlighted with a dashed red box, is also considered to analyze the intrinsic sea surface variability.

Table 1. Main features of the S1 SAR dataset.

Data ID	Acquisition Date	Figure	ROIs Wind Speed (m/s)	ROIs	Reference
1	10/8/2017	Figure 2a	2–3	Slick-free sea surface, oil slick, look-alike	[41]
2	8/10/2018	Figure 2b	<3 & 4–6	Slick-free sea surface, oil slick, look-alike	[42,43]
3	8/3/2017	Figure 2c	5	Slick-free sea surface, oil slick	[44]
4	11/3/2017	Figure 2d	<3	Slick-free sea surface, oil slick, look-alike	[44]
5	20/7/2019	Figure 2e	<3	Slick-free sea surface, look-alike	[45]
6	1/4/2018	Figure 2f	6–7	Slick-free sea surface, oil slick	[46,47]

4. Experiments

In this section, the sensitivity of the AACF to the low-backscattering areas highlighted in Figure 2 is analyzed.

4.1. Experimental Settings

The AACF over each ROI is estimated by averaging at least 10 AACFs evaluated according to the methodology described in Section 2. To allow a fair intercomparison of the estimated AACFs, they are normalized to their maximum value and the pedestal is set to zero, as it will be shown in Section 4.2, where the color coding is in accordance to the ROIs of Figure 2. In all the subsequent experiments, reference is made to slick-free sea surface ROIs (that will be labeled in blue as "Sea ref") at almost the same incidence angle of the low-backscattering areas, see dashed blue boxes in Figure 2. In each SAR image, an extra slick-free sea surface ROI (that will be labeled in red as "Sea"), see the dashed red boxes of Figure 2, is also selected to analyze the intrinsic sea surface variability. In the following, we refer to the slick-free sea surface AACFs as AACF_s and to the AACFs estimated within the low-backscattering areas as AACF_lb. In all subsequent SAR image analysis, the average contrast Δ, i.e., the difference between the slick-free sea surface and the low-backscattering area NRCS values in dB scale, is evaluated as follows:

$$\Delta(\text{dB}) = 10\log_{10}(\sigma^0_{\text{VV}})_\text{s} - 10\log_{10}(\sigma^0_{\text{VV}})_\text{lb}, \tag{5}$$

where the subscripts "s" and "lb" stand for slick-free sea surface and low-backscattering area, respectively.

To quantify the deviation of the AACF_lb with respect to the AACF_s, the Euclidean distance, d_E, and the percentage relative difference, D_rel, are introduced. The Euclidean distance is defined as:

$$d_\text{E} = \sqrt{\sum_{i=1}^{m}(lb_i - s_i)^2}, \tag{6}$$

where lb and s refer to AACF_lb and AACF_s, respectively, while i is the index that refers to the AACF samples of the selected low-backscattering and slick-free sea surface ROIs, each consisting of m = 1000 pixels. The percentage relative difference is defined as:

$$D_\text{rel}(\%) = \frac{(d_\text{E})_\text{lb} - (d_\text{E})_\text{s}}{(d_\text{E})_\text{lb}} \times 100, \tag{7}$$

where $(d_\text{E})_\text{lb}$ is the Euclidean distance between the low-backscattering ROI and the reference sea surface one, while $(d_\text{E})_\text{s}$ is the Euclidean distance between the two slick-free surface ROIs. This metric, as it is defined, is thought to analyze the deviation of the AACF estimated over low-backscattering areas from the corresponding slick-free sea surface reference one in order to filter out the intrinsic sea surface variability, i.e., induced by NRCS variability with respect to the azimuth angle (wind direction relative to the radar's azimuth look direction). All the above-mentioned quantitative parameters are listed in Table 2. Before proceeding further with the AACF analysis, since we are dealing with low-backscattering areas, a discussion on the effects of additive noise onto the backscattered signal is due. When dealing with S1 SAR data, the worst case nominal noise equivalent sigma zero (NESZ) is −22 dB. Hence, a received signal whose intensity is lower than NESZ may be uninformative. To perform a more accurate analysis, the SNR is used. The latter is not evaluated using the provided (worst case) NESZ but is estimated from the data according to [48,49]:

$$\mathrm{NESZ} = \frac{\eta}{A^2}, \tag{8}$$

where η and A are the noise calibration parameter and the NRCS calibration factor, respectively, which are pixel-dependent parameters provided in the ESA annotated metadata through a look-up-table. Then, the SNR is evaluated as follows:

$$\mathrm{SNR\ (dB)} = 10\log_{10}(\sigma^0_{\mathrm{VV}}) - 10\log_{10}(\mathrm{NESZ}). \tag{9}$$

4.2. Discussion

The SNR images evaluated over the whole SAR dataset are shown, in dB scale, in Figure 3, where land is masked in white. The average SNR values evaluated over the ROIs highlighted in Figure 2 are listed in Table 2.

Considering the reference slick-free sea surface ROIs (dashed blue boxes), the average SNR is always larger than about 7 dB, witnessing that the signal scattered off slick-free sea surface lies well above the NESZ along the whole SAR dataset. Considering the low-backscattering sea areas, i.e., both oil slicks and look-alikes, the average SNR estimated over the corresponding ROIs significantly varies along the SAR dataset depending on the damping properties and incidence angle. Oil slicks (dashed green boxes), call for an average SNR that lies in the range of 1.5 dB–4.2 dB, while the look-alikes (dashed orange boxes), result in general in lower SNR values falling in the range of 0.5 dB–2.8 dB. Note that these values can be either lower or larger than the oil slick ones. Hence, for the purposes of the spectral analysis, it can be concluded that reference slick-free sea surface samples are noise-free and oil slick samples can be considered to have an average SNR large enough to not affect significantly the spectral analysis, while look-alike samples are partly contaminated by noise and, therefore, particular attention must be paid in their spectral analysis.

The first experiment refers to the SAR image ID 1, see Table 1. The two low-backscattering ROIs are characterized by a great Δ value, i.e., >6 dB. The corresponding AACFs evaluated over the four ROIs are plotted in Figure 4a. The qualitative analysis clearly shows that the two $\mathrm{AACF_s}$ are narrow, while the two $\mathrm{AACF_{lb}}$ are wider. Although both $\mathrm{AACF_{lb}}$ show a distinct behavior with respect to $\mathrm{AACF_s}$, the oil-covered one shows the largest broadening due to the oil damping properties and the reduction of the energy transfer from wind to the sea waves. It can be also noted that both ROIs call for a large enough average SNR, i.e., >1.5 dB. The low-wind ROI, whose SNR exhibits large spatial variability, calls for the largest average SNR (2.8 dB). The low-backscattering ROIs are characterized by d_{E} values showing a significant deviation from the reference $\mathrm{AACF_s}$, i.e., 3.31 and 1.80 for the oil-covered and the low-wind ROI, respectively. The intrinsic sea variability, measured by computing the Euclidean distance d_{E} between the "Sea ref" and "Sea" AACFs, is equal to 0.30, witnessing a good overlapping of the corresponding $\mathrm{AACF_s}$. Note that a similar result is obtained by randomly changing the position of the two slick-free sea surface ROIs. In conclusion, the d_{E} values relevant to the low-wind and oil-covered ROIs are about 6 and 11 times greater than 0.30, respectively. The D_{rel} values are 90.9% and 83.3% over the oil-covered and low-wind ROIs, respectively.

Table 2. Experimental results relevant to the ROIs selected from the SAR dataset shown in Figure 2.

SAR Scene / Parameter	ID 1, See Figure 2a			ID 2, See Figure 2b			ID 3, See Figure 2c		
ROI	S	O	L	S	O	L	S	O	L
Incidence angle (°)	32			33–41			35		
σ^0_{VV} (dB)	−17.4 ± 1.3	−23.7 ± 1.4	−23.9 ± 1.3	−17.4 ± 1.4	−22.2 ± 2.2	−24.7 ± 1.3	−18.2 ± 1.5	−23.8 ± 1.3	–
SNR (dB)	7.7	1.6	2.8	9.0	4.2	0.7	6.8	1.5	–
Δ (dB)	–	6.3	6.5	–	4.8	7.3	–	5.6	–
d_E	0.30	3.31	1.80	0.01	0.55	0.14	0.08	1.15	–
D_{rel} (%)	–	90.9	83.3	–	98.2	92.9	–	93.0	–
SAR Scene / Parameter	**ID 4, see Figure 2d**			**ID 5, see Figure 2e**			**ID 6, see Figure 2f**		
ROI	S	O	L	S	O	L	S	O	L
Incidence angle (°)	35–40			38			43		
σ^0_{VV} (dB)	−21.7 ± 1.4	−28.5 ± 1.5	−25.7 ± 1.8	−18.4 ± 1.4	–	−27.2 ± 1.4	−20.9 ± 1.4	−27.9 ± 1.4	–
SNR (dB)	8.9	1.7	0.5	8.8	–	0.8	7.6	3.1	–
Δ (dB)	–	6.8	4.0	–	–	8.9	–	7.0	–
d_E	0.14	2.68	0.38	0.07	–	0.14	0.14	1.39	–
D_{rel} (%)	–	94.8	63.2	–	–	50.0	–	89.9	–

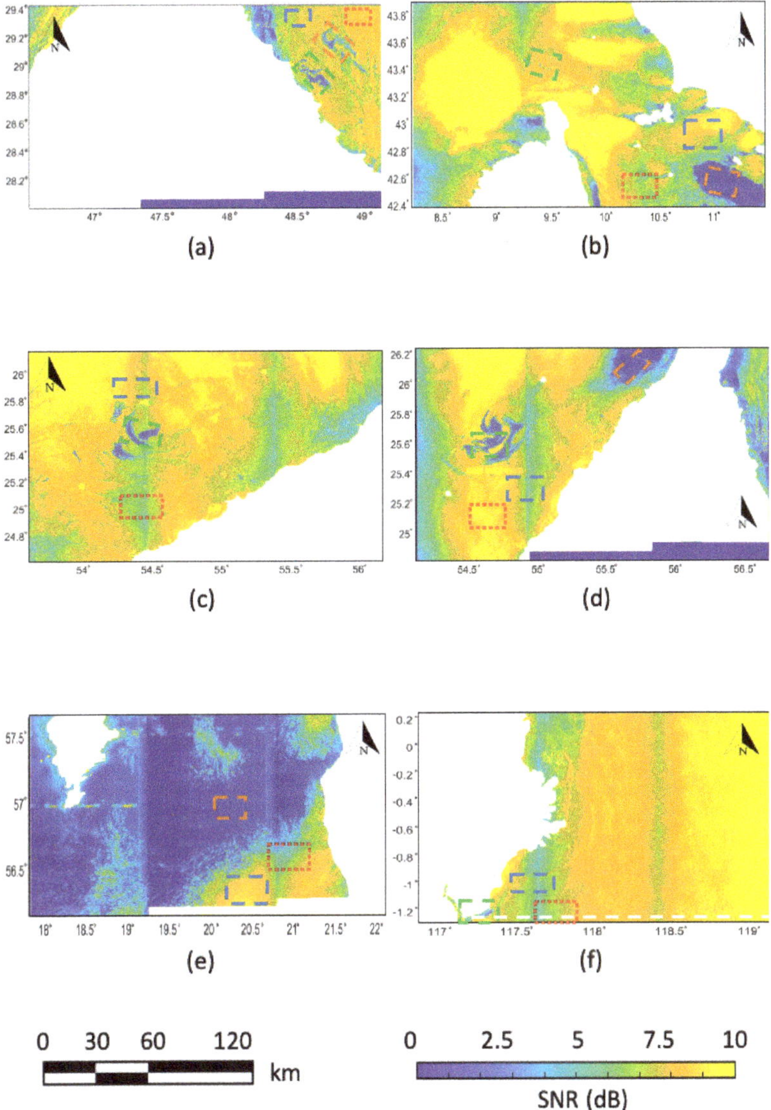

Figure 3. SNR images, in dB scale, evaluated over the SAR dataset shown in Figure 2. (**a**–**f**) refer to the SAR scenes labeled as data ID 1–6 in Table 1. Land is masked in white.

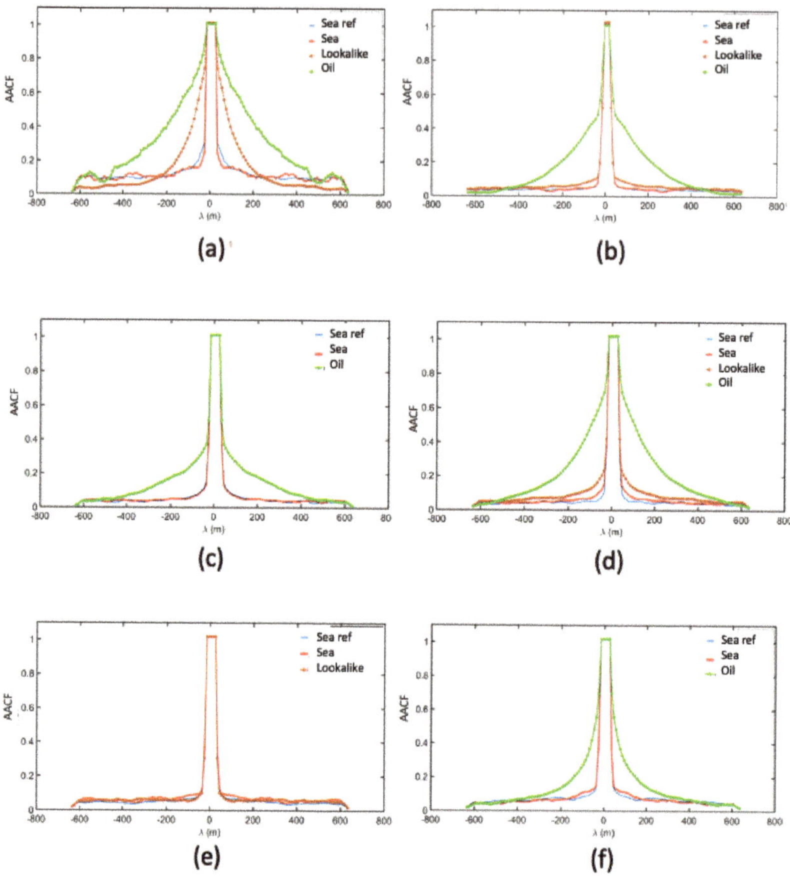

Figure 4. AACFs evaluated over the ROIs highlighted in Figure 2. (**a–f**) refer to the SAR scenes labeled as data ID 1–6 in Table 1. Note that each AACF is normalized to its maximum value.

The second experiment refers to the SAR image ID 2, see Table 1. The two low-backscattering ROIs show a great Δ value, i.e., about 5 dB and 7 dB for the oil-covered and low-wind ROI, respectively. The four AACFs are plotted in Figure 4b. The two $AACF_{lb}$ are wider than the two $AACF_s$, with the oil $AACF_{lb}$ resulting in the largest broadening. d_E is equal to 0.55 and 0.14 for the oil-covered and the low-wind ROI, respectively. This occurs even if the oil-covered ROI is characterized by an average SNR significantly larger than the low-wind ROI (4.2 dB versus 0.7 dB, respectively), witnessing that the broadening is mainly due to the different spectral properties of the low-backscattering sea areas rather than to the additive noise. The intrinsic variability of the slick-free sea surface results in d_E = 0.01, witnessing a very small $AACF_s$ variability. The d_E values relevant to $AACF_{lb}$ are at least one order of magnitude larger than the ones relevant to $AACF_s$. The D_{rel} values are 98.2% and 92.9% over the oil-covered and low-wind ROIs, respectively, showing a remarkable sensitivity of the AACF to the low-backscattering ROIs. In fact, $AACF_{lb}$ significantly deviates from $AACF_s$.

The third experiment refers to the SAR image ID 3 and 4, see Table 1. The two scenes refer to the same oil slick observed in two different dates and under different sea state conditions. The SAR image of Figure 2d also includes a low-wind ROI. Again, all the low-backscattering ROIs are characterized by a great Δ value, >4 dB. The corresponding AACF evaluated over the considered ROIs are plotted in Figure 4c,d. The $AACF_s$ are narrower than the $AACF_{lb}$, with the oil $AACF_{lb}$ showing the largest broadening. Even in this case, the largest broadening of the oil-covered ROI cannot be attributed to the additive noise since the oil-covered ROI calls for an average SNR which is more than three times larger than the low-wind one (1.7 dB versus 0.5 dB, respectively). For all the low-backscattering ROIs, d_E values are in the range of 0.38–2.68, always larger than the intrinsic sea variability that results in d_E equal to 0.08 and 0.14 for SAR image ID 3 and 4, respectively. The d_E values relevant to the oil and the low-wind $AACF_{lb}$ are more than one order of magnitude and just three times larger than the ones relevant to $AACF_s$, respectively. These results are consistent with the D_{rel} values which are about 94% and 63% for the oil-covered and low-wind ROIs, respectively.

The fourth experiment refers to the SAR image ID 5, see Table 1. The low-backscattering ROI shows a significantly great contrast, i.e., Δ about 9 dB. The three AACFs are plotted in Figure 4e, where it can be noted that $AACF_s$ and $AACF_{lb}$ overlap despite the remarkable contrast. Although it seems that the sensitivity of the AACF to this low-backscattering area is negligible, this is most likely due to the very low wind conditions that apply over the whole sea area. As a result, the longer-wave part of the spectrum does not change significantly from the slick-free sea surface to the low-backscattering ROI. It is interesting to note that $AACF_s$ and $AACF_{lb}$ call for almost the same width even if the average SNR of the low-backscattering ROI is more than 10 times lower than reference slick-free sea surface one (0.8 dB versus 8.8 dB, respectively), witnessing that additive noise is not the factor driving the AACF broadening. However, although $AACF_s$ and $AACF_{lb}$ appear overlapped, they result in $d_E = 0.14$, i.e., twice the intrinsic sea variability ($d_E = 0.07$). The corresponding D_{rel} value is 50.0% that, even though it is the smallest value among the whole SAR dataset, still represents a remarkable deviation.

The fifth experiment refers to the SAR image ID 6, see Table 1. Even in this case, the oil-covered ROI shows a pronounced contrast, i.e., $\Delta = 7$ dB. The three AACFs are plotted in Figure 4f. The oil $AACF_{lb}$ clearly deviates from the $AACF_s$. This broadening results in $d_E = 1.39$, 10 times larger than the intrinsic sea variability ($d_E = 0.14$). In this case, the oil-covered ROI calls for $D_{rel} = 89.9\%$.

The experimental results suggest that the NRCS and the AACF call for a different sensitivity to the incidence angle. Hence, a deeper investigation is due. The behavior of Δ and d_E with respect to the incidence angle is analyzed with reference to the scene depicted in Figure 2f. The two metrics are evaluated fixing the oil-covered ROI while moving the slick-free see surface ROI along the range transect to span the available range of incidence angles. As suggested in [48,50,51], for the incidence angles of interest (31°–43°), the slick-free sea surface NRCS variability is expected to be large, i.e., about 10 dB, while the oil NRCS variability is expected to be much smaller. Hence, Δ is expected to vary significantly with θ. The experimental results are depicted in Figure 5, where a dB scale is used. One can note, as expected, that Δ significantly varies with the incidence angle. This variability is practically negligible when d_E is considered.

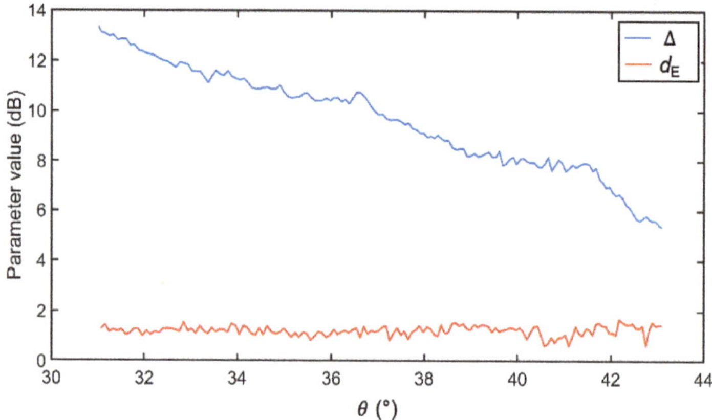

Figure 5. Behavior of the Δ and d_E parameters versus the incidence angle θ evaluated along with the dashed white range transect highlighted in Figure 2f.

5. Conclusions

In this study, a spectral analysis of low-backscattering sea areas of both anthropogenic and natural origin was addressed. This is of great interest for the marine pollution community. The rationale of the proposed analysis relies on the capability of low-backscattering sea features, including oil slicks, algal blooms, and low-wind regions, to modify the full-range sea surface spectrum. Hence, the sensitivity of the SAR image autocorrelation function evaluated along the azimuth direction, namely AACF, was investigated.

The AACF estimated over low-backscattering areas was analyzed with reference to a slick-free sea surface AACF. The deviation of the low-backscattering AACF from the slick-free sea surface one was quantified according to two objective metrics, i.e., the Euclidean distance d_E and the percentage relative difference D_{rel}. A comparison was also made with respect to the intrinsic sea variability, i.e., the AACF deviation between two different slick-free sea surface areas.

Experiments, undertaken on 6 S1 VV-polarized SAR images, collected in IW dual-polarimetric imaging mode, where known oil slicks and low-backscattering sea areas due to natural phenomena are observed under low-to-moderate wind conditions (2 m/s–7 m/s) in a broad range of incidence angles (≈30°–46°), showed that:

- The AACF is sensitive to different low-backscattering areas, with d_E and D_{rel} values which are at least twice and 50% larger, respectively, than the intrinsic sea surface variability;
- Among the low-backscattering sea areas, the oil slicks exhibit the largest AACF deviation with respect to the reference slick-free sea surface, with a maximum of $d_E = 3.31$ and $D_{rel} = 98.2\%$;
- The additive noise does not play a key role in broadening the AACF;
- The AACF is practically independent on the incidence angle while the backscattering contrast depends on it.

Author Contributions: Conceptualization, M.M.; methodology, F.N. and A.B.; software, V.C.; validation, V.C., A.B. and F.N.; formal analysis, V.C., A.B. and F.N.; data curation, V.C. and A.B.; writing–original draft preparation, A.B. and M.M.; writing–review and editing, F.N. and M.M; supervision, M.M. All authors have read and agreed to the published version of the manuscript.

Funding: This research was funded by the European Space Agency within the frame of ESA-MOST (Ministry of Science and Technology) Dragon 5 Cooperation (Monitoring harsh coastal environments and ocean surveillance using radar remote sensing sensors, Project ID 57979).

Acknowledgments: The authors thank the European Space Agency for providing Sentinel-1 SAR data free of charge through the Copernicus Scientific Hub. This paper was written during the COVID-19 pandemic in Italy and is dedicated to all people working in the healthcare system.

Conflicts of Interest: The authors declare no conflict of interest.

Abbreviations

The following abbreviations are used in this manuscript:

AACF	azimuth autocorrelation function
ECMWF	European centre for medium-range weather forecasting
ESA	European space agency
FFT	fast Fourier transform
ID	identifier
IFFT	inverse fast Fourier transform
IW	interferometric wide
NESZ	noise equivalent sigma zero
NRCS	normalized radar cross section
PSD	power spectral density
ROI	region of interest
S1	Sentinel-1
SAR	synthetic aperture radar
SNR	signal-to-noise ratio
TOPSAR	terrain observation with progressive scans SAR
VH	vertical transmit horizontal receive
VV	vertical transmit vertical receive

References

1. Visbeck, M. Ocean Science Research Is Key for a Sustainable Future. *Nat. Commun.* **2018**, *9*, 1–4. [CrossRef]
2. Li, X.; Guo, H.; Chen, K.-S.; Yang, X. *Advances in SAR Remote Sensing of Oceans*; Taylor & Francis Group: Boca Raton, FL, USA; CRC Press: Boca Raton, FL, USA, 2018; 344p.
3. Alpers, W.; Holt, W.B.; Zeng, K. Oil Spill Detection by Imaging Radars: Challenges and Pitfalls. *Remote Sens. Environ.* **2017**, *201*, 133–147. [CrossRef]
4. Migliaccio, M.; Huang, L.; Buono, A. SAR Speckle Dependence on Ocean Surface Wind Field. *IEEE Trans. Geosci. Remote Sens.* **2019**, *57*, 5447–5455. [CrossRef]
5. Velotto, D.; Bentes, C.; Tings, B.; Lehner, S. First Comparison of Sentinel-1 and TerraSAR-X Data in the Framework of Maritime Targets Detection: South Italy Case. *IEEE J. Ocean. Eng.* **2016**, *41*, 993–1006. [CrossRef]
6. Shen, H.; Perrie, W.; He, Y.; Liu, G. Wind Speed Retrieval From VH Dual-Polarization RADARSAT-2 SAR Images. *IEEE Trans. Geosci. Remote Sens.* **2014**, *52*, 5820–5826. [CrossRef]
7. Hanowski, P. *Copernicus Sentinel Data Access 2018 Annual Report*; COPE-SERCO-RP-19-0389; Frascati, Italy, 2019.
8. Vachon, P.W.; Wolfe, J. *GMES Sentinel-1 Analysis of Marine Applications Potential (AMAP)*; Defence Research and Development: Ottawa, ON, Canada, 2008.
9. Malenovsky, Z.; Rott, H.; Cihlar, J.; Schaepman, M.E.; García-Santos, G.; Fernandes, R.; Berger, M. Sentinels for Science: Potential of Sentinel-1, -2, and -3 Missions for Scientific Observations of Ocean, Cryosphere, and Land. *Remote Sens. Environ.* **2012**, *120*, 91–101. [CrossRef]
10. Misra, A.; Balaji, R. Simple Approaches to Oil Spill Detection Using Sentinel Application Platform (SNAP)-Ocean Application Tools and Texture Analysis: A Comparative Study. *J. Indian Soc. Remote Sens.* **2017**, *45*, 1065–1075. [CrossRef]
11. Trivero, P.; Fiscella, B.; Gomez, F.; Pavese, P. SAR Detection and Characterization of Sea Surface Slicks. *Int. J. Remote Sens.* **1998**, *19*, 543–548. [CrossRef]
12. Bertacca, M.; Berizzi, F.; Mese, E.D. A FARIMA-Based Technique for Oil Slick and Low-Wind Areas Discrimination in Sea SAR Imagery. *IEEE Trans. Geosci. Remote Sens.* **2005**, *43*, 2484–2493. [CrossRef]
13. Topouzelis, K. Oil Spill Detection by SAR Images: Dark Formation Detection, Feature Extraction and Classification Algorithms. *Sensors* **2008**, *8*, 6642–6659. [CrossRef]
14. Nunziata, F.; Buono, A.; Migliaccio, M. COSMO-SkyMed Synthetic Aperture Radar Data to Observe the Deepwater Horizon Oil Spill. *Sustainability* **2018**, *10*, 3599. [CrossRef]
15. Lang, H.; Zhang, X.; Xi, Y.; Zhang, X.; Li, W. Dark-Spot Segmentation for Oil Spill Detection Based on Multifeature Fusion Classification in Single-pol Synthetic Aperture Radar Imagery. *J. Appl. Remote Sens.* **2017**, *11*, 015006. [CrossRef]
16. Song, H.; Huang, B.; Zhang, K. A Globally Statistical Active Contour Model for Segmentation of Oil Slick in SAR Imagery. *IEEE J. Selected Topics Appl. Earth Obs. Remote Sens.* **2013**, *6*, 2402–2409. [CrossRef]

17. Girard-Ardhuin, F.; Mercier, G.; Collard, F.; Garello, R. Operational Oil-Slick Characterization by SAR Imagery and Synergistic Data. *IEEE J. Ocean. Eng.* **2005**, *30*, 487–495. [CrossRef]
18. Buono, A.; Nunziata, F.; de Macedo, C.R.; Velotto, D.; Migliaccio, M. A Sensitivity Analysis of the Standard Deviation of the Co-polarized Phase Difference for Sea Oil Slick Observation. *IEEE Trans. Geosci. Remote Sens.* **2019**, *57*, 2022–2030. [CrossRef]
19. Skrunes, S.; Brekke, C.; Eltoft, T. Characterization of Marine Surface Slicks by Radarsat-2 Multipolarization Features. *IEEE Trans. Geosci. Remote Sens.* **2014**, *52*, 5302–5319. [CrossRef]
20. Li, H.; Perrie, W.; He, Y.; Wu, J.; Luo, X. Analysis of the Polarimetric SAR Scattering Properties of Oil-Covered Waters. *IEEE J. Selected Topics Appl. Earth Obs. Remote Sens.* **2015**, *8*, 3751–3759. [CrossRef]
21. Yin, J.; Yang, J.; Zhou, Z.-S.; Song, J. The Extended Bragg Scattering Model-Based Method for Ship and Oil-Spill Observation Using Compact Polarimetric SAR. *IEEE J. Selected Topics Appl. Earth Obs. Remote Sens.* **2015**, *8*, 3760–3772. [CrossRef]
22. Clemente-Colón, P.; Yan, X.-H. Low-Backscatter Ocean Features in Synthetic Aperture Radar Imagery. *Johns Hopkins APL Tech. Digest* **2000**, *21*, 116–121.
23. Meng, H.; Wang, X.; Chong, J.; Wei, X.; Kong, W. Doppler Spectrum-Based NRCS Estimation Method for Low-Scattering Areas in Ocean SAR Images. *Remote Sens.* **2017**, *9*, 219. [CrossRef]
24. Raney, R.K. Theory and Measure of Certain Image Norms in SAR. *IEEE Trans. Geosci. Remote Sens.* **1985**, *23*, 343–348. [CrossRef]
25. Nunziata, F.; Migliaccio, M. On the COSMO-SkyMed PingPong Mode to Observe Metallic Targets at Sea. *IEEE J. Ocean. Eng.* **2013**, *38*, 71–79. [CrossRef]
26. Nunziata, F.; Migliaccio, M.; Li, X.; Ding, X. Coastline Extraction Using Dual-Polarimetric COSMO-SkyMed PingPong Mode SAR Data. *IEEE Geosci. Remote Sens. Lett.* **2014**, *11*, 104–108. [CrossRef]
27. Elfouhaily, T.; Guérin, C.A. A Critical Survey of Approximate Scattering Wave Theories From Random Rough Surfaces. *Waves Random Med.* **2004**, *14*, R1–R40. [CrossRef]
28. Nunziata, F.; Sobieski, P.; Migliaccio, M. The Two-Scale BPM Scattering Model for Sea Biogenic Slicks Contrast. *IEEE Trans. Geosci. Remote Sens.* **2009**, *47*, 1949–1956. [CrossRef]
29. Lemaire, D.; Sobieski, P.; Craeye, C.; Guissard, A. Two-scale models for rough surface scattering: Comparison between the boundary perturbation method and the integral equation method. *Radio Sci.* **2002**, *37*, 1001. [CrossRef]
30. Gade, M.; Alpers, W.; Hühnerfuss, H.; Wismann, V.R.; Lange, A. On the Reduction of the Radar Backscatter by Oceanic Surface Films: Scatterometer Measurements and Their Theoretical Interpretation. *Remote Sens. Environ* **1998**, *66*, 52–70. [CrossRef]
31. Boukabara, S.A.; Eymard, L.; Guillou, C.; Lemaire, D.; Sobieski, P.; Guissard, A. Development of a modified two-scale electromagnetic model simulating both active and passive microwave measurements: Comparison to data remotely sensed over the ocean. *Radio Sci.* **2002**, *37*, 16-1–16-11. [CrossRef]
32. Guissard, A.; Sobieski, P.; Baufays, C. A unified approach to bistatic scattering for active and passive remote sensing of rough ocean surfaces. *Trends Geophys. Res.* **1992**, *1*, 43–68.
33. Grieco, G.; Lin, W.; Migliaccio, M.; Nirchio, F.; Portabella, M. Dependency of the Sentinel-1 Azimuth Wavelength Cut-off on Significant Wave Height and Wind Speed. *Int. J. Remote Sens.* **2016**, *37*, 5086–5104. [CrossRef]
34. Lombardini, P.; Fiscella, B.; Trivero, P.; Cappa, C.; Garrett, W. Modulation of the Spectra of Short Gravity Waves by Sea Surface Films: Slick Detection and Characterization With a Microwave Probe. *J. Atmos. Ocean. Technol.* **1989**, *6*, 882–890. [CrossRef]
35. Hasselmann, K.; Hasselmann, S. On the Nonlinear Mapping of an Ocean Wave Spectrum into a Synthetic Aperture Radar Image Spectrum and Its Inversion. *J. Geophys. Res. Oceans* **1991**, *96*, 10713–10729. [CrossRef]
36. Vachon, P.W.; Raney, R.K. Ocean Waves and Optimal SAR Processing: Don't Adjust the Focus! *IEEE Trans. Geosci. Remote Sens.* **1992**, *30*, 627–629. [CrossRef]
37. Vachon, P.W.; Krogstad, H.E.; Paterson, J.S. Airborne and Spaceborne Synthetic Aperture Radar Observations of Ocean Waves. *Atmos. Ocean* **1994**, *32*, 83–112. [CrossRef]
38. Kerbaol, V.; Chapron, B.; Vachon, P.W. Analysis of ERS-1/2 Synthetic Aperture Radar Wave Mode Imagettes. *J. Geophys. Res.* **1998**, *103*, 7833–7846. [CrossRef]
39. Corcione, V.; Grieco, G.; Portabella, M.; Nunziata, F.; Migliaccio, M. A Novel Azimuth Cutoff Implementation to Retrieve Sea Surface Wind Speed From SAR Imagery. *IEEE Trans. Geosci. Remote Sens.* **2019**, *57*, 3331–3340. [CrossRef]
40. Mera, D.; Cotos, J.M.; Varela-Pet, J.; Garcia-Pineda, O. Adaptive Thresholding Algorithm Based on SAR Images and Wind Data to Segment Oil Spills Along the Northwest Coast of the Iberian Peninsula. *Marine Poll. Bullet.* **2012**, *64*, 2090–2069. [CrossRef] [PubMed]
41. Amos, J. Satellite Imagery Reveals Scope of Last Week's Oil Spill in Kuwait. Skytruth Report. 2017. Available online: https://skytruth.org/2017/08/satellite-imagery-reveals-scope-of-last-weeks-massive-oil-spill-in-kuwait/ (accessed on 11 March 2020).
42. ESI. Major Oil Spill in the Mediterranean. European Space Imaging Press. 2018. Available online: https://www.euspaceimaging.com/major-oil-spill-in-the-mediterranean/ (accessed on 11 March 2020).
43. ESA. Applications: Mediterranean Slick. European Space Agency Press. 2018. Available online: https://www.esa.int/ESA_Multimedia/Images/2018/10/Mediterranean_slick (accessed on 11 March 2020).
44. Thomas, C. Oil Spill in the Persian Gulf. Shytruth Report; 2017. Available online: https://skytruth.org/2017/03/oil-spill-in-the-persian-gulf/ (accessed on 15 July 2019).

45. ESA. Applications: Earth From Space—Gotland Baltic Blooms. European Space Agency Press. 2019. Available online: http://www.esa.int/ESA_Multimedia/Videos/2019/12/Earth_from_Space_Gotland_Baltic_blo-\oms (accessed on 11 March 2020).
46. Koto, J.; Putrawidjaja, M. Subsea Pipeline Damaged in Balikpapan Bay Caused by Anchor Load. *J. Subsea Offshore Sci. Eng.* **2018**, *14*, 6–12.
47. Setiani, P.; Ramdani, F. Oil Spill Mapping Using Multi-sensor Sentinel Data in Balikpapan Bay, Indonesia. In Proceedings of the 2018 4th IEEE International Symposium on Geoinformatics (ISyG), Malang, Indonesia, 17–18 November 2018; pp. 1–4.
48. Gouveia, N.D.A.; Alves, F.C.; Pereira, L.D.O. Pre-processing of Sentinel-1 C-Band SAR Images Based on Incidence Angle Correction for Dark Target Detection. *Remote Sens. Lett.* **2019**, *10*, 939–948. [CrossRef]
49. ESA. Sentinel-1 Product Specification. *Sentinel-1 User Handbook*, 2016. Available online: https://sentinels.copernicus.eu/documents/247904/685163/Sentinel-1_User_Handbook (accessed on 10 September 2020).
50. Nunziata, F.; de Macedo, C.R.; Buono, A.; Velotto, D.; Migliaccio, M. On the Analysis of a Time Series of X-band TerraSAR-X SAR Imagery Over Oil Seepages. *Int. J. Remote Sens.* **2019**, *40*, 3623–3646. [CrossRef]
51. Topouzelis, K.; Singha, S.; Kitsiou, D. Incidence angle Normalization of Wide Swath SAR Data for Oceanographic Applications. *Open Geosci.* **2016**, *8*, 450–464. [CrossRef]

Article

Vertical Migration of the Along-Slope Counter-Flow and Its Relation with the Kuroshio Intrusion off Northeastern Taiwan

Yuanshou He [1,2], Po Hu [1,3,4], Yuqi Yin [1,3,4], Ze Liu [1,3,4], Yahao Liu [1,3,4], Yijun Hou [1,2,3,4,*] and Yuanzhi Zhang [5]

1 CAS Key Laboratory of Ocean Circulation and Waves, Institute of Oceanology, Chinese Academy of Sciences, Qingdao 266071, China; heyuanshou14@mails.ucas.edu.cn (Y.H.); hupo@qdio.ac.cn (P.H.); yinyuqi@qdio.ac.cn (Y.Y.); liuze@qdio.ac.cn (Z.L.); yhliu@qdio.ac.cn (Y.L.)

2 School of Earth and Planetary Sciences, University of Chinese Academy of Sciences, Beijing 100049, China

3 Laboratory for Ocean and Climate Dynamics, Pilot National Laboratory for Marine Science and Technology, Qingdao 266237, China

4 Center for Ocean Mega-Science, Chinese Academy of Sciences, Qingdao 266071, China

5 Nanjing University of Information Science and Technology, Pukou District, Nanjing 210044, China; yuanzhizhang@cuhk.edu.hk

* Correspondence: yjhou@qdio.ac.cn

Received: 24 September 2019; Accepted: 8 November 2019; Published: 9 November 2019

Abstract: Based on satellite and analysis data and in situ observations acquired during May 23, 2017 to May 19, 2018, the spatiotemporal variations of the along-slope counter-flow off northeastern Taiwan were investigated. It was observed that the along-slope counter-flow in the subsurface layer was uplifted and lowered significantly during the study period. The counter-flow was significantly uplifted (lowered) while the sea surface was during an interval of positive (negative) geostrophic velocity anomaly (GVA) curl. The vertical migration of the counter-flow was also found closely linked with the Kuroshio intrusion (KI) to the northeast of Taiwan. The depths of both the upper boundary and the axis of the counter-flow were found proportional to the KI variance along the western continental slope off northeastern Taiwan. More importantly, it was established that the variation of the KI to the northeast of Taiwan had better correlation with the counter-flow than the Kuroshio derived from altimetry data. Thus, further study of the variation and mechanism of the along-slope counter-flow is needed to improve the understanding and prediction of the KI in the area of northeastern Taiwan, as well as the biochemical systems and marine economy in the East China Sea in the future.

Keywords: ocean modeling; counter-flow; vertical migration; Kuroshio intrusion; marine economy

---

1. Introduction

The Kuroshio is the strongest western boundary current in the Pacific Ocean, transporting warm, salty and nutrient-rich water from the seas off eastern Philippines northward to the seas off eastern Japan [1,2]. The Kuroshio has significant influence on the marginal seas, atmosphere and climate while traveling northward along the continental shelf west of the Pacific Ocean [3–7]. It is a unique and significant phenomenon that the Kuroshio current intrudes onto the East China Sea shelf off northeastern Taiwan [5,8–10]. Recent studies indicate that the seas to the northeast of Taiwan are the source regions of the Kuroshio branch currents on the East China Sea shelf [11,12], which have considerable influence on the regional circulation [13–15], chemical hydrography [16,17], and biological systems [18–20]. Therefore, it is essential to investigate the detailed flow structures and their variations in the region off northeastern Taiwan. Despite the northward-flowing Kuroshio water, a unique along-slope counter-flow exists in the subsurface layer below the depth of 150 m [21,22]. As illustrated in Figure 1, this counter-flow is directed towards the southwest along the steep continental slope from the North Mien-Hwa Canyon (NMHC) to the Mien-Hwa Canyon (MHC) [22–24] before flowing southward into the I-Lan Bay [25].

The southwestward flow was first observed during a hydrographic survey [26] and the existence of a counter-flow in the subsurface layer was first proposed by Chuang and Wu [27]. The year-round existence of the counter-flow was later confirmed by further observations [22,23] and numerical simulations [21,28,29]. Based on multiple historical observations, the counter-flow was initially considered part of a cyclonic eddy in the subsurface layer off northeastern Taiwan [23,24,30,31], and the cyclonic eddy was found to be closely related with the upwelling systems off northeastern Taiwan [25,29–31]. Observational studies have indicated that this counter-flow is a quasi-steady phenomenon that exhibits considerable seasonal and intraseasonal variations [22–24]. Earlier cruise observations revealed that the along-slope counter-flow extends to the surface layer above the depth of 50 m during summer months and descends to depths below 150 m during winter months [23]. In addition, substantial intraseasonal variability has also been reported based on previous in situ observations [22,23,32,33] and numerical simulations [21].

However, the uplift and lowering of the counter-flow in the subsurface layer remain confusing and unresolved. Furthermore, flowing along the western continental slope (D1-D2, Figure 1), where the Kuroshio branches intrude onto the East China Sea shelf [12,34], to the northeast of Taiwan and showing considerable variation in the extent of its vertical migration, the southwestward counter-flow should be linked closely with the variation of the Kuroshio intrusion (KI); however, the relationship between the along-slope counter-flow and the KI off northeastern Taiwan remains unclear. Above all, it is important that the specifics of the variations of the counter-flow be revealed because this could help improve the understanding of the local flow structure and KI variances off northeastern Taiwan.

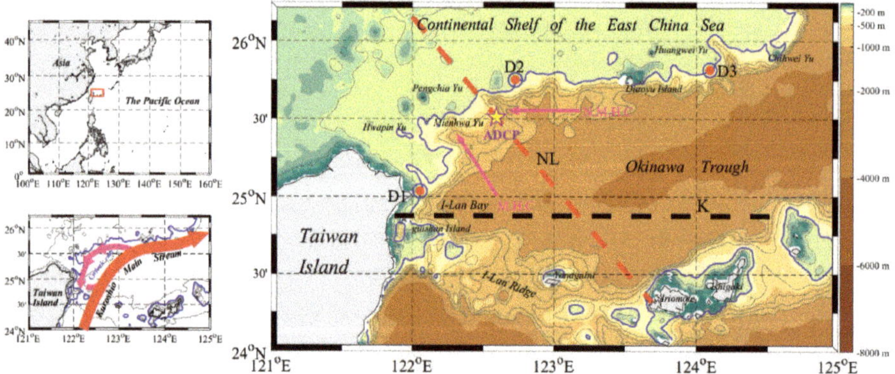

Figure 1. Location of the study area, the bathymetry off northeastern Taiwan, location of the mooring (25.51°N, 122.59°E; acoustic Doppler current profiler (ADCP) marked by yellow star), sections used in data analysis and a sketch of the horizontal flow pattern below the water depth of 200 m. The gray solid lines are isobaths of ETOPO1 [35], the blue solid line denotes the 200-m isobath and the red dots (D1 (25.03°N, 122.03°E), D2 (25.75°N, 122.72°E) and D3 (25.83°N, 124.10°E)) indicate segment points along the 200-m isobath. The red dashed line (NL) and the black dashed line (K) denote sections used in the data analysis. The panel in the lower-left corner shows the horizontal flow patterns off northeastern Taiwan in the subsurface layer below the water depth of 200 m. The deep red arrow denotes the main Kuroshio Current and the magenta arrows denote the counter-flow and the cyclonic eddy in the subsurface layer. The panel in the upper-left corner shows the location of the study area (red box).

2. Materials and Methods

2.1. Materials

2.1.1. Study Area

The East China Sea is one of the marginal seas west of the North Pacific Ocean, and seas off northeast Taiwan located at the southern East China Sea (Figure 1). The bathymetry as well as the flow structure in this area is complex. The sea water depth in the Okinawa Trough is deeper than 1000 m, while the sea water depth of the continental shelf is shallower than 200 m. Sea valleys (NMHC and MHC) are found on the steep continental slope off northeastern Taiwan. The Kuroshio current flows northward into the study area through the I-Lan ridge, and then collides with the continental slope off northeastern Taiwan, resulting in the significant Kuroshio water intrusion onto the East China Sea shelf. Although the Kuroshio waters intrude onto the shelf across the entire slope off northeastern Taiwan, the Kuroshio branches on the East China sea shelf were usually considered to intrude westward or northwestward onto the continental shelf mainly through the western continental slope (D1–D2) both during the summer months [9,12] and during the winter months [9,34].

2.1.2. In Situ Observations

In situ observations were carried out on the continental slope between the MHC and NMHC (Figure 1), where the angle of the local isobaths is approximately 30° from north. An acoustic Doppler current profiler (ADCP) mooring was deployed at a water depth of about 500.2 m with a standard deviation of 0.7 m, and first bin depth of 483.5 m from 23 May, 2017 to 19 September, 2017. The ADCP was again deployed at a water depth of 495.4 m with a standard deviation of 0.6 m and first bin depth of 478.6 m from 19 September, 2017 to 19 May, 2018; the bathymetry measured by ship at the in situ measurement sites was 621.0 m. The ADCP provided horizontal velocity records for 62 layers of the water column in 8 m vertical bins with a 1 h sampling interval. The uppermost six bins were excluded from the analysis because the data were contaminated by sidelobe reflection. Focusing on the circulation off northeastern Taiwan, a 36 h low-pass filter was applied to the remaining 56 vertical bins to remove tidal signals and other high-frequency fluctuations.

2.1.3. Satellite Altimeter Data

The all-satellite merged absolute dynamic topography (ADT), geostrophic velocity (GV), sea level anomaly (SLA) and geostrophic velocity anomaly (GVA) data from the Archiving, Validation, and Interpretation of Satellite Oceanographic (AVISO) dataset were used to provide geostrophic velocities off northeastern Taiwan. The AVISO dataset were derived from 15 altimeter missions: the TOPEX/Poseidon and Jason series; ERS-1, ERS-2, and ENVISAT; and Geosat Follow-On, Cryosat-2, Saral/AltiKa, Sentinel-3A, Sentinel-3B, and Haiyang-2A, CFOSAT. The resolution of the AVISO dataset is sufficient to resolve mesoscale eddy activity and mesoscale patterns off northeastern Taiwan [36] with 1-d temporal resolution and 0.25° spatial resolution. The daily satellite altimeter data are available from the Copernicus Marine Environment Monitoring Service (CMEMS) [37], the version of the datasets used in this study is "Global Ocean Gridded L4 Sea Surface Heights and Derived Variables Reprocessed (1993-ongoing)".

2.1.4. MODIS SST Data

The sea surface temperature (SST) level 3 datasets derived from the Moderate-resolution Imaging Spectroradiometer (MODIS) [38] observations were used to compare the surface temperature distribution patterns with the satellite altimeter data in the study. The first MODIS Flight Instrument, ProtoFlight Model or PFM, is integrated on the Terra (EOS AM-1) spacecraft. Terra successfully launched on 18 December, 1999. The second MODIS flight instrument, Flight Model 1 or FM1, is integrated on the Aqua (EOS PM-1) spacecraft; it was successfully launched on 4 May, 2002. These

MODIS instruments offer an unprecedented look at terrestrial, atmospheric, and ocean phenomenology for a wide and diverse community of users throughout the world. The weekly (8-d) daytime SST data with 4 km spatial resolution were used in this study.

2.1.5. The Analysis Data

The realistic ocean analysis datasets generated by the data assimilative global Hybrid Coordinate Ocean Model (HYCOM) [39] were also used to reveal the spatiotemporal patterns of the counter-flow in the region to the northeast of Taiwan, the version of the datasets used in this study is "GOFS 3.1 Global Analysis". The HYCOM data applies the Navy Coupled Ocean Data Assimilation (NCODA) system, which assimilates available satellite altimeter, sea surface wind stress, sea surface temperature observations, in situ sea surface temperatures, vertical temperature and salinity profiles from expendable bathythermographs (XBTs), Argo floats, and moored buoys. The three vertical diffusion mixing sub-models of the HYCOM are capable of resolving both geostrophic shear and ageostrophic wind-driven shear in the upper ocean [40]. The daily data are available with 0.08° horizontal resolution and 40 vertical z-levels, which are considered suitable for revealing accurate variation of the counter-flow and providing reliable detailed flow fields. The analysis data were validated, as shown in Figures 2 and 3.

2.2. Methods

The counter-flow is an along-slope current in the subsurface layer [21,23], and the direction of the local isobaths is about 30° from north (Figure 1). The cross-shelf shoreward direction is 150° (0° towards east, and 90° towards north), and the southwestward direction in Figure 2b,d are 240°. The along-isobath southwestward velocities reveals the along-slope counter-flow, and distinguish it from the surface velocities. To reveal and make quantitative estimation of the vertical migration of the observed along-slope counter-flow, we applied a formula to calculate the axis depth (D_a) of the counter-flow at the in situ site. The method was also used in the calculation of Kuroshio axis position [41]. The analysis velocities shallower than the sixth uppermost bin depth of the observation velocities were also excluded in the calculation.

$$D_a = \frac{\int_{z_B}^{z_U} v_{sw}(z) \cdot z dz}{\int_{z_B}^{z_U} v_{sw}(z) dz}, \tag{1}$$

where v_{sw} denotes the observed southwestward counter-flow velocity, z denotes water depth, and z_U and z_B denote the upper boundary and bottom depth of v_{sw}, respectively.

To make quantitative estimation of the Kuroshio mainstream, the Kuroshio intensity (INT) [41] derived from the altimeter data along the section K (Figure 1) during the observation period were calculated as follows.

$$\mathrm{INT} = \int_{x_W}^{x_E} v_g(x) dx, \tag{2}$$

where v_g denotes the normal geostrophic velocity of section K derived from the satellite altimeter data; x denotes distance from the western integral limits of section K, the x_W and x_E. are the western and eastern integral limits, respectively.

To make quantitative estimation the Kuroshio cross-shelf intrusion off northeastern Taiwan, the integral KI off northeastern Taiwan derived from the analysis data and integral surface Kuroshio intrusion (SKI) off northeastern Taiwan derived from the satellite altimeter data were given below.

$$\mathrm{KI} = \int_{S_1}^{S_n} v(s) ds, \tag{3}$$

$$\mathrm{SKI} = \int_{x_w}^{x_E} v_g(x)dx, \tag{4}$$

where $v(s)$ denotes cross-isobath (200-m isobath) component of horizontal velocity, s denote area of vertical grid cell from bottom to the surface along the 200-m isobath section; v_g indicates the normal geostrophic velocity of section D1–D2 and section D2–D3 derived from the satellite altimeter data; x denotes distance from the western integral limits of section D1–D2 and section D2–D3, the x_W and x_E are the western and eastern integral limits, respectively.

The least square regression method [42] was applied to review the linear trend of the Kuroshio Current, the along-slope counter-flow with the Kuroshio intrusion intensity off northeastern Taiwan.

$$\hat{y}_i = a + bx_i, \tag{5}$$

$$min \sum_{i=1}^{n} \delta_i^2 = min \sum (y_i - \hat{y}_i)^2, \tag{6}$$

$$R = \frac{Cov(x,y)}{\sqrt{var(x)var(y)}}, \tag{7}$$

where the x_i is the independent variable and y_i is the dependent variable, $\hat{y}_i$ is the fitting dependent variable, and δ_i is the error or residual. a and b are the linear regression coefficient satisfied the minimum δ_i^2. R is the related coefficient. $Cov(x,y)$ is the covariance of variable x and y, and $var(x)$, $var(y)$ are variance of variable x and y, respectively.

In addition, to reveal the surface cyclonic or anti-cyclonic GVA field variations, the GVA curl at the in situ site were calculated as follow.

$$V_{ga}curl = \frac{\partial v_{ga_y}}{\partial x} - \frac{\partial v_{ga_x}}{\partial y}, \tag{8}$$

where v_{ga} indicates the geostrophic velocity anomaly; x and y are the zonal and meridional direction.

3. Results

3.1. Vertical Migration of the Counter-Flow

The one year's in situ observations confirmed that the depths of the upper boundary and axis of the counter-flow experienced substantial fluctuations (Figures 2 and 3). The depth of the bottom of the counter-flow at the in situ observation sites was deeper than the depth of the deployed ADCP and therefore it could not be determined in this study. The depths of the upper boundary and axis of the counter-flow rose during the summer months (May–October) and fell during the winter months (November–April); the transition times were at the end of April and at the end of October (Figure 3). During the observation period, the mean depth of the upper boundary of the counter-flow was 141.9 (± 84.4) m and the mean depth of the axis was 307.4 (± 51.8) m. Specifically, during May–October 2017, the mean depths of the upper boundary and axis of the counter-flow were 102.0 (± 70.5) m and 269.6 (± 41.3) m, respectively. During November 2017 to April 2018, the mean depths of the upper boundary and axis of the counter-flow were 182.3 (± 77.9) m and 339.6 (± 37.4) m, respectively.

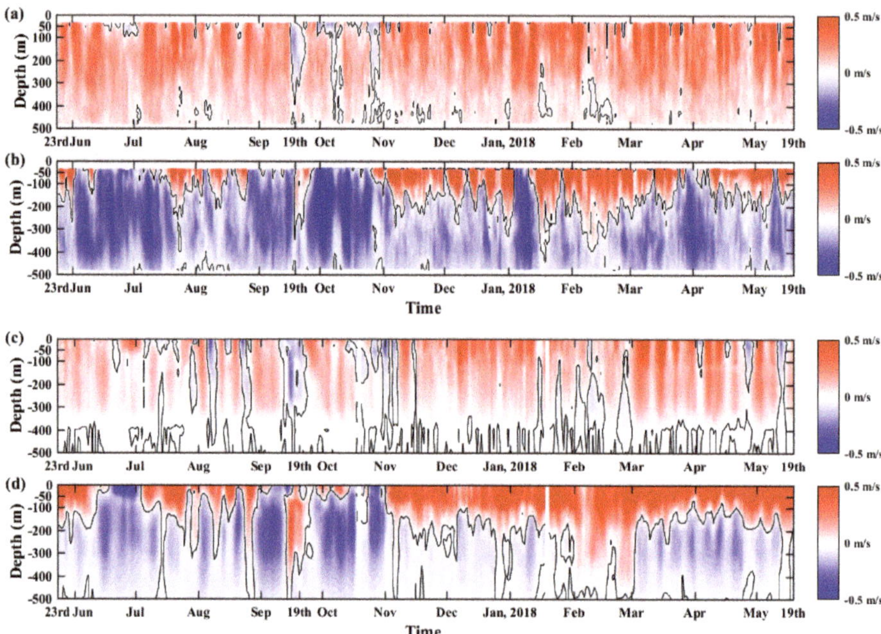

Figure 2. Current velocity distribution as a function of depth and time derived from (**a**,**b**) in situ observations and (**c**,**d**) analysis data. Panels a and c represent cross-shelf velocities, where red (blue) color indicates shoreward (seaward) velocity; panels b and d represent along-shelf velocities, where red (blue) color indicates northeastward (southwestward) velocity. Black contour denotes 0 m/s.

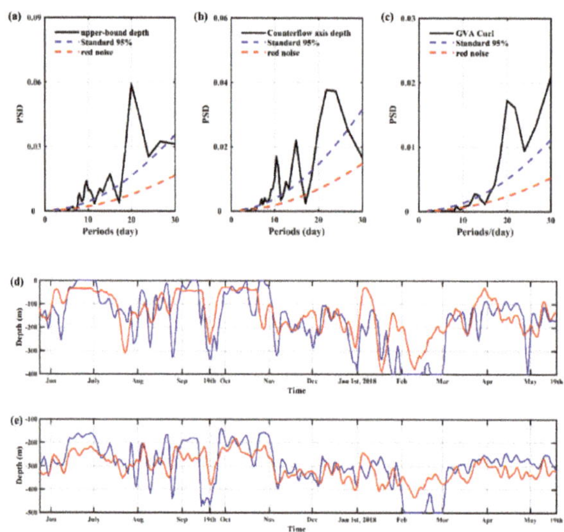

Figure 3. The power spectrum of the observed (**a**) counter-flow upper boundary depth, (**b**) counter-flow axis depth and (**c**) geostrophic velocity anomaly (GVA) curl at the in situ site, and time series of the observed (**d**) counter-flow upper boundary depth and (**e**) counter-flow axis depth. A 5-d filter was applied to the primary data. The red (blue) solid line was derived from in situ observations (analysis data).

In addition to the seasonal pattern, the depths of the counter-flow upper boundary and axis also rose and fell frequently within periods of tens of days (Figure 3); the 5-d smoothed daily time series showed near periodicity of 10, 15 and 20 d (Figure 3a,b). The near 10-d periodicity could be attributed to Kuroshio baroclinic instability waves, which are a characteristic of the Kuroshio current in the East China Sea [43,44]. The signal observed by James [43] in the East China Sea was 11-day, and they also pointed out that the continental shelf depth and core location attributed as well to their effects on the "stiffness" of the systems, and the model result reproduced was near 12-d. The near 15-d periodicity could be attributed to the lunisolar synodic fortnightly component of the tidal signal, which was not excluded in the 5-d low-pass filtering process. As for the near 20-d periodicity, the same signals were revealed in the daily GVA curl time series at the in situ site (Figure 3c). Furthermore, intraseasonal variations of the vertical migration of the counter-flow were also revealed in the GVA curl time series. The sea surface was during an interval of positive (negative) GVA curl while the counter-flow was significant uplifted (lowered). For instance, the counter-flow was uplifted significantly during June 2017 during an interval of positive GVA curl (Figures 2 and 4), while the counter-flow was lowered significantly during July 2017 during an interval of negative GVA curl. There are a total of five intervals of positive GVA curl and four intervals of negative GVA during the analysis period. Additionally, as revealed in Figure 4, the counter-flow could also be uplifted during winter months with significant positive GVA curl in the sea surface. A previous study [21] indicated that the flow field in the seas northeast of Taiwan fluctuates in a wide range of timescales, for the intraseasonal scale, the local structure was strongly influenced by the intraseasonal forces, such as westward-propagating mesoscale eddies east of Taiwan [36,45]. The significant counter-flow uplifted case during the winter months indicates that the intraseasonal forces imposed on the counter-flow off northeastern Taiwan is also significant during winter months.

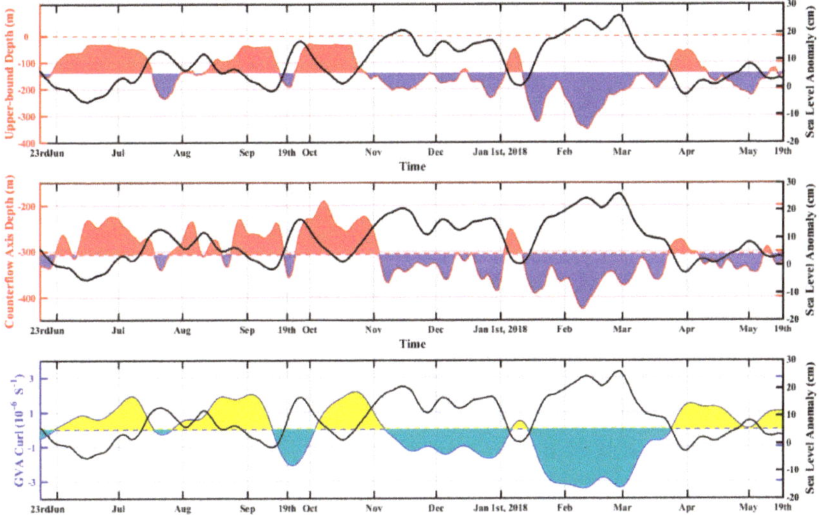

Figure 4. The daily sea level anomaly (SLA) (black lines), GVA curl (blue lines), and depths of the upper boundary and axis of the counter-flow (red lines). A 9-d low-pass filter was applied to the primary data. The SLA and GVA curl at the in situ site were derived from satellite altimeter data, and the depths of the upper boundary and axis of the counter-flow were derived from in situ observations. The SLA was defined as a deviation from mean sea level for the analysis period, and the seasonal variation was removed.

3.2. Horizontal and Vertical Patterns

The horizontal GV field, SST field, GVA field and the vertical velocity field during an uplifted case and a lowered case are presented in Figures 5 and 6, respectively. During June (July) 2017, a significant cyclonic (anti-cyclonic) GVA field covered the western continental slope off northeastern Taiwan, and the SST distribution variations (Figures 5b and 6b) fit with the surface GVA field variations. The vertical distribution of normal horizontal velocity along section NL (Figures 5d and 6d) indicates that the surface cyclonic or anti-cyclonic GVA fields near the western slope were dominant in the surface layer above the thermocline, while the along-slope counter-flow was dominant in the subsurface layer below the thermocline. Previous studies [46–48] have also highlighted that the positions of cyclonic and anticyclonic eddies in the surface layer off northeastern Taiwan can shift substantially. The horizontal distribution of the surface GVA field (Figures 5c and 6c) also helps us to distinguish the surface flow structures from the along-slope counter-flow in the subsurface layer. The variation and duration of each of the surface processes over the western continental slope to the northeast of Taiwan were revealed by the GVA curl, and it was found that the vertical migration of the counter-flow varies with the surface GVA curl (Figure 4).

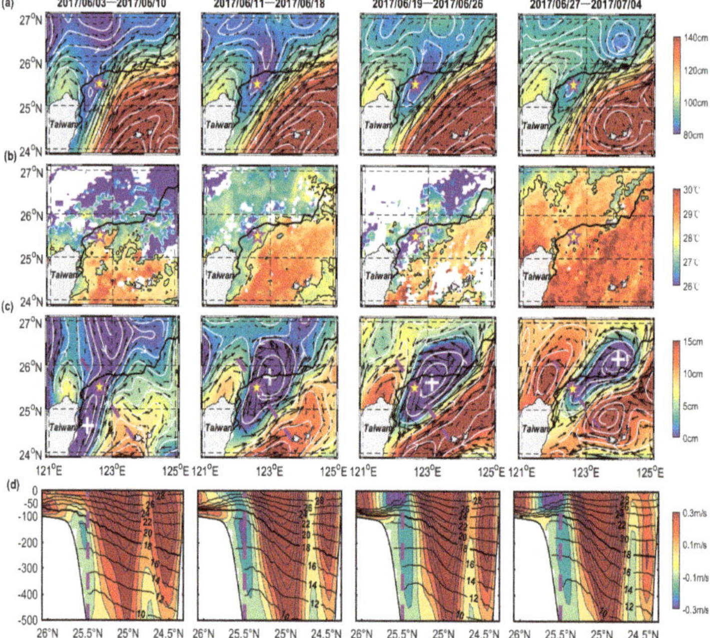

Figure 5. The horizontal absolute dynamic topography (ADT), geostrophic velocity (GV), sea level anomaly (SLA), geostrophic velocity anomaly (GVA) and sea surface Temperature (SST) distribution off northeastern Taiwan and the normal velocity distribution along section NL during an uplift case. (**a**) The horizontal ADT (colors) and GV (black arrows) distributions derived from satellite altimeter data. (**b**) The horizontal SST (colors) distributions derived from MODIS data, the white contours denote the 26.5 °C isotherm, and the black contours denote the 28 °C isotherm. (**c**) The horizontal SLA (colors) and GVA (black arrows) distributions derived from satellite altimeter data; the black line denotes the 200-m isobath, the in situ site is marked by a yellow star and the cyclonic GVA field center to the northeast of Taiwan is marked by a white '+' symbol. (**d**) Vertical distribution of normal velocity (colors) along section NL (Figure 1) derived from the analysis data; the black solid lines denote isotherms, the black bold solid line denotes the 18 °C isotherm, and the purple dashed line denotes the ADCP mooring site.

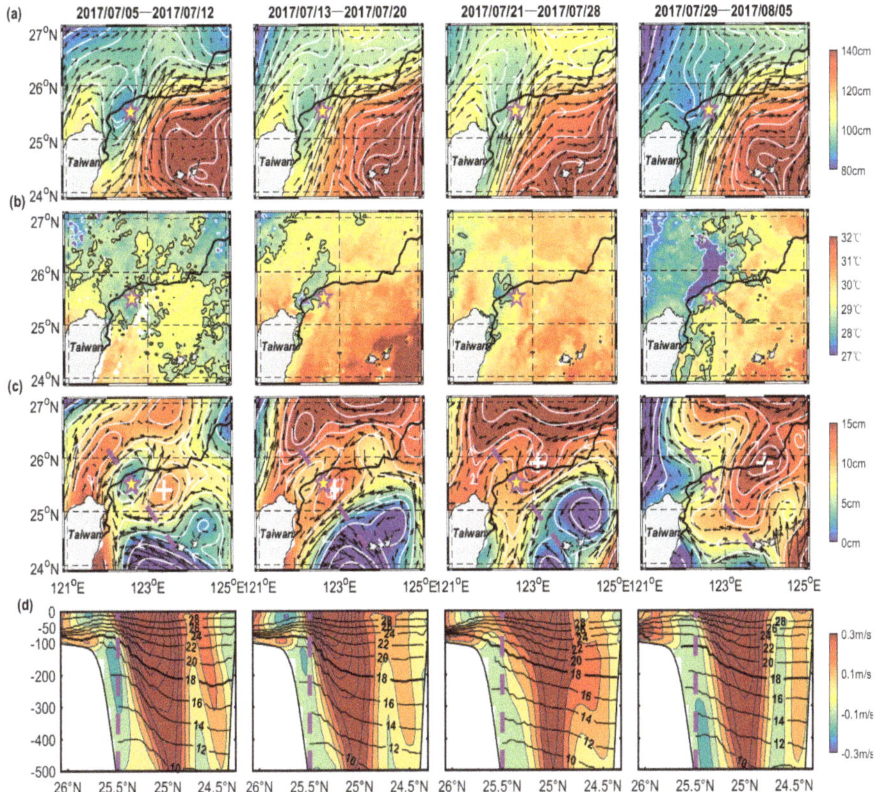

Figure 6. The horizontal absolute dynamic topography (ADT), geostrophic velocity (GV), sea level anomaly (SLA), geostrophic velocity anomaly (GVA) and sea surface temperature (SST) distribution off northeastern Taiwan and the normal velocity distribution along section NL during a lowered case. (**a**) The horizontal ADT (colors) and GV (black arrows) distributions derived from satellite altimeter data. (**b**) The horizontal SST (colors) distributions derived from MODIS data, the white contours denote the 27.5 °C isotherm, and the black contours denote the 29 °C isotherm. (**c**) The horizontal SLA (colors) and GVA (black arrows) distributions derived from satellite altimeter data; the black line denotes the 200-m isobath, the in situ site is marked by a yellow star and the anti-cyclonic GVA field center to the northeast of Taiwan is marked by a white '+' symbol. (**d**) Vertical distribution of normal velocity (colors) along section NL (Figure 1) derived from the analysis data; the black solid lines denote isotherms, the black bold solid line denotes the 18°C isotherm, and the purple dashed line denotes the ADCP mooring site.

4. Discussion

The Kuroshio Current transports warm, salty and nutrient-rich waters to the marginal seas west of the Pacific Ocean [3,16,17,49,50], and marine organisms and species in the seas are substantially sensitive to temperature, salinity, nitrate and phosphate [18–20,51,52]. Thus, the Kuroshio intrusion water onto the shelf is important for the biochemical systems and ecological environment in the East China Sea [50–53]. Previous studies indicate that the Kuroshio intrusion across the continental slope off northeastern Taiwan is closely related with the upwelling systems and the marine fishery off northeastern Taiwan [12,30,31,54]. Specifically, the nutrient-rich surface upwelling waters northeast Taiwan was supplied by the subsurface Kuroshio intrusion waters [55,56]. What is more, the near-shore Kuroshio branch current (NKBC) intrudes to the inner side of the East China Sea shelf and reaches the seas off eastern Zhejiang and Changjiang River estuary [54,57]. The Kuroshio subsurface water was observed in the upwelling systems off the Changjiang river estuary [54,58]. The off-shore branch current also transports nutrient-rich waters to the offshore regions in the East China Sea [49]. Both the phytoplankton, zooplankton and fish diversity in the East China Sea were significantly influenced by the Kuroshio bottom branches [19,58,59]. In addition, the nutrient supplement of Kuroshio intrusion waters is not only important for the prediction of the fisheries, but also important for the harmful algal blooms and red tides in the coastal area of the East China Sea [59,60].

The continental slope to the northeast of Taiwan is the source region of the Kuroshio branch currents on the continental shelf. Although the Kuroshio water intrude onto the shelf across the entire slope off northeastern Taiwan, and the cross-shelf intrusion across the eastern slope is also strong (Figure 7), previous studies indicate that the waters of the Kuroshio branch currents mainly intrude onto the East China Sea shelf through the western slope (D1-D2, Figure 1) of northeastern Taiwan [9,13,34,61]. Strong westward intrusion velocities were observed all year round in the subsurface layer of the MHC Chanel (Figure 1) [62,63]. During the winter months, the strong anticyclonic Kuroshio branch current intrude onto the East China Sea shelf close to the coast of northeastern Taiwan [9,34]. During the summer months, the surface velocity across the western slope is weak, while, the Kuroshio intrusion in the subsurface layer is strong [9]. The Kuroshio horizontal velocities in the subsurface layer were considered to be colliding with the western continental shelf in westward or northwestward direction, and the interior horizontal velocities $\vec{u_k}$ rotates clockwise with depth following the topography beta spiral [12,47].

Generally, The KI across the continental slope to the northeast of Taiwan shows significant seasonal variation [9,64]. The KI volume transport across the western continental slope becomes weak (strong) during summer (winter) months, although the subsurface Kuroshio intrusion is relatively strong, whereas the KI across the eastern continental slope becomes strong (weak) during summer (winter) months (Figure 7). The intraseasonal variation of the KI across the western and eastern continental slope is also in negative phase (Figure 7). Previous studies [36,65] have indicated that intraseasonal variation of the KI can be attributed to mesoscale eddies off eastern Taiwan, cyclonic (anticyclonic) mesoscale eddies off eastern Taiwan induce a strong (weak) KI across the western continental slope to the northeast of Taiwan, while the Kuroshio Current volume transport east of Taiwan is weakened (enhanced) [45,66]. Therefore, the KI to the northeast of Taiwan is strongly modulated by the Kuroshio Current off northeastern Taiwan. The INT time series and the SLA variations east of Taiwan during the analysis period was shown in Figure 8. The cyclonic mesoscale eddies were found east of Taiwan during July 2017, last third (LT) of October to LT of November 2017, LT of December 2017 to LT of February 2018, and LT of April to May 2018, whereas, anti-cyclonic mesoscale eddies east of Taiwan were found during Juny 2017, August to LT of September 2017, LT of November to LT of December 2017 and LT of February to middle third (MT) of March 2018. The INT time series was in response to the SLA east of Taiwan during the analysis period, namely, an anti-cyclonic (cyclonic) mesoscale eddy east of Taiwan induce a PE (NE) of the Kuroshio intensity. However, it worth noting that during LT of December 2017 to LT of February 2018, the INT revealed a slight increase while a significant cyclonic mesoscale eddy propagating westward east of Taiwan (Table 1), this is different from the rules above. Generally, the KI and surface GVA field northeast of Taiwan was in response to the Kuroshio and mesoscale eddies east of Taiwan, namely, an anti-cyclonic (cyclonic) mesoscale eddy east of Taiwan induce NE (PE) of KI through the western slope and cyclonic (anti-cyclonic) GVA field over the western slope northeast of Taiwan. However, it worth noting that there are significant cases different from this rules. The GVA field variation northeast of Taiwan is more complex, for instance, during case L2 (U4). The GVA field over the western slope were anti-cyclonic (cyclonic) while an anti-cyclonic (cyclonic) mesoscale eddy was found east of Taiwan (Table 1). These abnormal cases indicate that other undetermined seasonal and intraseasonal factors strongly influence the flow field off northeastern Taiwan.

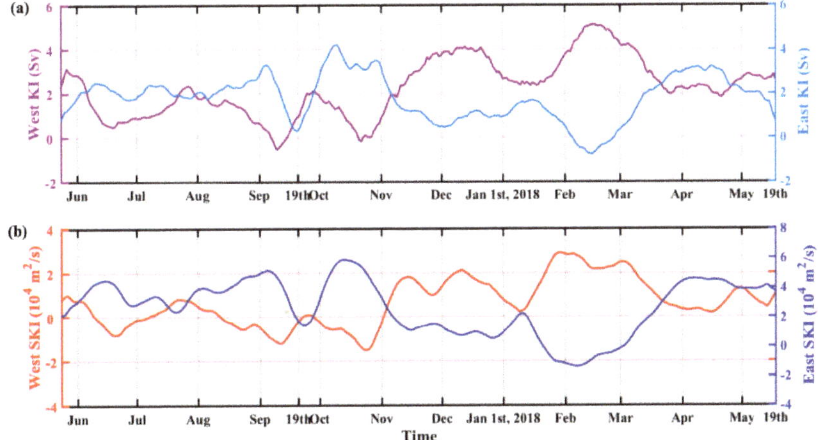

Figure 7. The KI across the western continental slope (D1–D2 in Figure 1) and across the eastern continental slope (D2–D3 in Figure 1). (**a**) The magnitude of the KI was derived from analysis data and calculated based on the vertical integral of the volume transport; (**b**) the magnitude of the SKI was derived from altimeter data and calculated based on the horizontal integral of the geostrophic velocity. A 15-d low-pass filter was applied to the primary data.

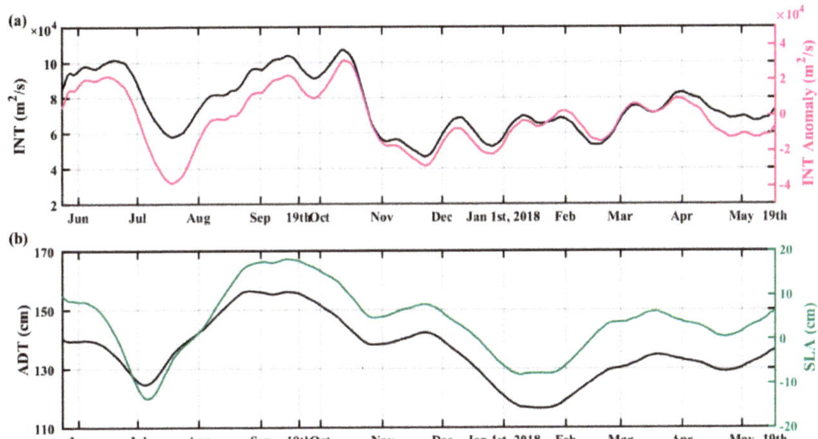

Figure 8. (**a**) The Kuroshio intensity (INT) and Kuroshio intensity anomaly through the K section derived from the altimeter data. The Kuroshio intensity anomaly was defined as a deviation from mean INT for the analysis period, and the seasonal variation was removed. (**b**) The ADT was the mean ADT derived from sea area of 22°N–24°N and 122°E–124°E east of Taiwan during the analysis period; the SLA was defined as a deviation from mean ADT for the analysis period, and the seasonal variation was removed. A 15-d low-pass filter was applied to the primary data.

Table 1. The information on different cases of the counter-flow and Kuroshio intrusion.

Case	Period	GVA Curl	Counter Flow Occurred (Yes, no)		Integral Kuroshio Intrusion (KI) (Sv)		Integral Surface Kuroshio Intrusion (SKI) (10^4 m^2/s)		INT (10^4 m^2/s)	Mesoscale Eddy East of Taiwan
			<100 m	subsurface	West (D1–D2)	East (D2–D3)	West (D1–D2)	East (D2–D3)		
U1	1 June–12 July, 2017	+	Yes	Uplifted	NE	PE	NE	PE	PE	Anti-cyclonic (June)
U2	29 July–16 September, 2017	+	Yes	Uplifted	NE	PE	NE	PE	PE	Anti-cyclonic (August–LT of Oct)
U3	3–26 October, 2017	+	Yes	Uplifted	NE	PE	NE	PE	PE	Anti-cyclonic (August–LT of October)
U4	3–11 January, 2018	+	Yes	Uplifted	NE	PE	NE	PE	PE	Cyclonic (LT of December–LT of February)
U5	24 March–7 April, 2018	+	Yes	Uplifted	NE	PE	NE	PE	PE	Anti-cyclonic (LT of February–MT of April)
L1	13–28 July, 2017	-	No	Lowered	PE	NE	PE	NE	NE	Cyclonic (July)
L2	17 September–2 October, 2017	-	No	Lowered	PE	NE	PE	NE	PE (slight decrease)	Anti-cyclonic (August–LT of October)
L3	27 October–2 January, 2017	-	No	Lowered	PE	NE	PE	NE	NE PE	Cyclonic (LT of October–LT of November) Anti-cyclonic (LT of November–LT of December)
L4	12 January–23 March, 2018	-	No	Lowered	PE	NE	PE	NE	PE PE	Cyclonic (LT of December–LT of February) Anti-cyclonic (LT of February–MT of April)

The PE (NE) denotes Positive (Negative) Extremums, symbol "-" and "+" indicates "Negative values" and "Positive values LT denotes "The Last Third", MT denotes "The Middle Third".

It is worth noting that previous observations [22–25], and observations in this study, as well as the simulations [21] supported the year-round existence of the strong along-slope counter-flow off northeastern Taiwan below the water depth of 150 m. The counter-flow flows southwestward along the western continental slope (D1–D2, Figure 1) in the subsurface layer, and more importantly, the counter-flow is a quasi-steady phenomenon in the subsurface layer that shows significant variation in its vertical scope. Therefore, it is essential to reveal the relationship between the along-slope counter-flow and the KI off northeastern Taiwan.

The least square regression method was applied to show the linear regression of the counterflow depths with the Kuroshio intrusion intensity off northeastern Taiwan (Figures 9 and 10). The related coefficient R of the standardized INT across section K (Figure 1) with the Western KI (D1–D2) was –0.643, while the R of the standardized depths of the upper boundary and axis of the counter-flow at the in situ site with the western KI were –0.750 and –0.791, respectively (Figure 9a). The West SKI between D1 and D2 derived from altimetry data was also used as validation in Figure 9b. The related coefficient R of the standardized INT across section K with the SKI was -0.678, respectively, while the R of the standardized depths of the upper boundary and axis of the counter-flow at the in situ site with the western SKI were –0.815 and –0.852, respectively (Figure 9b).

The related coefficient R of the standardized INT across section K (Figure 1) with the Eastern KI was 0.494, while the R of the standardized depths of the upper boundary and axis of the counter-flow at the in situ site with the East KI were 0.696 and 0.703, respectively (Figure 10a). The East SKI between D2 and D3 derived from altimetry data was also used as validation in Figure 10b. The related coefficient R of the standardized INT across section K with the East SKI was 0.624, respectively, while the R of the standardized depths of the upper boundary and axis of the counter-flow at the in situ site with the East SKI were 0.768 and 0.750, respectively (Figure 10b).

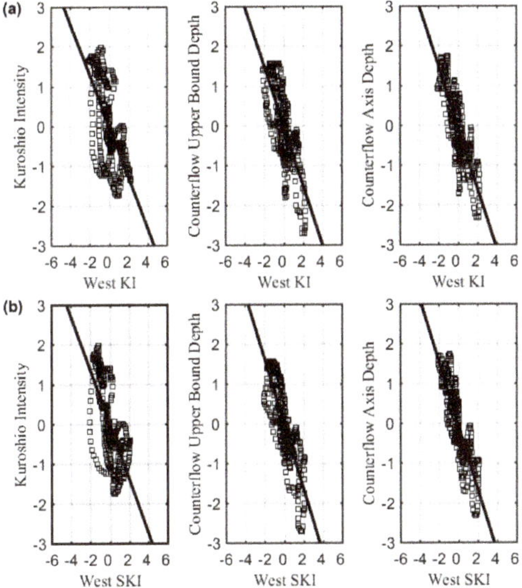

Figure 9. Scatter plots of standardized Kuroshio Intensity (INT), counter-flow upper boundary depth and counter-flow axis depth with the KI across the western continental slope (D1–D2, Figure 1) The INT was derived from satellite data along section K (Figure 1). The depths of the upper boundary and axis of the counter-flow were derived from in situ observations. (**a**) The West KI was derived from analysis data; (**b**) the West SKI was derived from satellite altimeter data. A 15-d low-pass filter was applied to the primary data. Locations of each section, segment points and in situ measurement sites are shown in Figure 1.

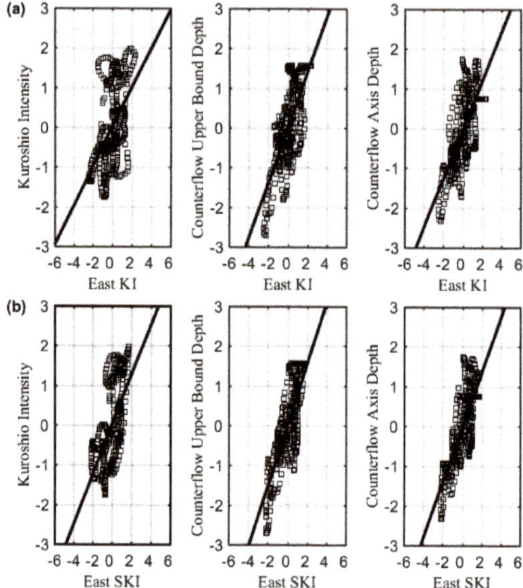

Figure 10. Scatter plots of standardized Kuroshio intensity (INT), counter-flow upper boundary depth and counter-flow axis depth with the KI across the Eastern continental slope (D2–D3, Figure 1). The INT were derived from satellite data along section K (Figure 1). The depths of the upper boundary and axis of the counter-flow were derived from in situ observations. (**a**) The East KI was derived from analysis data; (**b**) the East SKI was derived from satellite altimeter data. A 15-d low-pass filter was applied to the primary data. Locations of each section, segment points and in situ measurement sites are shown in Figure 1.

The results indicate that the depths of the upper boundary and axis of the counter-flow at the in situ site were proportional to the variance of the West KI and West SKI (Figure 9). The East KI and East SKI also showed high linear relation with the counter-flow depths (Figure 10). Moreover, the variation of the KI to the northeast of Taiwan had better correlation with the counter-flow than the Kuroshio derived from altimetry data. Although the INT derived from altimetry data exhibited linear trends with the KI, and the altimeter data are updated regularly on the open platform, the variable is inadequate for using as a linear index for the KI to the northeast of Taiwan.

The quasi-steady counter-flow is directed southwest along the western continental slope to the northeast of Taiwan in the subsurface layer and is frequently uplifted and lowered. This study has found that the vertical migration of the counter-flow was in well response to the local surface GVA curl, and that the vertical migration of the counter-flow exhibited reasonable linear correlation with the KI across the western continental slope off northeastern Taiwan, where the Kuroshio branch currents intrude onto the East China Sea shelf both during the summer and winter months [9,12,34]. Thus, further study of the variation and mechanism of the counter-flow is needed to improve the understanding of the KI to the northeast of Taiwan, and the counterflow variations would be helpful for oceanographers to make a better prediction of the KI off northeastern Taiwan, the KI and Kuroshio branch currents on the East China Sea shelf as well as the fisheries in the seas off northeastern Taiwan, eastern Zhejiang, and in the East China Sea.

5. Conclusions

Based on one year's sustained mooring observations, in conjunction with satellite and validated analysis data, this study revealed the vertical migration characteristics of the along-slope counter-flow to the northeast of Taiwan. The mean depths of the upper boundary and axis of the counter-flow at the in situ site were 141.9 (± 84.4) m and 307.4 (± 51.8) m, respectively. Specifically, during the summer half year, the counter-flow was uplifted with mean upper boundary depth of 102.0 (± 70.5) m and mean axis depth of 269.6 (± 41.3) m. During the winter half year, the counter-flow was lowered with mean upper boundary depth of 182.3 (± 77.9) m and mean axis depth of 339.6 (± 37.4) m. In addition to the seasonal pattern, the depths of the upper boundary and axis of the counter-flow also rose and fell frequently over periods of tens of days, e.g., with near 10-d and 20-d periodicity. The 20-d signal was revealed in the GVA curl time series, more importantly, the intraseasonal variation of the vertical migration of the counter-flow was also revealed in the sea surface GVA curl time series. There are five intervals of positive GVA curl and four intervals of negative GVA during the analysis period, and the counter-flow was significant uplifted during an interval of positive (negative) GVA curl in the sea surface. The sea surface GVA curl near the western slope well revealed the variations of the along-slope counter-flow in the subsurface layer. Additionally, the observations in this study also indicates that the strong intraseasonal forces imposed on the counter-flow can uplift the along-slope counter-flow to the surface layer during the winter months.

The depths of the upper boundary and axis of the counter-flow were also found closely linked with the KI off northeastern Taiwan, i.e., as the counter-flow became closer to the sea surface, the KI across the western continental slope (D1–D2, Figure 1) became weaker. The depths of the upper boundary and axis of the counter-flow were found to be proportional to the magnitude of the KI across the western continental slope to the northeast of Taiwan. Moreover, the variation of the KI to the northeast of Taiwan showed better correlation with the counter-flow than the Kuroshio derived from altimetry data. Thus, further study of the variation and mechanism of the along-slope counter-flow is needed to improve the understanding of the KI off northeastern Taiwan, and, a step further, the prediction of the biochemical systems and marine economy in the East China Sea.

Author Contributions: Y.H. (Yijun Hou), Y.H. (Yuanshou He) jointly conducted the research, data collecting, processing, analysis, and manuscript writing. Y.H. (Yuanshou He) worked on the data analysis, images preparation and drafting the manuscript; Z.L. (Ze Liu) and Y.L. (Yahao Liu) conducted the data collecting, and processing. Y.H. (Yijun Hou) and P.H. (Po Hu) designed and advised the research project and provided project support; Y.Y. (Yuqi Yin) and Y.Z. (Yuanzhi Zhang) assisted data analysis, reviewing and writing editing.

Funding: The research was funded by the National Natural Science Foundation of China (41630967 and 41476018), and the Global Change and Air-Sea Interaction Project (GASI-IPOVAI-01-06), and the National Natural Science Foundation of China (41776020, 41421005 and U1606402), and the Natural Scientific Foundation of Jiangsu Province (BK20181413).

Acknowledgments: The authors thank the crew of the R/V Xiang Yang Hong 18 No. NORC2017–02 and NORC2018–01 for the collection of the in situ observation data and the in situ flow field data available by Marine Data Science Center, IOCAS at Qingdao (https://doi.org/10.12157/QB.ADCP.TW.2017-2018. We want to thank the CAS Key Laboratory of Ocean Circulation and Waves, Institute of Oceanology, Chinese Academy of Sciences for their support. Furthermore, the critical and constructive comments and suggestions of the reviewers are very helpful and valuable for improving this paper. We would also like to express our gratitude to Xinhua Zhao, Fang Hu, Kai Wang and Bing Yang who helped with data collection. We thank Liwen Bianji, Edanz Goroup China for editing the English text of a draft of this manuscript.

Conflicts of Interest: The authors declare no conflict of interest.

References

1. Andres, M.; Jan, S.; Sanford, T.B.; Mensah, V.; Centurioni, L.R.; Book, J.W. Mean Structure and variability of the Kuroshio from northeastern Taiwan to southwestern Japan. *Oceanography* **2015**, *28*, 84–95. [CrossRef]
2. Hsueh, Y. The Kuroshio in the East China Sea. *J. Mar. Syst.* **2000**, *24*, 131–139. [CrossRef]
3. Hu, D.X.; Wu, L.X.; Cai, W.J.; Gupta, A.S.; Ganachaud, A.; Qui, B.; Gordon, A.L.; Lin, X.P.; Chen, Z.H.; Hu, S.J.; et al. Pacific western boundary currents and their roles in climate. *Nature* **2015**, *522*, 299–308. [CrossRef] [PubMed]
4. Isobe, A. Recent Advances in Ocean-Circulation Research on the Yellow Sea and East China Sea Shelves. *J. Oceanogr.* **2008**, *64*, 569–584. [CrossRef]
5. Wu, C.R.; Wang, Y.L.; Lin, Y.F.; Chao, S.Y. Intrusion of the Kuroshio into the South and East China Seas. *Sci. Rep.* **2017**, *7*, 7895. [CrossRef]

6. Nitani, H. Variation of the Kuroshio South of Japan. *J. Oceanogr. Soc. Jpn.* **1975**, *31*, 154–173.
7. Matsuno, T.; Lee, J.S.; Yanao, S. Influence of the Kuroshio on the water properties in the shelf. *Ocean Sci. Discuss.* **2009**, *6*, 741–764. [CrossRef]
8. Gawarkiewicz, G.; Jan, S.; Lermusiaux, P.F.J.; Mcclean, J.L.; Centurioni, L.; Duda, T.F.; Wang, J.; Yang, Y.J.; Sanford, T.; Lien, R.C.; et al. Circulation and intrusions northeast of Taiwan: Chasing and predicting uncertainty in the cold dome. *Oceanography* **2011**, *24*, 111–121. [CrossRef]
9. Liu, X.H.; Dong, C.M.; Chen, D.K.; Su, J.L. The Pattern and Variability of Winter Kuroshio Intrusion Northeast of Taiwan. *J. Geophys. Res. Oceans* **2014**, *119*, 5380–5394. [CrossRef]
10. Lee, J.S.; Matsuno, T. Intrusion of Kuroshio Water onto the Continental shelf of the East China Sea. *J. Oceanogr.* **2007**, *63*, 309–325. [CrossRef]
11. Yang, D.Z.; Yin, B.S.; Liu, Z.L.; Bai, T.; Qi, J.F.; Chen, H.Y. Numerical study on the pattern and origins of Kuroshio branches in the bottom water of southern East China Sea in summer. *J. Geophys. Res. Oceans* **2012**, *117*. [CrossRef]
12. Yang, D.Z.; Huang, R.X.; Yin, B.S.; Feng, X.R.; Chen, H.Y.; Qi, J.F.; Xu, L.J.; Shi, Y.L.; Cui, X.; Gao, G.D.; et al. Topographic Beta Spiral and Onshore Intrusion of the Kuroshio current. *Geophys. Res. Lett.* **2018**, *45*, 287–296. [CrossRef]
13. Yang, D.Z.; Yin, B.S.; Chai, F.; Feng, X.R.; Xue, H.J.; Gao, G.D. The Onshore Intrusion of Kuroshio Subsurface Water from February to July and a Mechanism for the Intrusion Variation. *Prog. Oceanogr.* **2018**, *167*, 97–115. [CrossRef]
14. Lie, H.J.; Cho, C.H. Recent advances in understanding the circulation and hydrography of the East China Sea. *Fish. Oceanogr.* **2002**, *11*, 318–328. [CrossRef]
15. Lie, H.J.; Cho, C.H. Progress in Oceanography Seasonal circulation patterns of the Yellow and East China Seas derived from satellite-tracked drifter trajectories and hydrographic observations. *Prog. Oceanogr.* **2016**, *146*, 121–141. [CrossRef]
16. Zhou, P.; Song, X.X.; Yuan, Y.Q.; Wang, W.T.; Cao, X.C.; Yu, Z.M. Intrusion pattern of the Kuroshio Subsurface Water onto the East China Sea continental shelf traced by dissolved inorganic iodine species during the spring and autumn of 2014. *Mar. Chem.* **2017**, *196*, 24–34. [CrossRef]
17. Guo, X.Y.; Zhu, X.; Long, Y.; Huang, D.J. Spatial variations in the Kuroshio nutrient transport from the East China Sea to south of Japan. *Biogeosciences* **2013**, *10*, 6403–6417. [CrossRef]
18. Shiah, F.K.; Liu, K.K.; Kao, S.J.; Gong, G.C. The coupling of bacterial production and hydrography in the southern East China Sea: Spatial patterns in spring and fall. *Cont. Shelf Res.* **2000**, *20*, 459–477. [CrossRef]
19. Xu, Y.; Ma, L.; Sun, Y.; Li, X.Z.; Wang, H.F.; Zhang, H.M. Spatial variation of demersal fish diversity and distribution in the East China Sea: Impact of the bottom branches of the Kuroshio current. *J. Sea Res.* **2019**, *144*, 22–32. [CrossRef]
20. Zhao, Y.; Yu, R.C.; Kong, F.Z.; Wei, C.J.; Liu, Z.; Geng, H.X.; Dai, L.; Zhou, Z.X.; Zhang, Q.C.; Zhou, M.J. Distribution Patterns of Picosized and Nanosized Phytoplankton Assemblages in the East China Sea and the Yellow Sea: Implications on the Impacts of Kuroshio Intrusion. *J. Geophys. Res. Oceans* **2019**, *124*, 1262–1276. [CrossRef]
21. Wu, C.R.; Lu, H.F.; Chao, S.Y. A numerical study on the formation of upwelling off northeast Taiwan. *J. Geophys. Res. Oceans* **2008**, *113*. [CrossRef]
22. Chuang, W.S.; Li, H.W.; Tang, T.Y.; Wu, C.K. Observations of the Countercurrent on the Inshore Side of the Kuroshio Northeast of Taiwan. *J. Oceanogr.* **1993**, *49*, 581–592. [CrossRef]
23. Tang, T.Y.; Tai, J.H.; Yang, Y.J. The Flow Pattern North of Taiwan and the Migration of the Kuroshio. *Cont. Shelf Res.* **2000**, *20*, 349–371. [CrossRef]
24. Wong, G.T.F.; Chao, S.Y.; Li, Y.H.; Shiah, F.K. The Kuroshio Edge Exchange Processes (KEEP) Study—An Introduction to Hypotheses and Highlights. *Cont. Shelf Res.* **2000**, *20*, 335–347. [CrossRef]
25. Hsu, P.C.; Zheng, Q.A.; Lu, C.Y.; Cheng, K.H.; Lee, H.J.; Ho, C.R. Interaction of coastal Countercurrent in I-Lan Bay with the Kuroshio Northeast of Taiwan. *Cont. Shelf Res.* **2018**, *171*, 30–41. [CrossRef]
26. Miyaji, K.; Inoue, N. Characteristics of the flow of the Kuroshio in the vicinity of Senkaku Islands. *Bull. Seikai Reg. Fish. Res. Lab.* **1983**, *60*, 50–70, (In Japanese with English Abstract).
27. Chuang, W.S.; Wu, C.K. Slope-Current fluctuations northeast of Taiwan winter 1990. *J. Oceanogr. Soc. Jpn.* **1991**, *47*, 185–193. [CrossRef]

28. Hsueh, Y.; Chern, C.S.; Wang, J. Blocking of the Kuroshio by the Continental Shelf Northeast of Taiwan. *J. Geophys. Res.* **1993**, *98*, 12351–12359. [CrossRef]
29. Shen, M.L.; Tseng, Y.H.; Jan, S. The formation and dynamics of the cold-dome off northeastern Taiwan. *J. Mar. Syst.* **2011**, *86*, 10–27. [CrossRef]
30. Jan, S.; Chen, C.C.; Tsai, Y.L.; Yang, Y.J.; Wang, J.; Chern, C.S.; Gawarkiewicz, G.; Lien, R.C.; Centurioni, L.; Kuo, J.Y. Mean Structure and Variability of the Cold Dome Northeast of Taiwan. *Oceanography* **2011**, *24*, 100–109. [CrossRef]
31. Xu, F.H.; Yao, Yuan.; Oey, L.; Lin, Y.L. Impacts of pre-existing ocean cyclonic circulation on sea surface chlorophyll-a concentrations off northeastern Taiwan following episodic typhoon passages. *J. Geophys. Res. Oceans* **2017**, *122*, 6482–6497. [CrossRef]
32. Tang, T.Y.; Hsueh, Y.; Yang, Y.J.; Ma, J.C. Continental Slope Flow Northeast of Taiwan. *J. Phys. Oceanogr.* **1999**, *29*, 1353–1362. [CrossRef]
33. Tang, T.Y.; Yang, Y.J. Low frequency current variability on the shelf break northeast of Taiwan. *J. Oceanogr.* **1993**, *49*, 193–210. [CrossRef]
34. Oey, L.Y.; Hsin, Y.C.; Wu, C.R. Why Does the Kuroshio Northeast of Taiwan Shift Shelfward in Winter? *Ocean Dyn.* **2010**, *60*, 413–426. [CrossRef]
35. Amante, C.; Eakins, B.W. ETOPO1 1 Arc-Minute Global Relief Model: Procedures, Data Sources and Analysis. NOAA Technical Memorandum NESDIS NGDC-24; National Geophysical Data Center, NOAA. Available online: https://www.ngdc.noaa.gov/mgg/global/global.html (accessed on 1 September 2009). [CrossRef]
36. Yin, Y.Q.; Lin, X.P.; He, R.Y.; Hou, Y.J. Impact of Mesoscale Eddies on Kuroshio Intrusion Variability Northeast of Taiwan. *J. Geophys. Res. Oceans* **2017**, *122*, 3021–3040. [CrossRef]
37. SSALTO/DUACS Experimental Product Handbook. Available online: https://www.aviso.altimetry.fr/en/data/product-information/aviso-user-handbooks.html (accessed on 1 September 2019).
38. NASA Goddard Space Flight Center, Ocean Ecology Laboratory, Ocean Biology Processing Group. MODIS-Terra Ocean Color Data. NASA Goddard Space Flight Center, Ocean Ecology Laboratory, Ocean Biology Processing Group, 2014. Available online: https://modis.gsfc.nasa.gov/data/dataprod/mod28.php (accessed on 28 July 2015). [CrossRef]
39. Wallcraft, A.J. HYCOM and Navy ESPC Future High Performance Computing Needs. Available online: https://www.hycom.org/hycom/documentation (accessed on 1 November 2017).
40. Chassignet, E.P.; Hurlburt, H.E.; Martin, O.M.; Halliwell, G.R.; Hogan, P.J.; Wallcraft, A.J.; Baraille, R.; Bleck, R. The HYCOM (HYbrid Coordinate Ocean Model) data assimilative system. *J. Mar. Syst.* **2007**, *65*, 60–83. [CrossRef]
41. Hsin, Y.C.; Qui, B.; Chiang, T.L.; Wu, C.R. Seasonal to interannual variations in the intensity and central position of the surface Kuroshio east of Taiwan. *J. Geophys. Res. Oceans* **2013**, *118*, 4305–4316. [CrossRef]
42. Geladi, P.; Kowalshi, B.R. Partial Least-Squares Regression: A Tutorial. *Anal. Chim. Acta* **1986**, *185*, 1–17. [CrossRef]
43. James, C.; Wimbush, M.; Ichikawa, H. Kuroshio Meanders in the East China Sea. *J. Phys. Oceanogr.* **1999**, *29*, 259–272. [CrossRef]
44. Johns, W.E.; Lee, T.N.; Zhang, D.X.; Zantopp, R. The Kuroshio East of Taiwan: Moored Transport Observations from the WOCE PCM-1 Array. *J. Phys. Oceanogr.* **2001**, *31*, 1031–1053. [CrossRef]
45. Lee, I.H.; Ko, D.S.; Wang, Y.H.; Wang, Y.H.; Centurioni, L.; Wang, D.P. The mesoscale eddies and Kuroshio transport in the western North Pacific east of Taiwan from 8-year (2003–2010) model reanalysis. *Ocean Dyn.* **2013**, *63*, 1027–1040. [CrossRef]
46. Cheng, Y.H.; Ho, C.R.; Zheng, Z.W.; Lee, Y.H.; Kuo, N.J. An Algorithm for Cold Patch Detection in the Sea off Northeast Taiwan Using Multi-Sensor Data. *Sensors* **2009**, *9*, 5521–5533. [CrossRef] [PubMed]
47. Yin, W.; Huang, D.J. Short-Term Variations in the Surface Upwelling off Northeastern Taiwan Observed via Satellite Data. *J. Geophys. Res. Oceans* **2019**, *124*, 939–954. [CrossRef]
48. Qin, D.D.; Wang, J.H.; Liu, Y.; Dong, C.M. Eddy Analysis in the Eastern China Sea Using Altimetry Data. *Front. Earth Sci.* **2015**, *9*, 709–721. [CrossRef]
49. Wang, W.T.; Yu, Z.M.; Song, X.X.; Yuan, Y.Q.; Wu, Z.X.; Zhou, P.; Cao, X.H. Intrusion Pattern of the Offshore Kuroshio Branch Current and Its Effects on Nutrient Contributions in the East China Sea. *J. Geophys. Res. Oceans* **2018**, *123*, 2116–2128. [CrossRef]

50. Gong, G.C.; Shiah, F.K.; Liu, K.K.; Wen, Y.H.; Liang, M.H. Spatial and temporal variation of chlorophyll a, primary productivity and chemical hydrography in the southern East China Sea. *Cont. Shelf Res.* **2000**, *20*, 411–436. [CrossRef]
51. Lin, Y.H.; Nakamura, Y. Distribution of planktonic copepods in the Kuroshio area of the East China Sea to the southwest of Kyushu in Japan. *Acta Oceanol. Sin.* **1993**, *12*, 445–456.
52. Shiozaki, T.; Kondo, Y.; Yuasa, D.; Takeda, S. Distribution of major diazotrophs in the surface water of the Kuroshio from northeastern Taiwan to south of mainland Japan. *J. Plankton Res.* **2018**, *40*, 407–419. [CrossRef]
53. Shiozaki, T.; Takeda, S.; Itoh, S.; Kodama, T.; Liu, X.; Hashihama, F.; Furuya, K. Why is Trichodesmium aboundant in the Kuroshio? *Biogeosciences* **2015**, *12*, 6931–6943. [CrossRef]
54. Umezawa, Y.; Yamaguchi, A.; Ishizaka, J.; Hasegawa, T.; Yoshimizu, C.; Tayasu, I.; Yoshimura, H.; Morii, Y.; Aoshima, T.; Yamawaki, N. Seasonal shifts in the contributions of the Changjiang River and the Kuroshio Current to nitrate dynamics in the continental shelf of the northern East China Sea based on a nitrate dual isotopic composition approach. *Biogeosciences* **2014**, *11*, 1297–1317. [CrossRef]
55. Liu, K.K.; Gong, G.C.; Lin, S.; Yang, C.Y.; Wei, C.L.; Pai, S.C.; Wu, C.K. The year-round upwelling at the shelf break near the northern tip of Taiwan as evidenced by chemical hydrography. *Terr. Atmos. Oceanic Sci.* **1992**, *3*, 243–275. [CrossRef]
56. Chang, Y.L.; Wu, C.R.; Oey, L.Y. Bimodal behavior of the seasonal upwelling off the northeastern coast of Taiwan. *J. Geophys. Res. Oceans* **2009**, *114*. [CrossRef]
57. Xu, L.J.; Yang, D.Z.; Benthuysen, J.A.; Yin, B.S. Key Dynamical Factors Driving the Kuroshio Subsurface Water to Reach the Zhejiang Coastal Area. *J. Geophys. Res. Oceans* **2018**, *123*, 9061–9081. [CrossRef]
58. Tseng, Y.F.; Lin, J.; Dai, M.; Kao, S.J. Joint effect of freshwater plume and coastal upwelling on phytoplankton growth off the Changjiang River. *Biogeosciences* **2014**, *11*, 409–423. [CrossRef]
59. Yu, R.C.; Zhang, Q.C.; Kong, F.Z.; Zhou, Z.X.; Chen, Z.F.; Zhao, Y.; Geng, H.X.; Dai, L.; Yan, T.; Zhou, M.J. Status, impacts and long-term changes of harmful algal blooms in the sea area adjacent to the Changjiang river estuary. *Oceanol. Limn. Sin.* **2017**, *48*, 1178–1186, (In Chinese with English Abstract).
60. Wang, J.H.; Wu, J.Y. Occurrence and potential risks of harmful algal blooms in the East China Sea. *Sci. Total Environ.* **2009**, *407*, 4012–4021. [CrossRef]
61. Liang, X.S.; Su, J.L. A two-layer model for the circulation of the East China Sea. *Donghai Mar. Sci.* **1994**, *12*, 1–20, (In Chinese with English Abstract).
62. Guo, B.H. Cold water and upwelling in the area around some capes of Taiwan island. *Adv. Mar. Sci.* **1986**, *3*, 3, (In Chinese with English Abstract).
63. Guo, B.H.; Lin, K. Characteristics of Circulation in the Continental shelf Area of the East China Sea in Winter. *Acta Oceanol. Sin. V* **1987**, *6*, 51–60.
64. Liu, X.H.; Chen, D.K.; Dong, C.M.; He, H.L. Variation of the Kuroshio Intrusion Pathways Northeast of Taiwan Using the Lagrangian Method. *Sci. China Earth Sci.* **2016**, *59*, 268–280. [CrossRef]
65. Vélez-Belchí, P.; Centurioni, L.R.; Lee, D.K.; Jan, S.; Niiler, P.P. Eddy induced Kuroshio intrusions onto the continental shelf of the East China Sea. *J. Mar. Res.* **2013**, *71*, 83–107. [CrossRef]
66. Yin, Y.Q.; Lin, X.P.; Hou, Y.J. Seasonality of the Kuroshio Intensity East of Taiwan Modulated by Mesoscale Eddies. *J. Mar. Syst.* **2019**, *193*, 84–93. [CrossRef]

Article

Sea Echoes for Airborne HF/VHF Radar: Mathematical Model and Simulation

Fan Ding [1], Chen Zhao [1,*], Zezong Chen [1,2] and Jian Li [1]

1 School of Electronic Information, Wuhan University, Wuhan 430072, China; dingf242@whu.edu.cn (F.D.); chenzz@whu.edu.cn (Z.C.); ivanlijian@whu.edu.cn (J.L.)

2 Collaborative Innovation Center for Geospatial Technology, Wuhan University, Wuhan 430072, China

* Correspondence: zhaoc@whu.edu.cn

Received: 17 October 2020; Accepted: 12 November 2020; Published: 15 November 2020

Abstract: Currently, shore-based HF radars are widely used for coastal observations, and airborne radars are utilized for monitoring the ocean with a relatively large coverage offshore. In order to take the advantage of airborne radars, the theoretical mechanism of airborne HF/VHF radar for ocean surface observation has been studied in this paper. First, we describe the ocean surface wave height with the linear and nonlinear parts in a reasonable mathematical form and adopt the small perturbation method (SPM) to compute the HF/VHF radio scattered field induced by the sea surface. Second, the normalized radar cross section (NRCS) of the ocean surface is derived by tackling the field scattered from the random sea as a stochastic process. Third, the NRCS is simulated using the SPM under different sea states, at various radar operating frequencies and incident angles, and then the influences of these factors on radar sea echoes are investigated. At last, a comparison of NRCS using the SPM and the generalized function method (GFM) is done and analyzed. The mathematical model links the sea echoes and the ocean wave height spectrum, and it also offers a theoretical basis for designing a potential airborne HF/VHF radar for ocean surface remote sensing.

Keywords: airborne HF/VHF radar; sea echo; mathematical model; radar cross section

1. Introduction

The sea echoes of HF or VHF ocean radars contain rich information about the sea surface since the length of the HF/VHF radio wave is very close to the wave length of gravity wave at the ocean surface [1]. On the one hand, shore-based HF radars are important components of coastal operational monitoring systems [2–4]. Many countries have utilized shore-based HF radars to obtain ocean current, wind and wave fields [5,6]. The maximum detection range of shore-based HF radars can reach 250 km, with the time resolution ranging from 10 min to 1 h and the spatial resolution varying from 300 m to 5 km. On the other hand, along with the development of electronic technology, airborne radars have been widely used for ocean remote sensing [7–9]. The size of radar is becoming smaller, and the cost of developing airborne radars is lower. In addition, an airborne VHF radar has been developed for forest remote sensing [10]. All these developments make it possible to design and develop an airborne HF/VHF radar to monitor the sea surface. The objective of this paper is to investigate the interaction mechanism of HF/VHF electromagnetic waves scattering from the ocean's surface, and this should provide a theoretical basis for designing novel airborne HF/VHF radars for ocean remote sensing.

Many scholars have analyzed the interaction mechanism between HF/VHF electromagnetic waves and ocean waves for shore-based HF radar. Barrick [11] adopted the small perturbation method (SPM) to compute the scattered field from the time-varying sea surface and derived the normalized radar cross section (NRCS) of the sea surface for monostatic HF radar. Subsequently, Johnstone [12] and Anderson [13] respectively extended Barrick's work to the configuration of shore-based bistatic HF radar. For the SPM, Hisaki [14] considered the effect of finite illumination area to derive the NRCS for shore-based monostatic HF radar. More recently, Hardman et al. [15] also presented the shore-based bistatic NRCS utilizing the SPM. Besides that, Srivastava and Walsh [16,17] proposed a generalized function method (GFM) to analyze the scattered field from the sea surface and derived the NRCS for the shore-based monostatic HF radar. Afterwards, the GFM was also extended to derive the NRCS for the shore-based bistatic HF radar [18–20]. It is noted that Silva et al. [21] modified the usual way of the GFM and derived a more general NRCS of the sea surface with arbitrary sea states.

For airborne HF/VHF radar, Bernhardt et al. [22,23] proposed the concept of HF Ground-Ionosphere-Ocean-Space (GIOS), and conducted experiments to observe the sea surface. Anderson [24] proposed the airborne passive HF radar which can be used to monitor the sea. Later, Chen et al. [25,26] theoretically analyzed the sea echoes of the shore-to-air bistatic HF radar. Meanwhile, Voronovich and Zavorotny [27] proved the possibility of extracting the wave height spectrum using airborne HF/VHF radars.

However, the theoretical study on the airborne HF/VHF radar is still in an initial stage. Voronovich and Zavorotny [27] analyzed the first-order interaction between HF/VHF radio waves and ocean waves, but they omitted the second-order information which is much more complicated than the first-order interaction and crucial for investigating the interaction mechanism between the HF/VHF radio waves and ocean surface waves. This paper analyzes the first- and second-order interactions which occur in the scattering of HF/VHF electromagnetic waves from the sea surface. First, a big square area of the sea surface is considered as the scattering patch. Taking into account the randomness of the sea surface, the ocean surface wave height is represented as the superposition of linear and nonlinear wave heights. Next, the SPM is employed to derive the scattered field from the sea surface. Then we obtain the NRCS of the sea surface for airborne HF/VHF radars. Finally, the theoretical NRCS of the sea surface is simulated with various parameters, such as sea states and radar operating frequencies.

This paper is organized as follows. Section 2 gives the description of the calculation of the scattered field. In Section 3, the NRCS of the sea surface is derived. Section 4 consists of the simulation of the NRCS and the analysis of the simulated sea echoes. Discussion and conclusions are presented in Sections 5 and 6, respectively.

2. Description of the Scattering Problem

2.1. The Review of the Description of Wave Heights

As shown in Figure 1, the geometry of the scattering patch is established using a three-dimensional Cartesian coordinate system, and the center of the scattering patch is set as the origin. The scattering patch is assumed to be a square area with a very large side length of L. The x axis is assumed as the projection direction of radar beam at the sea surface and the y axis is at the sea surface and perpendicular to the x axis. The z axis is vertical to the sea surface.

Then the sea surface wave height $z = f(x,y,t)$ can be expressed by Fourier series as:

$$z = f(x,y,t) = f^{(1)}(x,y,t) + f^{(2)}(x,y,t), \tag{1}$$

$$f^{(1)}(x,y,t) = \sum_{m,n,l} p_1(m,n,l)e^{iamx+iany-iwlt}, \tag{2}$$

$$f^{(2)}(x,y,t) = \sum_{m,n,l} p_2(m,n,l)e^{iamx+iany-iwlt}, \tag{3}$$

where m, n and l are integers between $-\infty$ and $+\infty$, $a = 2\pi/L$ and $w = 2\pi/T$; t denotes time; T is the temporal period of the Fourier expansion; $f^{(1)}(x,y,t)$ and $f^{(2)}(x,y,t)$ are Fourier series which denote linear and nonlinear wave heights [28], respectively; the superscripts 1 and 2 denote the first-order and second-order terms in the perturbational analysis, respectively; $p_1(m,n,l)$ and $p_2(m,n,l)$ are the Fourier coefficients of the linear and nonlinear wave heights, respectively.

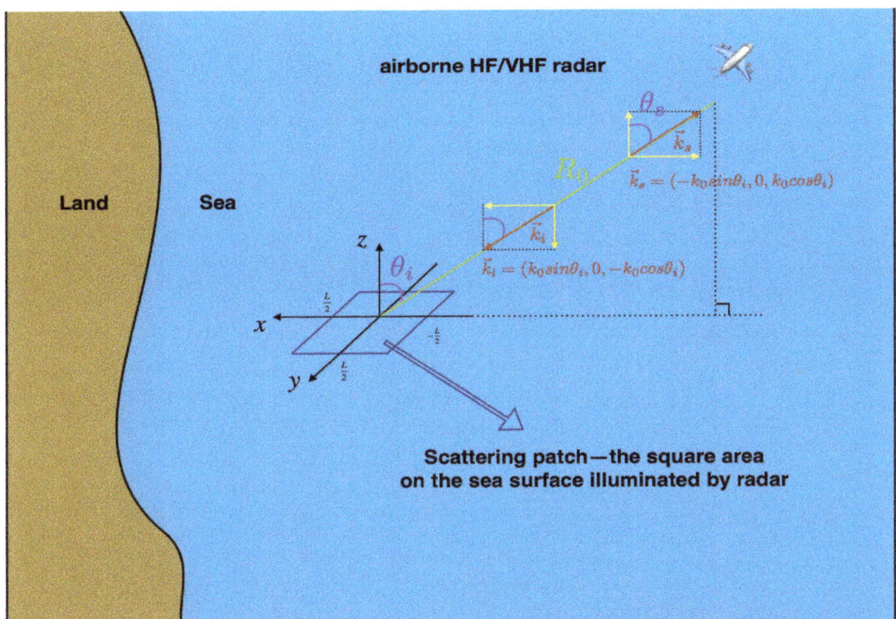

Figure 1. The geometry of the scattering patch. The square scattering patch is represented by the parallelogram whose sides are navy blue straight lines. The radar is located in the far zone of the scattering patch ($R_0 \gg L$). Backscattering is considered, i.e., $\theta_i = \theta_s$. Here the x-z plane is perpendicular to the sea surface and contains the center of the scattering patch and the point of the radar position. $\vec{k_i}$ and $\vec{k_s}$ represent the wave vectors of the incident and scattered fields, respectively.

When perturbational analysis is utilized to solve the hydrodynamic equations, it is found that $p_2(m,n,l)$ can be expressed using $p_1(m,n,l)$ [28]:

$$p_2(m,n,l) = \sum_{m',n',l'} \Gamma(\vec{k''},\omega'',\vec{k'},\omega') p_1(m',n',l') p_1(m-m',n-n',l-l'), \tag{4}$$

$$(\vec{k''},\omega'') = (am',an',wl'), \tag{5}$$

$$(\vec{k'},\omega') = (am-am',an-an',wl-wl'), \tag{6}$$

where m', n' and l' are integers between $-\infty$ and $+\infty$; g is the gravitational acceleration; $\vec{k''}$ and $\vec{k'}$ represent two ocean wave vectors; ω'' and ω' are the angular frequencies corresponding to $\vec{k''}$ and $\vec{k'}$, respectively. If $\vec{k''} = -\vec{k'}$ and $\omega'' = -\omega'$, $\Gamma(\vec{k''},\omega'',\vec{k'},\omega') = 0$; otherwise

$$\Gamma(\vec{k''},\omega'',\vec{k'},\omega') = \frac{1}{2}\left[|\vec{k''}| + |\vec{k'}| + \frac{\omega_0''\omega_0'}{g}\left(1 - \frac{\vec{k''}\cdot\vec{k'}}{|\vec{k''}||\vec{k'}|}\right)\frac{g|\vec{k''}+\vec{k'}| + (\omega_0''+\omega_0')^2}{g|\vec{k''}+\vec{k'}| - (\omega_0''+\omega_0')^2}\right], \tag{7}$$

where $|\vec{k''}|$ and $|\vec{k'}|$ are the lengths of $\vec{k''}$ and $\vec{k'}$, respectively. The angular frequencies $\omega_0'' = \pm\sqrt{g|\vec{k''}|}$ and $\omega_0' = \pm\sqrt{g|\vec{k'}|}$ are given by the dispersion relationship of the gravity waves in deep water.

2.2. Statistical Characteristics of the Scattering Patch

According to [29], the Fourier coefficient $p_1(m,n,l)$ of linear wave height can be considered as a Gaussian random variable so that (1)–(3) can represent a real random sea surface. The mean of the random variable $p_1(m,n,l)$ is zero:

$$< p_1(m,n,l) >= 0, \tag{8}$$

where $< \cdots >$ denotes a statistical ensemble average. $p_2(m,n,l)$ is also a random variable, because $p_2(m,n,l)$ is determined by random variable $p_1(m,n,l)$.

The linear wave height $f^{(1)}(x,y,t)$ can be regarded as a stationary random process, so the power spectral density of the linear wave height is calculated as:

$$W(p,q,\omega) = \frac{1}{\pi^3} \iiint_{-\infty}^{+\infty} < f^{(1)}(x_1,y_1,t_1) f^{(1)*}(x_2,y_2,t_2) > e^{-ip\tau_x - iq\tau_y - i\omega\tau}\, d\tau_x\, d\tau_y\, d\tau, \tag{9}$$

where $\tau_x = x_1 - x_2$, $\tau_y = y_1 - y_2$, $\tau = t_2 - t_1$ and $f^{(1)*}(x_2,y_2,t_2)$ means the complex conjugation of $f^{(1)}(x_2,y_2,t_2)$. $W(p,q,\omega)$ is called the spatial-temporal spectrum of ocean waves. p and q denote the components of a ocean wave vector $\vec{k}$ along the x axis and y axis, respectively. ω is the angular frequency corresponding to $\vec{k}$. After calculation, the relationship between $W(p,q,\omega)$ and $p_1(m,n,l)$ is

$$W(p,q,\omega) = \frac{L^2 T}{\pi^3} < p_1(m,n,l), p_1^*(m,n,l) >, \tag{10}$$

where $(\vec{k},\omega) = (p,q,\omega) = (am, an, wl)$. The spatial-temporal spectrum $W(p,q,\omega)$ also can be expressed as

$$W(\vec{k},\omega) = 4S(\vec{k})\delta(\omega - \sqrt{g|\vec{k}|}) + 4S(-\vec{k})\delta(\omega + \sqrt{g|\vec{k}|}), \tag{11}$$

where $\delta(\cdot)$ is the Dirac delta function, and $S(\vec{k})$ is the directional wavenumber spectrum.

2.3. The Incident and Scattered Fields Near the Sea Surface

Now we assume the incident field arriving at the scattering patch is vertically polarized with an incidence angle of θ_i. The wave vector $\vec{k_i} = (k_0 \sin\theta_i, 0, -k_0\cos\theta_i)$ of the incident electromagtic wave is shown in Figure 1. Then the incident plane wave near the scattering patch, $\vec{E^i}$, can be expressed as:

$$\vec{E^i} = E_0(\cos\theta_i \hat{x} + \sin\theta_i \hat{z}) e^{ik_0 \sin\theta_i x - ik_0\cos\theta_i z - i\omega_c t} = E_x^i \hat{x} + E_y^i \hat{y} + E_z^i \hat{z}, \tag{12}$$

where E_0 is the magnitude of the electric field intensity of the incident field; ω_c and k_0 are the angular frequency and wavenumber corresponding to the radio frequency f_c in the free space, respectively; $\hat{x}$, $\hat{y}$ and $\hat{z}$ are unit vectors along each coordinate axis; E_x^i, E_y^i and E_z^i are the components of $\vec{E^i}$ along the x, y and z axes, respectively.

In (1)–(3), the whole sea surface has been treated as a periodic repetition of the scattering patch. In this way, the scattered field near the scattering patch can be derived using the SPM, which is a classical way to calculate the scattered field generated by periodic rough surface.

The slightly rough sea surface within the scattering patch can be divided into two parts: one is the planar part of the surface, and the other is the rough part of the surface. Consequently, the scattered field near the scattering patch contains two parts: the field induced by the planar surface, $\vec{E^{sp}} = E_x^{sp}\hat{x} + E_y^{sp}\hat{y} + E_z^{sp}\hat{z}$, and the field caused by the rough surface, $\vec{E^{sr}} = E_x^{sr}\hat{x} + E_y^{sr}\hat{y} + E_z^{sr}\hat{z}$. With the assumption that the sea water is an ideal conductor and the incident field is a plane wave at a frequency of f_c, the total electric field intensity, $\vec{E^t} = E_x^t\hat{x} + E_y^t\hat{y} + E_z^t\hat{z}$, near the scattering patch, can be expressed as:

$$\vec{E^t} = \vec{E^i} + \vec{E^{sp}} + \vec{E^{sr}}. \tag{13}$$

The scattered field induced by the planar part is

$$\vec{E^{sp}} = E_0(-\cos\theta_i\hat{x} + \sin\theta_i\hat{z})e^{ik_0\sin\theta_i x + ik_0\cos\theta_i z - i\omega_c t}. \tag{14}$$

The components E_x^{sr}, E_y^{sr} and E_z^{sr} of the scattered field induced by the rough part are expressed as Fourier series:

$$\begin{cases} E_x^{sr} = \sum_{m,n,l} A(m,n,l)E(m,n,l), \\ E_y^{sr} = \sum_{m,n,l} B(m,n,l)E(m,n,l), \\ E_z^{sr} = \sum_{m,n,l} C(m,n,l)E(m,n,l), \end{cases} \tag{15}$$

where $A(m,n,l), B(m,n,l)$ and $C(m,n,l)$ are unkown Fourier coefficients. $E(m,n,l)$ is assumed as:

$$E(m,n,l) = E_0 e^{iamx + iany + ib(m,n)z - i(wl+\omega_c)t}, \tag{16}$$

and $b(m,n)$ is defined as

$$b(m,n) = \begin{cases} \sqrt{k_0^2 - a^2m^2 - a^2n^2}, & \text{if } m^2+n^2 < k_0^2/a^2 \\ i\sqrt{a^2m^2 + a^2n^2 - k_0^2}, & \text{if } m^2+n^2 > k_0^2/a^2 \end{cases}. \tag{17}$$

The coefficients $A(m,n,l), B(m,n,l)$ and $C(m,n,l)$ can be derived by expanding boundary conditions in perturbation parameter or smallness [30]. Here $f^{(1)}(x,y,t)$ and $f^{(2)}(x,y,t)$ are selected as the first- and second-order smallness, respectively.

Two boundary conditions must be satisfied. First, the tangential component of the total electric field intensity is zero at the interface between the sea water and the air, because the sea water is perfectly conducting. Second, the divergence of the total electric field intensity is zero, because the zone above the sea surface is sourceless. Then substituting the components E_x^t, E_y^t and E_z^t of the total electric field into these two boundary conditions gives the first- and second-order solutions of $A(m,n,l)$, $B(m,n,l)$ and $C(m,n,l)$. The results are presented in (18)–(25) where $av = \sin\theta_i$ with an integer v is assumed to facilitate the calculation. $A^{(1)}(m,n,l)$, $B^{(1)}(m,n,l)$ and $C^{(1)}(m,n,l)$ are the first-order solutions. $A^{(2)}(m,n,l)$, $B^{(2)}(m,n,l)$ and $C^{(2)}(m,n,l)$ are the second-order solutions. Substituting these Fourier coefficients into E_x^{sr}, E_y^{sr} and E_z^{sr}, the total electric field near the sea surface can be obtained.

Referring to Barrick's work [31], the first- and second-order terms of the scattered field are caused by the first- and second-order Bragg scattering, respectively.

$$\begin{cases} A(m,n,l) = A^{(1)}(m,n,l) + A^{(2)}(m,n,l) \\ B(m,n,l) = B^{(1)}(m,n,l) + B^{(2)}(m,n,l) \\ C(m,n,l) = C^{(1)}(m,n,l) + C^{(2)}(m,n,l) \end{cases} \tag{18}$$

$$A^{(1)}(m,n,l) = 2i(k_0 - am\sin\theta_i)p_1(m-v,n,l) \tag{19}$$

$$B^{(1)}(m,n,l) = -2ian\sin\theta_i p_1(m-v,n,l) \tag{20}$$

$$C^{(1)}(m,n,l) = \frac{1}{b(m,n)}2i\left[-k_0a(m-v) - b^2(m,n)\sin\theta_i\right]p_1(m-v,n,l) \tag{21}$$

$$A^{(2)}(m,n,l) = 2i(k_0 - am\sin\theta_i)p_2(m-v,n,l) + \\ \sum_{m',n',l'} 2\left[(k_0 - am\sin\theta_i)b^2(m',n') + k_0(am-am')(av-am')\right] Q(m,n,l,m',n',l') \tag{22}$$

$$B^{(2)}(m,n,l) = -2ian\sin\theta_i p_2(m-v,n,l) + \\ \sum_{m',n',l'} 2\left[k_0(an-an')(av-am') - an\sin\theta_i b^2(m',n')\right] Q(m,n,l,m',n',l') \tag{23}$$

$$C^{(2)}(m,n,l) = \frac{1}{b(m,n)} 2i\left[-k_0 a(m-v) - b^2(m,n)\sin\theta_i\right] p_2(m-v,n,l) + \\ \frac{1}{b(m,n)} \sum_{m',n',l'} 2\left\{\left[a^2(m^2+n^2)\sin\theta_i - amk_0\right] b^2(m',n')\right\} Q(m,n,l,m',n',l') + \\ \frac{1}{b(m,n)} \sum_{m',n',l'} 2\left\{a^3 k_0(m'-v)(m^2+n^2-mm'-nn')\right\} Q(m,n,l,m',n',l') \tag{24}$$

$$Q(m,n,l,m',n',l') = \frac{p_1(m'-v,n',l')p_1(m-m',n-n',l-l')}{b(m',n')} \tag{25}$$

2.4. The Scattered Field Far from the Scattering Patch

For airborne HF radars, the antennas are located in the far zone of the scattering patch. As shown in Figure 1, the far zone means that the distance R_0 between the radar antenna and the center of the scattering patch is much longer than the side length L of the scattering patch.

Here the Stratton–Chu integral is employed to calculate the scattered field in the far zone of the scattering patch [12,32]. For monostatic configuration, substituting the scattered field from the rough part of the scattering patch into the Stratton–Chu integral gives (26), which represents the scattered field $\vec{H^f}(R_0,t)$ at the receive antenna.

$$\vec{H^f}(R_0,r) = \frac{ie^{ik_0R_0}}{4\pi R_0}\int_{-L/2}^{L/2}\int_{-L/2}^{L/2}\{\vec{k_s}\times(\hat{n}\times\vec{H^{sr}}) - k_0\sqrt{\frac{\epsilon_0}{\mu_0}}(\hat{n}\times\vec{E^{sr}}) \\ +\sqrt{\frac{\epsilon_0}{\mu_0}}\vec{k_s}\cdot(\hat{n}\times\vec{E^{sr}})\frac{\vec{k_s}}{k_0}\}e^{-i\vec{k_s}\cdot\vec{r}}\,dx\,dy \tag{26}$$

In (26), $\vec{H^{sr}}$ denotes the magnetic field corresponding to $\vec{E^{sr}}$; $\vec{k_s} = (-k_0\sin\theta_i, 0, k_0\cos\theta_i)$ is the wave vector of the scatterd field; the square area $\{(x,y,z)| -L/2 \le x \le L/2, -L/2 \le y \le L/2, z=0\}$ is the integration interval of the Stratton–Chu integral; $\hat{n} = \hat{z}$ is the unit normal vector of the integration plane; $\vec{r} = (x,y,z)$ is the vector pointing from the center of the scattering patch to any point in the integration area; ϵ_0 and μ_0 are the electrical and magnetic permittivity of free space, respectively. For backscattering, the angle of reflection θ_s is identical to the angle of incidence θ_i, i.e., $\theta_i = \theta_s$. As mentioned in [14,15], when L is very big, i.e., $L \to +\infty$ is assumed, the integration interval of the Stratton–Chu integral can also be set as $\{(x,y,z)| -L/2 \le x \le L/2, -L/2 \le y \le L/2, z = f(x,y,t)\}$. The results of the NRCS are the same for these two cases of the integration interval.

As shown in Figure 1, the antenna locates at the point $(R_0, \theta_s = \theta_i, \phi_s = \pi)$ in the spherical coordinate system. The vertical polarization is considered herein. Thus the vertically polarized component of $\vec{H^f}(R_0,t)$ is the component of $\vec{H^f}(R_0,t)$ along the direction $\phi_s = \pi$ in the spherical coordinate system:

$$H^f_{\phi_s}(R_0,t) = \vec{H^f}(R_0,t)\cdot(-\hat{y}) = \sum_{m,n,l}\{\zeta(m,n,l)\frac{\sin[(am+k_0\sin\theta_i)L/2]\sin(anL/2)}{(am+k_0\sin\theta_i)an}\}, \tag{27}$$

$$\zeta(m,n,l) = \frac{ie^{ik_0R_0}}{\pi R_0}E_0\sqrt{\frac{\epsilon_0}{\mu_0}}\{[\cos\theta_i b(m,n) + k_0]\,A(m,n,l) - am\cos\theta_i C(m,n,l)\}e^{-i(wl+\omega_c)t}. \quad (28)$$

3. The NRCS of the Scattering Patch for Backscattering

3.1. The Power Spectral Density of the Scattered Field

The normalized power spectral density of the vertically polarized component $H^f_{\phi_s}(R_0,t)$ of the scattered field at the receive antenna can be calculated as follows:

1. Obtain the time autocorrelation function $R(\tau)$. The time autocorrelation function of $H^f_{\phi_s}(R_0,t)$ is defined as

$$R(\tau) = < H^f_{\phi_s}(R_0,t_1)H^{f*}_{\phi_s}(R_0,t_2) >, \quad (29)$$

where $\tau = t_2 - t_1$.

2. Estimate the power spectral density. Take the Fourier transform of $R(\tau)$ and estimate the power density spectrum $R(\omega'')$:

$$R(\omega'') = \frac{1}{\pi}\int R(\tau)e^{-i\omega''\tau}\,d\tau. \quad (30)$$

3. Calculate the normalized power spectral density. The normalized power density spectrum $\sigma(\omega'')$ is derived by:

$$\sigma(\omega'') = R(\omega'')\cdot\frac{4\pi R_0^2}{L^2H_0^2}, \quad (31)$$

where $H_0 = E_0\sqrt{\epsilon_0/\mu_0}$ is the magnitude of the magnetic field intensity corresponding to the magnitude of the electric field intensity of the incident field. $\sigma(\omega'')$ is also called the NRCS of the sea surface. The normalization is applied to derive the range-independent NRCS at the sea surface area.

For an airborne HF/VHF radar, $\sigma(\omega'')$ is a function of the incidence angle θ_i. For that $\omega_d = \omega'' - \omega_c$ is the Doppler frequency, $\sigma(\omega'')$ is rewritten as $\sigma(\omega_d,\theta_i)$ which is given in (32)–(34). The definitions of the coefficients and vectors in (32)–(34) are given in (35)–(41). Here the velocity of airplane is assumed to be constant within the coherent integration time and has been left out.

$$\sigma(\omega_d,\theta_i) = \sigma^{(1)}(\omega_d,\theta_i) + \sigma^{(2)}(\omega_d,\theta_i), \quad (32)$$

$$\sigma^{(1)}(\omega_d,\theta_i) = 2^4\pi k_0^4(1+\sin^2\theta_i)^2\sum_{m_1=\pm1}S(-2m_1\sin\theta_i\vec{k_0})\delta(\omega_d - m_1\omega_B), \quad (33)$$

$$\sigma^{(2)}(\omega_d,\theta_i) = 2^4\pi k_0^4\sum_{m_1,m_2=\pm1}\iint\left|(1+\sin^2\theta_i)\Gamma_H + \Gamma_{EM}\right|^2 S(m_1\vec{k_1}) S(m_2\vec{k_2})\delta(\omega_d - m_1\sqrt{gk_1} - m_2\sqrt{gk_2})\,dpdq, \quad (34)$$

$$\vec{k_0} = (k_0,0), \quad (35)$$

$$\omega_B = \sqrt{2gk_0\sin\theta_i}, \quad (36)$$

$$\vec{k_1} = (p - k_0 \sin\theta_i, q), \tag{37}$$

$$\vec{k_2} = (-k_0 \sin\theta_i - p, -q), \tag{38}$$

$$\vec{k_1} + \vec{k_2} = -2\sin\theta_i \vec{k_0}, \tag{39}$$

$$\Gamma_{EM} = \left[\frac{-(\vec{k_1}\cdot\hat{x})(\vec{k_2}\cdot\hat{x}) + (1+\sin^2\theta_i)[k_0^2\cos^2\theta_i + \vec{k_1}\cdot\vec{k_2}]}{\sqrt{k_0^2\cos^2\theta_i + \vec{k_1}\cdot\vec{k_2}} - k_0\Delta} \right], \tag{40}$$

$$\Gamma(\vec{k},\omega,\vec{k'},\omega') \underset{\vec{k''}=\vec{k_1},\vec{k'}=\vec{k_2}}{\overset{\omega_0''+\omega_0'=\omega_d}{\longrightarrow}} \Gamma_H = \frac{1}{2}\left[|\vec{k_1}| + |\vec{k_2}| + m_1 m_2 \sqrt{|\vec{k_1}||\vec{k_2}|}(1 - \frac{\vec{k_1}\cdot\vec{k_2}}{|\vec{k_1}||\vec{k_2}|})\frac{2gk_0 sin\theta_i + \omega_d^2}{2gk_0 sin\theta_i - \omega_d^2} \right]. \tag{41}$$

Both the first- and second-order scattered fields are considered; therefore, $\sigma(\omega_d, \theta_i)$ consists of two parts: $\sigma^{(1)}(\omega_d, \theta_i)$ and $\sigma^{(2)}(\omega_d, \theta_i)$, which are called the first- and second-order NRCS of the sea surface, respectively. $\sigma^{(1)}(\omega_d, \theta_i)$ comes from the first-order component of $\vec{E^{sr}}$, i.e., $A^{(1)}(m,n,l)$, $B^{(1)}(m,n,l)$ and $C^{(1)}(m,n,l)$, which are directly proportional to the $p_1(m-v,n,l)$. Hence, the first-order NRCS $\sigma^{(1)}(\omega_d, \theta_i)$ is only caused by the first-order Bragg scattering. Similarly, $\sigma^{(2)}(\omega_d, \theta_i)$ comes from the second-order components of $\vec{E^{sr}}$, i.e., $A^{(2)}(m,n,l)$, $B^{(2)}(m,n,l)$ and $C^{(2)}(m,n,l)$, which include $p_2(m-v,n,l)$ and $Q(m,n,l,m',n',l')$. The $\Gamma(\vec{k''},\omega'',\vec{k'},\omega')$ in $p_2(m-v,n,l)$ (given in (7)) becomes the hydrodynamic coupling coefficient Γ_H which is given in (41). According to Barrick's definitions [31], Γ_{EM} (given in (40)) is called the electromagnetic coupling coefficient which comes from the second-order scattered field composed of $Q(m,n,l,m',n',l')$. It can be seen that both $p_2(m-v,n,l)$ and $Q(m,n,l,m',n',l')$ are the products of $p_1(m',n',l')$ and $p_1(m-m',n-n',l-l')$. Thus the second-order NRCS $\sigma^{(2)}(\omega_d, \theta_i)$ is a result of the second-order Bragg scattering.

There is a singularity in the denominator of Γ_{EM} when $k_0^2\cos^2\theta_i + \vec{k_1}\cdot\vec{k_2}$ becomes zero. The assumption that the sea water is perfectly conducting causes the singularity. A term, $-k_0\Delta$, is added in the denominator of Γ_{EM} to eliminate this singularity [12,33]. Δ is the normalized surface impedance which is a complex constant, i.e., $\Delta = 0.011 - i(0.012)$. The added term $-k_0\Delta$ means the small energy loss of HF electromagnetic waves traveling along the actual sea surface which is good at conducting rather than perfectly conducting.

3.2. The Effectiveness of the NRCS

The NRCS of the sea surface has been derived using the SPM. Accordingly, the approximation made in the perturbational analysis must satisfy the condition:

$$k_0 h \cos\theta_i \ll 0.5 \text{ and } k_0 h \sin\theta_i \ll 0.5, \tag{42}$$

where h is the root mean square (RMS) wave height of the sea surface, k_0 is the wavenumber of the incident plane wave at a frequency of f_c and θ_i is the incident angle [28,34]. To ensure the correctness of the results from the perturbational analysis, a more rigorous condition is adopted in this work:

$$k_0 h \cos\theta_i \leq 0.2 \text{ and } k_0 h \sin\theta_i \leq 0.2. \tag{43}$$

Considering that $h_s = 4h$ (h_s is significant wave height), the NRCS $\sigma(\omega_d, \theta_i)$ is effective only if the following inequality is satisfied:

$$G(f_c) \leq 0.8, \tag{44}$$

where $G(f_c)$ is defined as:

$$G(f_c) = \begin{cases} k_0 h_s \cos\theta_i, & \text{if } 20^\circ \leq \theta_i \leq 45^\circ \\ k_0 h_s \sin\theta_i, & \text{if } 45^\circ < \theta_i \leq 90^\circ \end{cases}. \tag{45}$$

The scattered field induced by the rough part of the sea surface is taken into consideration herein. As a result, the NRCS $\sigma(\omega_d, \theta_i)$ is effective only when the angle of incidence θ_i satisfies $20^\circ \leq \theta_i \leq 90^\circ$ where the intensity of the scattered field from the plane part of the sea surface is much smaller than the intensity of the scattered field from the rough part of the sea surface [31]. Figure 2 shows the effective region of the NRCS $\sigma(\omega_d, \theta_i)$ in the $f_c - h_s$ plane with different θ_i values.

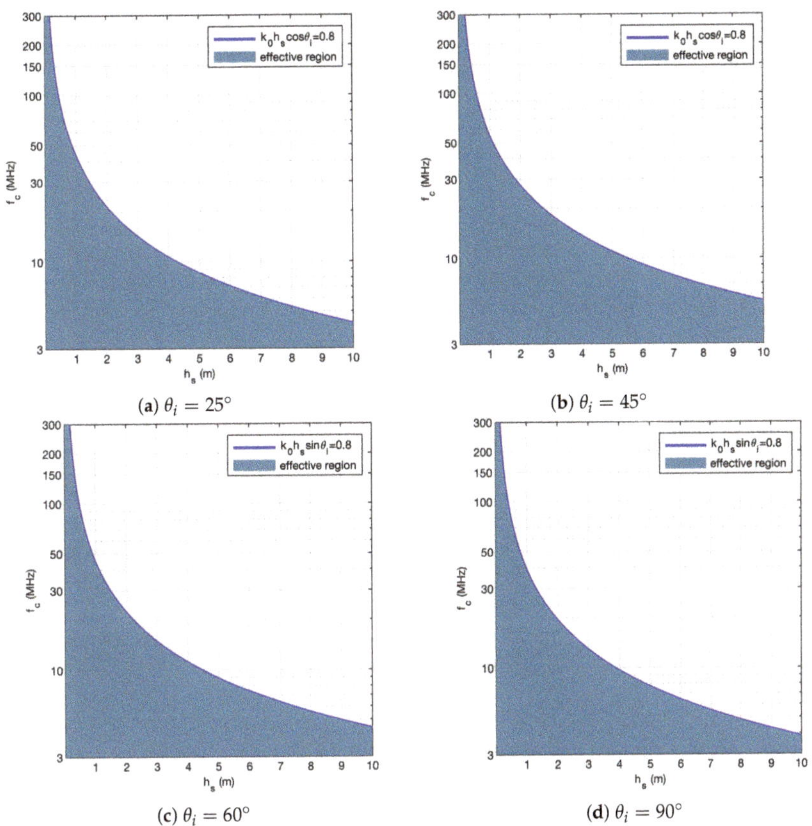

Figure 2. The effective region of the NRCS $\sigma(\omega_d, \theta_i)$. In (**a**) $\theta_i = 25^\circ$, (**b**) $\theta_i = 45^\circ$, (**c**) $\theta_i = 60^\circ$ and (**d**) $\theta_i = 90^\circ$, the area filled with dark blue is the effective region for each case.

4. The Simulation and Analysis of the Sea Echo

The interpretation of the NRCS of the sea surface is crucial to analyze the sea echoes. The NRCS of the sea surface, $\sigma(\omega_d, \theta_i)$, is interpreted as the theoretical prediction of the sea-echo Doppler spectrum. Consequently, the simulations of $\sigma^{(1)}(\omega_d, \theta_i)$ and $\sigma^{(2)}(\omega_d, \theta_i)$ are treated as the first- and second-order sea-echo Doppler spectra, respectively.

It can be found from the Formulas (32)–(34) that the directional wavenumber spectrum $S(\vec{k})$ is included in the theoretical sea-echo Doppler spectrum $\sigma(\omega_d, \theta_i)$. $S(\vec{k})$ is the product of a non-directional wave spectrum $S(k)$ and a directional distribution function $g(\alpha)$. It is assumed that only wind waves exist and they are fully developed. The Pierson–Moskowitz spectrum [35] and the cardioid distribution model [33] are assumed:

$$S(\vec{k}) = S(k)g(\alpha), \tag{46}$$

$$S(k) = \frac{4.05 \times 10^{-3}}{k^4} e^{-0.74(\frac{g}{kU^2})^2}, \tag{47}$$

$$g(\alpha) = \frac{\cos^4(\frac{\alpha-\alpha'}{2})}{\int_{-\pi}^{\pi} \cos^4(\frac{\alpha}{2})\, d\alpha}, \tag{48}$$

where U is the wind speed at 19.5 m above the sea surface and α' is the dominant wave direction which is the same with wind direction for wind-wave sea state. The relationship between U and h_s is:

$$h_s = 4\sqrt{\iint S(\vec{k})\, d\vec{k}} = 0.2\frac{U^2}{g}. \tag{49}$$

It can be seen that the theoretically predicted sea-echo Doppler spectrum $\sigma(\omega_d, \theta_i)$ is influenced by four factors: the dominant wave direction α', the incident angle θ_i, the radar frequency f_c and the sea state h_s. Here we investigate the effects of the latter three factors on $\sigma(\omega_d, \theta_i)$. For simplification, a normalized Doppler frequency is defined as $\eta = \omega_d/\omega_B$ in the simulation. It is easy to prove that when $\theta_i = 90°$, the theoretical sea-echo Doppler spectrum $\sigma(\omega_d, \theta_i)$ is reduced to the classical NRCS for shore-based HF radar.

4.1. Sea Echoes at Different Radar Frequencies and Sea States

When $\alpha' = 90°$, different values of f_c, θ_i and h_s are selected to simulate the Doppler spectrum. The simulated results for $\theta_i = 30°, 45°, 55°$ and $70°$ are given in Figures 3–6, respectively. Each sub-figure in Figures 3–6 corresponds to a combination of θ_i and h_s and shows the Doppler spectrum at six radar frequencies, i.e., $f_c = 3, 9, 15, 30, 45$ and 55 MHz. The first-order sea-echo Doppler spectrum $\sigma^{(1)}(\omega_d, \theta_i)$ is represented by the two peaks at $\eta = \pm 1$, and the continuous curves around these two peaks are the second-order sea-echo Doppler spectra $\sigma^{(2)}(\omega_d, \theta_i)$.

First, as shown in Figure 3, the symmetry characteristics of the simulated results (when $\alpha' = 90°$ and $\theta_i \neq 90°$) are the same as the simulated sea-echo Doppler spectra for shore-based HF radar (when $\alpha' = 90°$ and $\theta_i = 90°$) [33].

Second, it is noted that the first-order sea-echo Doppler spectrum seems a constant for each θ_i when the radar works in a higher frequency band, e.g., $f_c \geq 15$ MHz. It can be found from Figures 3a, 4a, 5a and 6a that the energy of the first-order peak is relatively smaller when radar frequency is low and sea state is calm, e.g., $f_c = 3$ MHz and $h_s = 0.7$ m. The reason for this is that the energy of ocean waves which cause the first-order Bragg scattering does not vary dramatically when f_c is high and h_s is large.

Finally, it can be seen from Figure 3a–d that the second-order spectrum increases in magnitude when f_c increases. However, when the values of f_c and h_s do not meet the condition given in (44), the second-order spectrum is even higher than the first-order peaks. As mentioned in [33], in this case, the theoretical Doppler spectrum predicted by the SPM is not accurate.

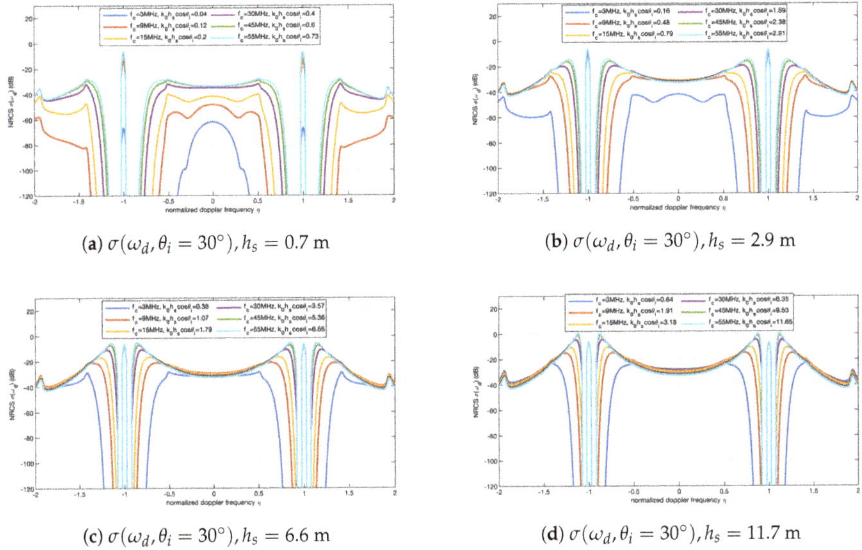

(**a**) $\sigma(\omega_d, \theta_i = 30°), h_s = 0.7$ m (**b**) $\sigma(\omega_d, \theta_i = 30°), h_s = 2.9$ m

(**c**) $\sigma(\omega_d, \theta_i = 30°), h_s = 6.6$ m (**d**) $\sigma(\omega_d, \theta_i = 30°), h_s = 11.7$ m

Figure 3. The Doppler spectra for $\theta_i = 30°$ with different f_c and h_s. (**a**) $\sigma(\omega_d, \theta_i = 30°), h_s = 0.7$ m; (**b**) $\sigma(\omega_d, \theta_i = 30°), h_s = 2.9$ m; (**c**) $\sigma(\omega_d, \theta_i = 30°), h_s = 6.6$ m; and (**d**) $\sigma(\omega_d, \theta_i = 30°), h_s = 11.7$ m. For each pair of θ_i and h_s, six radar frequencies were selected to simulate the spectra. The six values of f_c were 3, 9, 15, 30, 45 and 55 MHz. The dominant wave direction α' is 90°.

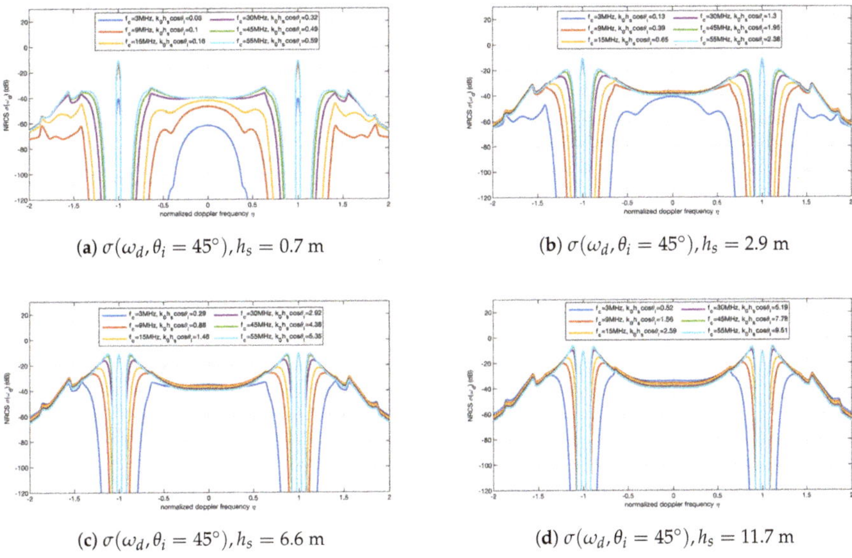

(**a**) $\sigma(\omega_d, \theta_i = 45°), h_s = 0.7$ m (**b**) $\sigma(\omega_d, \theta_i = 45°), h_s = 2.9$ m

(**c**) $\sigma(\omega_d, \theta_i = 45°), h_s = 6.6$ m (**d**) $\sigma(\omega_d, \theta_i = 45°), h_s = 11.7$ m

Figure 4. The Doppler spectra for $\theta_i = 45°$ with different f_c and h_s. (**a**) $\sigma(\omega_d, \theta_i = 45°), h_s = 0.7$ m; (**b**) $\sigma(\omega_d, \theta_i = 45°), h_s = 2.9$ m; (**c**) $\sigma(\omega_d, \theta_i = 45°), h_s = 6.6$ m; and (**d**) $\sigma(\omega_d, \theta_i = 45°), h_s = 11.7$ m. For each pair of θ_i and h_s, six radar frequencies were selected to simulate the spectra. The six values of f_c were 3, 9, 15, 30, 45 and 55 MHz. The dominant wave direction α' is 90°.

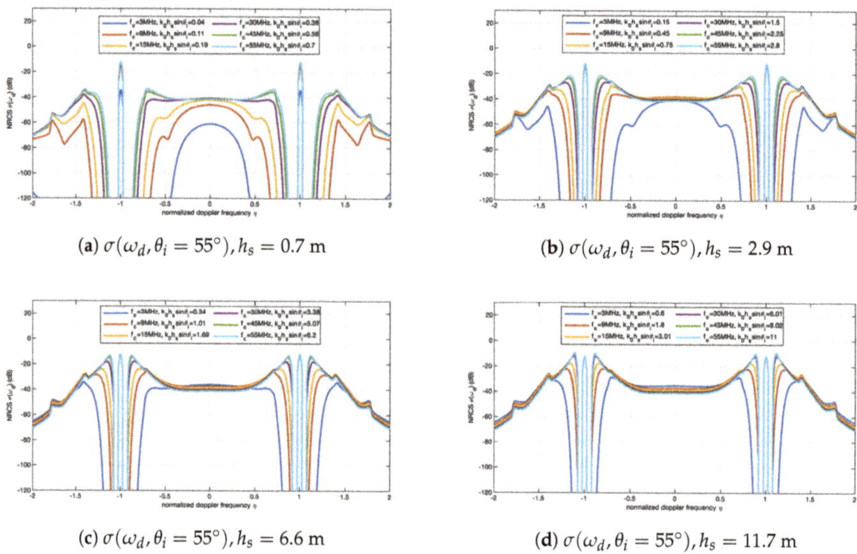

(**a**) $\sigma(\omega_d, \theta_i = 55°), h_s = 0.7$ m (**b**) $\sigma(\omega_d, \theta_i = 55°), h_s = 2.9$ m

(**c**) $\sigma(\omega_d, \theta_i = 55°), h_s = 6.6$ m (**d**) $\sigma(\omega_d, \theta_i = 55°), h_s = 11.7$ m

Figure 5. The Doppler spectra for $\theta_i = 55°$ with different f_c and h_s. (**a**) $\sigma(\omega_d, \theta_i = 55°), h_s = 0.7$ m; (**b**) $\sigma(\omega_d, \theta_i = 55°), h_s = 2.9$ m; (**c**) $\sigma(\omega_d, \theta_i = 55°), h_s = 6.6$ m; and (**d**) $\sigma(\omega_d, \theta_i = 55°), h_s = 11.7$ m. For each pair of θ_i and h_s, six radar frequencies were selected to simulate the spectra. The six values of f_c were 3, 9, 15, 30, 45 and 55 MHz. The dominant wave direction α' is 90°.

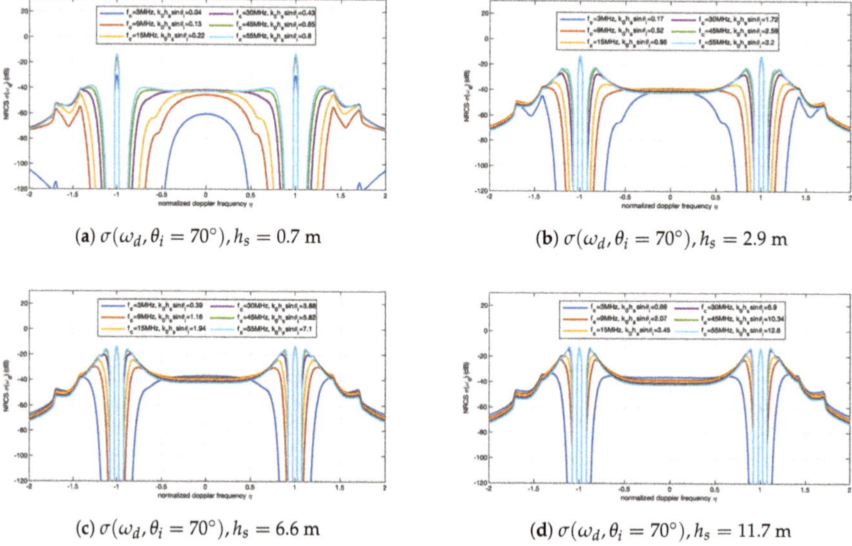

(**a**) $\sigma(\omega_d, \theta_i = 70°), h_s = 0.7$ m (**b**) $\sigma(\omega_d, \theta_i = 70°), h_s = 2.9$ m

(**c**) $\sigma(\omega_d, \theta_i = 70°), h_s = 6.6$ m (**d**) $\sigma(\omega_d, \theta_i = 70°), h_s = 11.7$ m

Figure 6. The Doppler spectra for $\theta_i = 70°$ with different f_c and h_s. (**a**) $\sigma(\omega_d, \theta_i = 70°), h_s = 0.7$ m; (**b**) $\sigma(\omega_d, \theta_i = 70°), h_s = 2.9$ m; (**c**) $\sigma(\omega_d, \theta_i = 70°), h_s = 6.6$ m; and (**d**) $\sigma(\omega_d, \theta_i = 70°), h_s = 11.7$ m. For each pair of θ_i and h_s, six radar frequencies were selected to simulate the spectra. The six values of f_c were 3, 9, 15, 30, 45 and 55 MHz. The dominant wave direction α' is 90°.

4.2. Sea Echoes for Different Incidence Angles

Comparing Figures 3a, 4a, 5a and 6a, it can be found that the first-order spectrum $\sigma^{(1)}(\omega_d, \theta_i)$, which is represented by the two highest peaks in the spectra, varies with θ_i. To make it clear, the values of $\sigma_0^{(1)} = \frac{1}{2}\int_{-\infty}^{\infty} \sigma^{(1)}(\omega_d, \theta_i)\, d\omega_d$ against incident angles are shown in Figure 7a, and it shows that the radar-received energy caused by the first-order Bragg scattering drops from -11 dB to -23 dB when θ_i increases from 20° to 90°.

As shown in Figure 7b, this descending trend also exists in the second-order Doppler spectra for different θ_i. For $|\eta| < 1$, the values of second-order spectra decrease nearly 10 dB when θ_i varies from 25° to 90°. In contrast, for $|\eta| > 1$, the magnitude of the second-order spectrum decreases even more than 10 dB. Figure 7c demonstrates the value of $\sigma_0^{(2)} = \frac{1}{2}\int_{-\infty}^{\infty} \sigma^{(2)}(\omega_d, \theta_i)\, d\omega_d$ against θ_i, and it clearly shows the decrease in the radar-received energy caused by the second-order Bragg scattering.

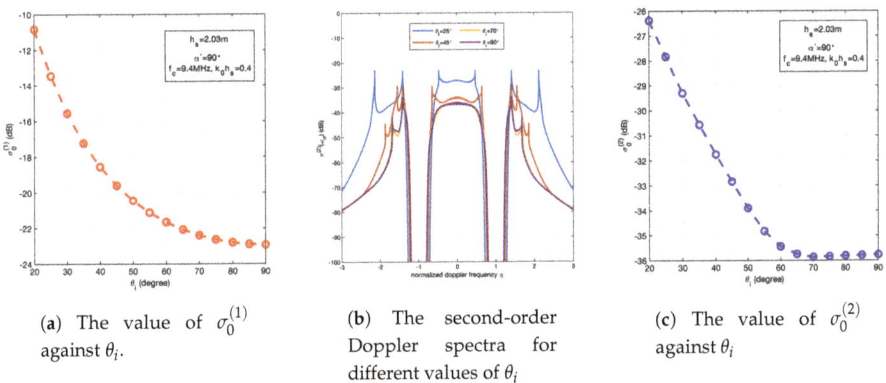

(**a**) The value of $\sigma_0^{(1)}$ against θ_i.

(**b**) The second-order Doppler spectra for different values of θ_i

(**c**) The value of $\sigma_0^{(2)}$ against θ_i

Figure 7. (**a**) The values of $\sigma_0^{(1)}$ for $20^\circ \leq \theta_i \leq 90^\circ$. (**b**) The second-order Doppler spectra $\sigma^{(2)}(\omega_d, \theta_i)$ for several incident angles θ_i, i.e., $\theta_i = 25^\circ, 45^\circ, 70^\circ$ and 90°. (**c**) The values of $\sigma_0^{(1)}$ for $20^\circ \leq \theta_i \leq 90^\circ$. $h_s = 2.03$ m, $\alpha' = 90^\circ$ and $f_c = 9.4$ MHz are assumed for (a), (b) and (c).

However, this descending trend is not significant for the near-grazing case, i.e., for $70^\circ \leq \theta_i \leq 90^\circ$. The value of $\sigma_0^{(1)}$ drops less than 1 dB when θ_i changes from 70° to 90°. The second-order Doppler spectrum for $\theta_i = 70^\circ$ is nearly identical to that for $\theta_i = 90^\circ$. Consequently, the values of $\sigma_0^{(2)}$ for $\theta_i = 70^\circ$ and 90° are nearly equal. There are two reasons for this phenomenon. One reason is that the values of the functions $\sin\theta_i$ and $\cos\theta_i$ vary slightly with θ_i changing from 70° to 90°, which causes a small variation in the length of the vector $-2\sin\theta_i\vec{k}_0$. The other one is that the ocean waves which cause the second-order Bragg scattering contain nearly equal energy for $70^\circ \leq \theta_i \leq 90^\circ$.

4.3. Sea Echoes for Different Sea States

In Section 4.1, it has been clearly seen that the first-order Doppler spectra do not vary as the sea state becomes higher. Here it is necessary to investigate the variation of the second-order spectrum when sea state is higher. As shown in Figure 8, the Doppler spectra under three different sea states ($h_s = 1.3$ m, 2.92 m and 4.56 m) for different angles of incidence ($\theta_i = 25^\circ, 55^\circ$ and 90°) were simulated while $\alpha' = 45^\circ$ and $f_c = 8$ MHz. It is obvious that the energy of the second-order Doppler spectrum becomes stronger along with the higher sea state.

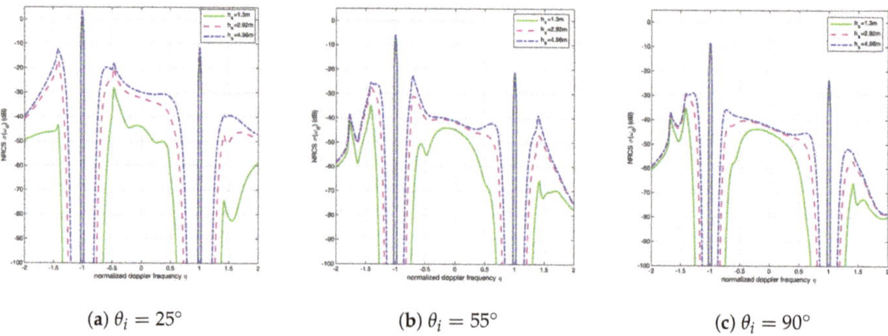

(**a**) $\theta_i = 25°$ (**b**) $\theta_i = 55°$ (**c**) $\theta_i = 90°$

Figure 8. The Doppler spectra for different values of h_s and θ_i. (**a**) $\theta_i = 25°$. (**b**) $\theta_i = 55°$. (**c**) $\theta_i = 90°$. $\alpha' = 45°$ and $f_c = 8$ MHz. The Doppler spectra were simulated for three distinct values of h_s, i.e., $h_s = 1.3$ m, 2.92 m and 4.56 m. These three values of h_s correspond to $U = 8$ m/s, 12 m/s and 15 m/s, respectively.

4.4. Comparison between SPM and GFM

For the case of shore-based monostatic HF radar, both the SPM [31] and GFM [16] have been utilized to derive the NRCS of the sea surface. For a comparison between these two methods, it is convenient to simulate the sea echoes derived by the two methods. Under the same condition as Figure 5 in [36], we simulated the model which was derived by using the SPM. The simulated result is shown in Figure 9.

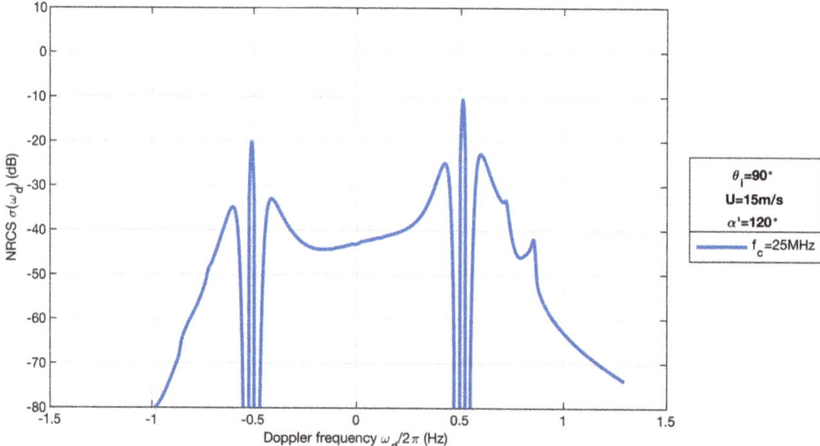

Figure 9. The simulated Doppler spectrum for monostatic radar ($\theta_i = 90°$). In order to compare the simulated result with Figure 5 in [36], $U = 15$ m/s, $f_c = 25$ MHz and $\alpha' = 120°$ were assumed.

It is seen that the Doppler spectra simulated by the two methods are similar in shape. Each result shows that the positive first-order peak is nearly 10 dB larger than the negative one. However the amplitudes of these two spectra are not equal. As mentioned in [36], these two methods are different although they have the same form. The significant difference is that the NRCS based on the GFM is affected by the range resolution of the radar while the NRCS derived using the SPM is not based on this parameter.

The above comparison shows a typical example of the NRCSs simulated using the GFM and the SPM. However, a recent work [21] seems to indicate that the derivation of the NRCS using the GFM has a wider application range in terms of approximation restrictions. The derivation of the NRCS using the SPM is on the basis of three assumptions: first, the sea water is a good conductor; second, the slope of ocean surface wave height is much smaller than 1; third, the product of the significant wave height and the radio wavenumber is small. The results in Figures 3–8 were obtained based on those conditions. Additionally, HF radar NRCS simulated using the SPM has been validated using real data for more than 50 years. In contrast, it is possible to remove the significant wave height restriction using the GFM as shown in [21]. In that work, the NRCS with arbitrary roughness scales has been obtained, but it has not been compared with real Doppler spectrum.

5. Discussion

Four factors, radar frequency f_c, the angle of incidence θ_i, the significant wave height h_s and the dominant wave direction α', which influence the shape and the magnitude of the sea-echo Doppler spectrum, have been investigated.

First, it was found that the first- and second-order spectra increase when radar frequency becomes higher. However, the Doppler spectrum becomes saturated when radar frequency is too high to meet the effective condition of the SPM. From the radar equation, we know that $SNR_o \propto P_t\sigma_0/L_p$, where SNR_o is the signal to noise ratio at the output of the radar receiver, P_t is the transmitted power of radar, $\sigma_0 = \frac{1}{2}\int_{-\infty}^{+\infty}\sigma(\omega_d,\theta)\,d\omega_d$ and L_p represents the propagation loss of radio waves. If radar frequency increases, both the L_p and σ_0 vary. Thus, it is much better to combine the σ_0 (derived in this paper) with a suitable L_p (which is not the focus of our work) to select radar frequency for designing an airborne HF/VHF radar for ocean remote sensing.

Second, the variation that occurs in the sea-echo Doppler spectrum when θ_i changes attracts our attention. It can be known from Figure 7a,c that σ_0 increases nearly 10 dB with θ_i changing from 90° to 20° ($\sigma_0 = \sigma_0^{(1)} + \sigma_0^{(2)}$). The σ_0 becomes large when the incident angle becomes small, and L_p is smaller when radio waves propagate in the air than when they propagate along the air–sea surface. Consequently, considering the same SNR_o for the airborne HF/VHF radar and the shore-based HF radar, P_t could be much smaller for airborne HF/VHF radars. It is convenient to design a relatively compact and low-power airborne HF/VHF radar.

Third, the energy of the sea echo increases when the sea state becomes higher, which is similar to the case of shore-based HF radar.

Finally, since the NRCS connects the sea echoes and the waveheight spectrum, it is possible to retrieve wave parameters from radar sea echoes by inversing the NRCS. In addition, sea surface current may also be extracted from the first-order echoes by determining the Doppler shift induced by current. The difference between $\theta_i \neq 90°$ and $\theta_i = 90°$ for current inversion is shown in Figure 10. If airborne and shore-based HF radars are located at the positions as the red points in the picture, the current velocity measured by the shore-based HF radar is $\vec{V}_x$, whereas the current measured by the airborne radar is $\vec{V}'$ which is a component of $\vec{V}_x$.

6. Conclusions

In this paper, the sea surface wave height has been expressed as the superposition of two Fourier series which represent linear and nonlinear wave heights. Then the SPM was adopted to get the scattered field from the sea surface. The scattered field has been calculated by taking into account both the first- and second-order Bragg scatterings between the sea surface waves and the electromagnetic waves. At last, theoretical models of the first- and second-order sea-echo Doppler spectra for the airborne HF/VHF radars have been derived. Besides that, the effectiveness region of the theoretical sea-echo Doppler spectrum $\sigma(\omega_d,\theta_i)$ was given.

There are continuous second-order spectra $\sigma^{(2)}(\omega_d,\theta_i = 90°)$ around the first-order Bragg peaks $\sigma^{(1)}(\omega_d,\theta_i = 90°)$ in the sea-echo Doppler spectra of the shore-based HF radar, and the continuous spectra have been used for wave parameter inversion in practice. Thus, the second-order terms in the SPM are not neglected in order to get the theoretical second-order sea-echo Doppler spectrum for the airborne HF/VHF radar. Both the first- and second-order spectra were simulated under different environment conditions to give a brief demonstration of the sea echo is received by radar. In addition, the results of the simulated sea echoes may provide a basic guide for designing an airborne HF/VHF radar to monitor the sea state in the future.

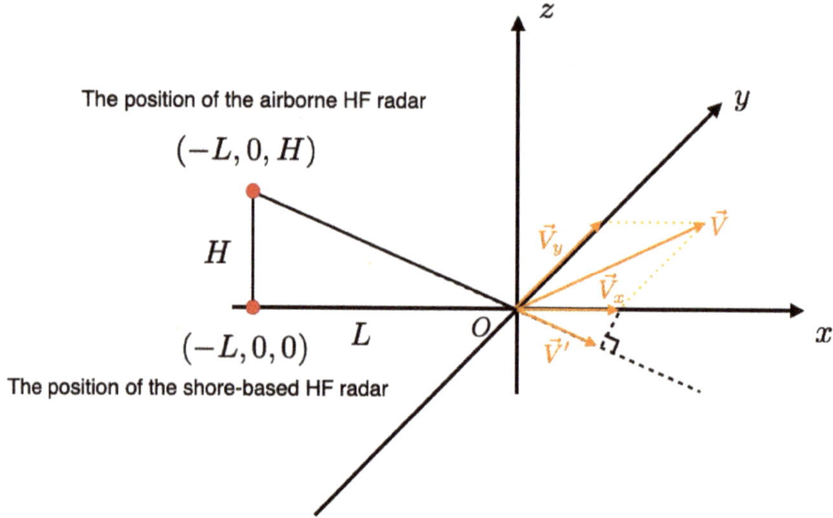

Figure 10. The difference between $\theta_i \neq 90°$ and $\theta_i = 90°$ for current inversion. A current with velocity vector $\vec{V}$ exists at the origin O. The airborne and shore-based HF radars are located at $(-L, 0, H)$ and $(-L, 0, 0)$, respectively.

Author Contributions: Conceptualization, F.D. and C.Z.; methodology, F.D. and C.Z.; software, F.D. and J.L.; validation, Z.C., C.Z. and J.L.; writing—original draft preparation, F.D.; supervision, C.Z.; project administration, C.Z. and Z.C.; funding acquisition, C.Z. and Z.C. All authors have read and agreed to the published version of the manuscript.

Funding: This work was supported in part by the National Natural Science Foundation of China under grant 61871296, grant 41506201 and grant 41376182; and in part by the National Key Research and Development Program of China under grant 2017YFF0206404 and grant 2016YFC1400504.

Conflicts of Interest: The authors declare no conflict of interest.

Abbreviations

The following abbreviations are used in this manuscript:

HF	high frequency
VHF	very high frequency
NRCS	normalized radar cross section
SPM	small perturbation method
GFM	generalized function method
GIOS	Ground-Ionosphere-Ocean-Space
RMS	root mean square

References

1. Crombie, D.D. Doppler spectrum of sea echo at 13.56 Mc./s. *Nature* **1955**, *175*, 681–682. [CrossRef]
2. Prandle, D.; Ryder, D. Measurement of surface currents in Liverpool Bay by high-frequency radar. *Nature* **1985**, *315*, 128–131. [CrossRef]
3. Georges, T.; Harlan, J.; Lematta, R. Large-scale mapping of ocean surface currents with dual over-the-horizon radars. *Nature* **1996**, *379*, 434–436. [CrossRef]
4. Zhao, C.; Chen, Z.; He, C.; Xie, F.; Chen, X. A Hybrid Beam-Forming and Direction-Finding Method for Wind Direction Sensing Based on HF Radar. *IEEE Trans. Geosci. Remote Sens.* **2018**, *56*, 6622–6629. [CrossRef]
5. Capodici, F.; Cosoli, S.; Ciraolo, G.; Nasello, C.; Maltese, A.; Poulain, P.M.; Drago, A.; Azzopardi, J.; Gauci, A. Validation of HF radar sea surface currents in the Malta-Sicily Channel. *Remote Sens. Environ.* **2019**, *225*, 65–76. [CrossRef]
6. Zhao, C.; Chen, Z.; Li, J.; Zhang, L.; Huang, W.; Gill, E.W. Wind Direction Estimation Using Small-Aperture HF Radar Based on a Circular Array. *IEEE Trans. Geosci. Remote Sens.* **2020**, *58*, 2745–2754. [CrossRef]
7. Jackson, G.; Fornaro, G.; Berardino, P.; Esposito, C.; Lanari, R.; Pauciullo, A.; Reale, D.; Zamparelli, V.; Perna, S. Experiments of sea surface currents estimation with space and airborne SAR systems. In Proceedings of the 2015 IEEE International Geoscience and Remote Sensing Symposium (IGARSS), Milan, Italy, 26–31 July 2015; pp. 373–376.
8. Forget, P.; Broche, P. Slicks, waves, and fronts observed in a sea coastal area by an X-band airborne synthetic aperture radar. *Remote Sen. Environ.* **1996**, *57*, 1–12. [CrossRef]
9. Martin, A.C.; Gommenginger, C. Towards wide-swath high-resolution mapping of total ocean surface current vectors from space: Airborne proof-of-concept and validation. *Remote Sens. Environ.* **2017**, *197*, 58–71. [CrossRef]
10. Israelsson, H.; Ulander, L.M.H.; Askne, J.L.H.; Fransson, J.E.S.; Frolind, P..; Gustavsson, A.; Hellsten, H. Retrieval of forest stem volume using VHF SAR. *IEEE Trans. Geosci. Remote Sens.* **1997**, *35*, 36–40. [CrossRef]
11. Barrick, D. First-order theory and analysis of MF/HF/VHF scatter from the sea. *IEEE Trans. Antennas Propag.* **1972**, *20*, 2–10. [CrossRef]
12. Johnstone, D.L. Second-Order Electromagnetic and Hydrodynamic Effects in High-Frequency Radio-Wave Scattering from the Sea. Ph.D. Thesis, Stanford Univeristy, Stanford, CA, USA, 1975.
13. Anderson, S.J. Directional wave spectrum measurement with multistatic HF surface wave radar. In Proceedings of the IGARSS 2000. IEEE 2000 International Geoscience and Remote Sensing Symposium. Taking the Pulse of the Planet: The Role of Remote Sensing in Managing the Environment. Proceedings (Cat. No.00CH37120), Honolulu, HI, USA, 24–28 July 2000; Volume 7, pp. 2946–2948.
14. Hisaki, Y.; Tokuda, M. VHF and HF sea echo Doppler spectrum for a finite illuminated area. *Radio Sci.* **2001**, *36*, 425–440. [CrossRef]
15. Hardman, R.L.; Wyatt, L.R.; Engleback, C.C. Measuring the Directional Ocean Spectrum from Simulated Bistatic HF Radar Data. *Remote Sens.* **2020**, *12*, 313. [CrossRef]
16. Srivastava, S.K. Scattering of High-Frequency Electromagnetic Waves from an Ocean Surface: An Alternative Approach Incorporating a Dipole Source. Ph.D. Thesis, Memorial University of Newfoundland, Saint John, NL, Canada, 1984.
17. Srivastava, S.; Walsh, J. An alternate of HF Scattering from an ocean surface. In Proceedings of the 1983 Antennas and Propagation Society International Symposium, Houston, TX, USA, 23–26 May 1983; Volume 21, pp. 680–683.
18. Gill, E.W.; Walsh, J. High-frequency bistatic cross sections of the ocean surface. *Radio Sci.* **2001**, *36*, 1459–1475. [CrossRef]
19. Huang, W. The Second-Order High Frequency Bistatic Radar Cross Section of the Ocean Surface for Patch Scatter. Master's Thesis, Memorial University of Newfoundland, Saint John, NL, Canda, 2004.
20. Ma, Y.; Gill, E.W.; Huang, W. Bistatic High-Frequency Radar Ocean Surface Cross Section Incorporating a Dual-Frequency Platform Motion Model. *IEEE J. Ocean. Eng.* **2018**, *43*, 205–210. [CrossRef]
21. Silva, M.T.; Huang, W.; Gill, E.W. High-Frequency Radar Cross-Section of the Ocean Surface with Arbitrary Roughness Scales: A Generalized Functions Approach. *IEEE Trans. Antennas Propag.* **2020**. [CrossRef]
22. Bernhardt, P.A.; Briczinski, S.J.; Siefring, C.L.; Barrick, D.E.; Bryant, J.; Howarth, A.; James, G.; Enno, G.; Yau, A. Large area sea mapping with Ground-Ionosphere-Ocean-Space (GIOS). In Proceedings of the OCEANS 2016 MTS/IEEE Monterey, Monterey, CA, USA, 19–23 September 2016; pp. 1–10.

23. Bernhardt, P.A.; Siefring, C.L.; Briczinski, S.C.; Vierinen, J.; Miller, E.; Howarth, A.; James, H.G.; Blincoe, E. Bistatic observations of the ocean surface with HF radar, satellite and airborne receivers. In Proceedings of the OCEANS 2017, Anchorage, AK, USA, 18–21 September 2017; pp. 1–5.
24. Anderson, S.J. Space-borne Passive HF Radar for Surveillance and Remote Sensing. In Proceedings of the Progress In Electromagnetics Research Symposium, St Petersburg, Russia, 22–25 May 2017; p. 1649.
25. Zhao, C.; Chen, Z. Model of shore-to-air bistatic HF radar for ocean observation. In Proceedings of the OCEANS 2018 MTS/IEEE Charleston, Charleston, SC, USA, 22–25 October 2018; pp. 1–4.
26. Chen, Z.; Li, J.; Zhao, C.; Ding, F.; Chen, X. The Scattering Coefficient for Shore-to-Air Bistatic High Frequency (HF) Radar Configurations as Applied to Ocean Observations. *Remote Sens.* **2019**, *11*, 2978. [CrossRef]
27. Voronovich, A.G.; Zavorotny, V.U. Measurement of Ocean Wave Directional Spectra Using Airborne HF/VHF Synthetic Aperture Radar: A Theoretical Evaluation. *IEEE Trans. Geosci. Remote Sens.* **2017**, *55*, 3169–3176. [CrossRef]
28. Weber, B.L.; Barrick, D.E. On the Nonlinear Theory for Gravity Waves on the Ocean's Surface. Part I: Derivations. *J. Phys. Oceanogr.* **1977**, *7*, 3–10. [CrossRef]
29. Barrick, D.E.; Weber, B.L. On the Nonlinear Theory for Gravity Waves on the Ocean's Surface. Part II: Interpretation and Applications. *J. Phys. Oceanogr.* **1977**, *7*, 11–21. [CrossRef]
30. Rice, S.O. Reflection of electromagnetic waves from slightly rough surfaces. *Commun. Pure Appl. Math.* **1951**, *4*, 351–378. [CrossRef]
31. Barrick, D.E. Remote Sensing of Sea State by Radar. In *Remote Sensing of the Troposphere*; Derr, V.E., Ed.; U.S. Govt. Printing Office: Washington, DC, USA, 1972; pp. 1–46.
32. Barrick, D.E. The interaction of HF/VHF radio waves with the sea surface and its implications. In *Electromagnetic of the Sea, AGARD Conference Proceedings*; AGARD: Neuilly sur Seine, France, 1970.
33. Lipa, B.J.; Barrick, D.E. Extraction of sea state from HF radar sea echo: Mathematical theory and modeling. *Radio Sci.* **1986**, *21*, 81–100 [CrossRef]
34. Barrick, D.E.; Peake, W.H. A review of scattering from surfaces with different roughness scales. *Radio Sci.* **1968**, *3*, 865–868. [CrossRef]
35. Katopodes, N.D. Air-Water Interface. In *Free-Surface Flow:Shallow Water Dynamics*; Elsevier: Amsterdam, The Netherlands, 2019; Chapter 2, pp. 44–121.
36. Gill, E.; Huang, W.; Walsh, J. On the Development of a Second-Order Bistatic Radar Cross Section of the Ocean Surface: A High-Frequency Result for a Finite Scattering Patch. *IEEE J. Ocean. Eng.* **2006**, *31*, 740–750. [CrossRef]

MDPI
St. Alban-Anlage 66
4052 Basel
Switzerland
Tel. +41 61 683 77 34
Fax +41 61 302 89 18
www.mdpi.com

Remote Sensing Editorial Office
E-mail: remotesensing@mdpi.com
www.mdpi.com/journal/remotesensing

www.ingramcontent.com/pod-product-compliance
Lightning Source LLC
LaVergne TN
LVHW080503180726
843515LV00016B/1384